Computer Simulation of Dynamic Systems

MAURICE F. ABURDENE
Bucknell University

wcb
Wm. C. Brown Publishers
Dubuque, Iowa

Library of Congress Catalog Card Number: 88–1210

ISBN 0–697–06753–X

Printed in the United States of America

10 9 8 7 6 5 4 3 2 1

To Laurie,
who worked long hours with me,
encouraged me, and juggled the duties of
wife, mother, and editor-in-chief with
dedication and a sense of humor

Contents

Preface *xi*

1 Introduction to Simulation 1

1.1 Introduction 1
1.2 The Elements of Simulation 2
1.3 Uses and Purposes of Simulation 6
1.4 Pitfalls of Simulation 7
1.5 Classification of Models 9
1.6 Simulation Languages 10
1.7 Summary 11
1.8 Review Problems 11
1.9 References for Further Reading 12

2 Simulation Examples 17

2.1 Introduction 17
2.2 Simulation of Stochastic Systems: The Salesperson Problem 17
2.3 Simulation of Dynamic Continuous Systems: The Information Diffusion Problem 22
2.4 Equilibrium of Dynamic Systems: The Supply and Demand Problem 27
2.5 Summary 31
2.6 Review Problems 31
2.7 Appendix: Simulation of the Salesperson Problem 33
2.8 Appendix: Simulation of the Information Diffusion Problem 39
2.9 Appendix: Simulation of the Supply and Demand Problem 43
2.10 References 44

3 System Description 45

3.1 Introduction 45
3.2 System Descriptions 45
3.3 Open and Closed Systems 60
3.4 Open-Loop and Feedback Systems 60
3.5 Dynamic and Static Systems 62
3.6 Stochastic and Deterministic Systems 62

3.7 Linear and Nonlinear Systems 63
3.8 Causal Loop Diagrams 64
3.9 Summary 64
3.10 Review Problems 65
3.11 References 68

4 Basic Concepts of Probability Theory 69

4.1 Introduction 69
4.2 Probability 69
4.3 The Concept of a Random Variable 70
4.4 Discrete and Continuous Random Variables 72
4.5 The Distribution Function 73
4.6 The Probability Density Function 74
4.7 The Expected Value 76
4.8 The Variance 77
4.9 Probability Distribution Functions 77
4.10 Sampling 86
4.11 Central Limit Theorem 87
4.12 Regression Analysis 92
4.13 Correlation Analysis 95
4.14 Summary 96
4.15 Review Problems 96
4.16 References 100

5 The Monte Carlo Method and Simulation 102

5.1 Introduction 102
5.2 Uniformly Distributed Random Number Generators 103
5.3 Testing Random Number Generators 104
5.4 The Chi-Square Test 105
5.5 Scatter Diagrams 107
5.6 Principles of Random Process Generators 108
5.7 Exponential Process Generator 109
5.8 Poisson Process Generator 110
5.9 Applications: Monte Carlo Simulation 112
5.10 Reduction of Variance 114
5.11 Confidence Intervals and Number of Simulations 116
5.12 Summary 117
5.13 Review Problems 117
5.14 Appendix: Chi-Square Distributions 120
5.15 Appendix: Random Number Tables 122
5.16 Appendix: Normal Distributions 124
5.17 References 127

6 Discrete-Event Simulation 129

6.1 Introduction 129
6.2 A Perspective on Simulation Languages 129
6.3 A Discrete-Event Simulation: Principles of Operation 131
6.4 World Views 137
6.5 Discrete-Event Simulation of a Distributed Computing System 141
6.6 Discrete-Event Simulation of a Computer Communication Network 145
6.7 Discrete-Event Simulation of a Five-Node Computer Network 152
6.8 Discrete-Event Simulation of a Manufacturing Process 158
6.9 Summary 165
6.10 Review Problems 166
6.11 Appendix: SIMAN Run Processor Release (Trace for First 60 Units of Time) 168
6.12 References 188

7 Discrete Systems and Markov Processes 190

7.1 Introduction 190
7.2 Mathematical Representation of Markov Processes 190
7.3 Examples of Markov Processes 193
7.4 Determining the Probability Transition Matrix 197
7.5 Sample Markov Simulation Algorithm 200
7.6 Summary 207
7.7 Review Problems 207
7.8 References 208

8 System Dynamics and Integration 210

8.1 Introduction 210
8.2 Integration and Exponentiation 211
8.3 Euler Integration Algorithm 215
8.4 Defense Expenditures Model 216
8.5 Second-Order System: An Arms Race Model 218
8.6 Equilibrium Conditions 219
8.7 Solution of the Arms Race Model 220
8.8 System Trajectories 224
8.9 Time-Varying Systems 226
8.10 Nonlinear Systems 226
8.11 Limiting Values 229
8.12 Time Delays in Continuous Systems 230
8.13 Choice of Integration Increment 234
8.14 Simulation of Large-Scale Social Systems 234
8.15 Sensitivity 236

8.16 Summary 236
8.17 Review Problems 237
8.18 Appendix: Proof of Exponential Relationship 240
8.19 Appendix: Second-Order System Response and State Variables 242
8.20 Appendix: Runge-Kutta Integration Method 244
8.21 References 244

9 Continuous-System Simulation Languages 246

9.1 Introduction 246
9.2 Principles of Operation 246
9.3 CSSL Examples 248
9.4 Computer-Aided Design and Simulation 265
9.5 Summary 269
9.6 Review Problems 269
9.7 Appendix: More Simulation Programs 271
9.8 References 292

10 Parameter Estimation 293

10.1 Introduction 293
10.2 Parameter Estimation of Growth and Decay Processes 293
10.3 Estimating the Logistics Curve Coefficient of Diffusion 295
10.4 Parameter Estimation of State Variable Systems 296
10.5 Parameter Estimation of Stochastic Systems 297
10.6 Summary 304
10.7 Review Problems 304
10.8 Appendix: Solution for Estimating the Logistics Curve Coefficient of Diffusion 306
10.9 References 308

11 Parameter Optimization 309

11.1 Introduction 309
11.2 Cost Functions 309
11.3 Single-Variable Optimization Methods 313
11.4 Multiparameter Optimization 320
11.5 Summary 320
11.6 Review Problems 320
11.7 References 322

12 Computer Simulation Methodology 323

12.1 Introduction 323
12.2 Simulation Methodology: Cardiovascular System Example 323
12.3 Major Problems in Simulation Design 336
12.4 Tools for Simulation Management and Methodology 336

12.5 Top-Down Design 338
12.6 Warnier Diagrams 340
12.7 Summary 341
12.8 Appendix: Parameter Identification Using Maximum Likeliho
Estimation 341
12.9 Appendix: The Ten Fatal Mistakes and How to Avoid Them
12.10 References 345

13 Verification and Validation 347

13.1 The End of the Simulation Process: Verification and Validat
13.2 References 349

Index 351

Preface

Simulation is an important tool in the understanding and design of physical systems, engineering systems, and social systems. Because of its importance and broad range, it has been the subject of numerous books of differing perspectives. The objective of *Computer Simulation of Dynamic Systems* is to provide a text that is easy to read and understand. Real-life examples are used to spark an interest in the vast potential that simulation holds. Most of the concepts are demonstrated through examples from engineering, the physical sciences, and the social sciences.

The text is dynamic in that it provides students with the fundamentals and allows teachers to be creative—that is, to use their experiences as well as research literature examples to bring the real world into the classroom through applications of text concepts.

Chapter 1 introduces the student to systems and simulation by presenting the elements of simulation, the uses and purposes of simulation, the simulation types, and the classification of models.

Chapter 2 introduces the simulation of probabilistic systems, whose outcome cannot be predicted with certainty, and deterministic systems, whose outcomes can be predicted. The purpose of Chapter 2 is to provide the student with confidence in the methods of simulation by using simple examples whose solutions can be obtained without the use of a computer. In fact, a coin or a die and very elementary math skills will suffice. The systems presented for simulation also have analytic solutions, thus providing the opportunity to compare simulation results with expected results.

Chapter 3 presents a framework for the construction of models of systems. Mathematical models of real systems are presented along with systems classifications, which include discrete, continuous, deterministic, stochastic (probabilistic), linear, and nonlinear classifications.

The purpose of Chapter 4 is to provide basic concepts of probability theory, which is needed to simulate stochastic systems such as queuing systems (waiting lines). The statistics presented in this chapter also acquaint us with the methods used to determine the number of simulation experiments needed to properly interpret the results of a simulation experiment.

Chapter 5 addresses stochastic simulations by focusing on the methods for generating random numbers for use in computer simulation. The principles of testing and evaluating random number generators also are presented.

The purpose of Chapter 6 is to introduce simulation languages and to give an understanding of the basic mechanisms of discrete event simulation languages. Other authors present this chapter much earlier, as some students seem anxious to use computer simulation languages and to be able to master new tools. However, I feel it is much easier to understand the mechanisms inherent in simulation languages when they are presented at this point. Each individual instructor should choose when to present this chapter.

Chapter 7 is devoted to Markov processes, an important class of stochastic systems. Markov processes are used in many fields, including sociology, psychology, engineering, and computer communication networks design. Students should have a fundamental understanding of the simulation of Markov processes after completing this chapter.

Chapter 8 addresses the simulation of continuous systems. Numerical integration is essential to the behavior of this important class of systems. Emphasis is placed on those described by second-order equations. Forrester's World Dynamic model also is presented.

Chapter 9 introduces continuous simulation languages. As in Chapter 6, it is at the discretion of the instructor to decide at what point this material will be best understood and most useful to the students.

The purpose of Chapter 10 is to provide methods for identifying parameters of simulation models. Parameter values are critical because they identify relationships between variables in the systems under study.

Chapter 11 discusses optimization methods. This topic is obviously a very crucial consideration in designing systems and determining policy.

The focus of Chapter 12 is on the methodology for performing large system simulations. The methodology is presented at this point, rather than in earlier chapters, because by this time the students have a better understanding of simulation methods. Otherwise, experience has shown that students view this as an academic exercise due to their lack of experience with simulation and, as a result, do not benefit from the presentation.

Chapter 13 is devoted to the important topics of verification and validation, which are the last two stages of the simulation process. In the verification stage, we ask "Is our program an accurate translation of our lumped model?" In the validation stage, the question is "Does the model reflect the real world system?" These two topics are often neglected by users, which leads to simulation analysis failures.

Depending on the focus of the course, teachers may wish to present the chapters in this book in a different order. To maximize its effectiveness, however, students should have one semester each of calculus, probability and statistics, and computer science. This background is helpful, but not absolutely required to read and understand the text. Students with a strong background in mathematics and statistics might not need to cover Chapter 4, but coverage of this chapter reinforces the concepts, and most students have appreciated the review. Students with both engineering and mathematics backgrounds could use Chapters 3 and 4 for review only, while those with no mathematics will be best served by a thorough reading of Chapters 1 and 9. An advantage of this book, from the teacher's perspective, is that it encourages students to follow up the references in the library for more detailed understanding of the models presented.

This text was written with no particular simulation language in mind, thus allowing students and teachers to make use of whatever language is available to them. For example, in a discrete simulation course, Chapters 1, 2, 4–7, 10–13 may be used in conjunction with one or more simulation language manuals for SLAM, SIMULA, SIMAN, GPSS, and SIMSCRIPT. Similarly, for a continuous simulation course, Chapters 1–5, and 8–13 are recommended and can be used with CSMP, DYNAMO, NDTRAN, and ACSL manuals. In the text, sample programs are presented in SLAM, SIMAN, GPSS, SIMSCRIPT, CSMP, DYNAMO, NDTRAN and ACSL. Simulation algorithms are presented in Pascal.

The text has grown out of class notes for a simulation theory course and has evolved over ten years of in-class use. Student input (through questions, comments, and class performance) has been essential in its revision and refinement. At Bucknell University, I have used these notes in simulation seminars with juniors, seniors, and first-year graduate students majoring in engineering, mathematics, and computer science. Class periods were devoted to students' presentations of the concepts from the notes, while laboratory periods were used for exercises and an introduction to simulation languages. These enterprising and inventive engineering and science students were fascinated by the application of methods in their fields to problems in the social sciences. At Swarthmore College, my classes included juniors and seniors studying engineering, economics, physics, and political science. These aspiring young social scientists and physical scientists found simulation to be a common ground between them, which increased their mutual understanding and communication. It also made for some very interesting and thought-provoking class discussions. At St. John's University, Dr. Raouf Farag used these notes for a mathematical modelling course for mathematics and computer science students.

Acknowledgments

As an undergraduate at the University of Connecticut, I had the good fortune to stumble upon a fantastic professor, who conveyed his excitement and enthusiasm for engineering and education to me in every course he taught, including simulation. Over the years, we grew to be good friends and colleagues. Though he has died, Ralph J. Kochenburger continues to be an inspiration to me, and I would like to remember him now at this very happy time in my life.

Writing this book has been a long and varied adventure, and I have received much help along the way. For their constructive and insightful comments, without which I would still be at my desk, I thank my reviewers: Edward J. Williams, Robert Trenary, Edgar B. Lewis, Marlin Mickle, Tom Hartley, Thomas Duvall, James Dutt, and Raouf Farag. For their unflagging support, I am indebted to my colleagues: Edward Mastascusa, Mohamed El-Sharkawy, Lee Hammarstrom,

Saleem Kassam, Daniel Hyde, Alex Nauda, David Finkel, David Bowler, Nelson Macken, Scott Dickson, Searl Dunn, Harold Sandler, Danielle Goldwater, and Louise Jones. This book has been developed over years of classroom use, and for their suggestions and criticisms, I thank my students, in particular: Talal Al-Aggad, George Thomas, Alan Snyder, Andrew Love, Paula Durlach, Lisa Keleman, Steve Branch, Mark Luders, Kawalinder Singh, Karl Schwarz, Arvind Betrabet, Radwan Shaban, Susan Managan, Wing Leung, Tim Ross, Stephen Love, Moses Sutherland, and Chris Berry. In addition, I wish to express my appreciation to Andrew Yett for encouraging me to continue developing the book and for his many helpful editorial comments. I owe a special thanks to Diane S. Goodling for her good nature and efficient typing of what seemed to be scores of drafts and revisions under deadlines. And lastly, I thank my children, Leila and Joseph, for their understanding and patience, during all those "lost" weekends. I will never be able to recapture those missed moments with them. Oh, what I wouldn't give for another game of soccer and a sax concert.

Maurice Felix Aburdene

1 Introduction to Simulation

1.1 INTRODUCTION

"Nature's laws and ordinances, whereby all things come to pass and change from one form to another, are everywhere and always the same; so that there should be one and the same method of understanding the nature of all things whatsoever, namely, through nature's universal laws and rules."[80] We do not yet understand all of these natural laws. Simulation can help us discover and understand the natural laws and rules that govern the behavior of physical systems and social systems.

We consider a system to be a collection of interrelated entities whose characteristics may often be related. For example, think of a game of shooting baskets between a father and son to be a system. The rules of our game state that

1. The son shoots first in each new game.
2. When one player misses a shot, the ball is given to the other player.
3. The winner of a game is the first player to score two consecutive baskets.

Therefore, if the son shoots and makes a basket, he shoots again. If he misses on his first try, his father takes a turn. If the father shoots and makes it, he shoots again. If he misses, the son gets his second turn. And so on.

The entities of the system are the father, the son, the basketball, and the basketball hoop. The characteristics of the father and son are age, size, skill, and experience. The characteristics of the basketball are its size and weight. The characteristics of the hoop are its diameter and how high it is placed. The quality of the game depends on these characteristics. The objective of studying this system might be to determine who wins the most games in a specified number of games.

Simulation is the process of developing a simplified model of a complex system and using the model to analyze and predict the behavior of the original system. Why simulate? The key reasons are that real-life systems are often difficult or impossible to analyze in all their complexity, and it is usually unnecessary to do so anyway. By carefully extracting from the real system the elements relevant to the stated requirements and ignoring the relatively insignificant ones (which is not as easy as it sounds), it is generally possible to develop a model that can be used to predict the behavior of the real system accurately.

You already know a lot about simulation because it is simply a natural way of thinking, of using analogy to unravel the workings of complex systems. You encounter the methods and results of simulation daily, for example in predictions on the economy and in simulation of car performance. You probably use the methods yourself, unaware that you are using the concepts of simulation. For example, upon entering the school cafeteria, you must choose from one of two lines to get your dinner. If you are really hungry, then you will quickly scan the lines to see which is moving faster and has fewer people. You may also know which individuals take longer to fill their plates or chat with the servers. Taking these and other factors into account, you will choose one line over the other, with the hope that your "simulation" will prove correct. How often have you been in this type of situation and made the right choice? This book is dedicated to a step-by-step development of methods for the simulation of different types of systems.

Simulation was first used extensively in the physical sciences and engineering, after being developed and perfected by the military and the aerospace industry. The simulation approach to the study and use of models has recently been in vogue in business, industry, and government, and in social science and behavioral science research. Widely publicized simulations in the social sciences include voting patterns of population groups, Forrester's "World" and "Urban" models, the Wharton School's econometric model of the U.S. economy, the Stanford University Graduate School of Business national economy model, and the "distributive justice" simulation of Gullahorn and Gullahorn. Simulations in science and engineering have covered theories of the moon's origin, ocean waves, models of the heart, semiconductor fabrication processes, distributed electrical energy systems, software systems, computer communication networks, local area networks, plane kinematic mechanisms, aircraft systems, weapons systems, nuclear attacks, customer queues, grocery store inventory, automated warehouses, and the design of computer systems and semiconductors.

The models developed in this text are mathematical, consisting of equations that relate internal variables of the system to one another and to external variables that act upon the system. These equations describe both the model and the real-life system and can be programmed into a computer. It is important to understand that the computer neither adds to nor subtracts from the model. In theory, the computer is not even a requirement for simulation. Its role is simply to process the equations with greater speed and efficiency and to "exercise" the model by stepping through it second by second, or by any time increment specified. We can change the inputs to the model, including initial conditions and parameters of the model. For example, when testing the performance of a computer system, we might want to see how the average waiting time differs when we change the inputs of the system from 100 to 150 interactive users.

1.2 THE ELEMENTS OF SIMULATION

Simulation is one of the most powerful tools available for the analysis and design of complex systems. The basic elements of the simulation process are[84,85]

1. *The real system.* The real system refers to a source of observable data. The system may be natural, such as planets and their orbits or the physiological system comprising the human body, or man-made, such as an automobile. The system may also involve both natural and man-made subsystems, such as the urban and world systems. The distinguishing characteristic of the system is that a segment of reality can be identified and measured. We shall have a lot more to say about systems in Chapter 3.

2. *The experimental frame.* This element describes a limited set of circumstances under which the real system is observed or tested.

3. *The base model.* This element is the hypothetical and complete explanation of the system. It is the "true" model of the system and is difficult to obtain. It is also called the reference model.

4. *The lumped model.* This is a simplification of the base model and is the modeller's image of the real system. We, as modellers, lump together components and then simplify their interactions accordingly.

5. *The computer.* This is a computational device that helps us solve the behavior of the lumped model.

6. *The algorithm.* This element is the step-by-step process for describing the equations of the lumped model to the computer. The execution of the algorithm simulates the behavior of the system. The computer implementation of the algorithm is referred to as a *program.* We will call thc program that is executed by the computer the *simulation model.* In developing the computer program, the following factors should be considered:

The computer's availability, memory size, and speed of execution

The availability of programming languages (for example, FORTRAN, Pascal, PL/1, BASIC) or simulation languages (for example, GPSS, ACSL, SIMULA)

The above elements of the simulation process have two basic functions: *model construction* and *model use*. These functions and their development are presented in later chapters. Figure 1.1 shows the steps involved in the development of a simulation model.

Consider the following real system. A resting ball of mass M is suspended by an elastic cord, as shown in Figure 1.2. The ball is pulled downward and bounces up and down. As modellers, we will consider the experimental frame to be only the ball's vertical motion. What does it take to simulate the ball's motion?

Our next step is to develop the base model. The base model is the comprehensive analysis of the system, with every factor that could affect the system taken into account. In complex systems, arriving at a complete base model is often impossible, because the way in which certain parts of the system work or interact is unknown, and perhaps is the very object of the simulation.

FIGURE 1.1 Steps in the Development of a Simulation Model

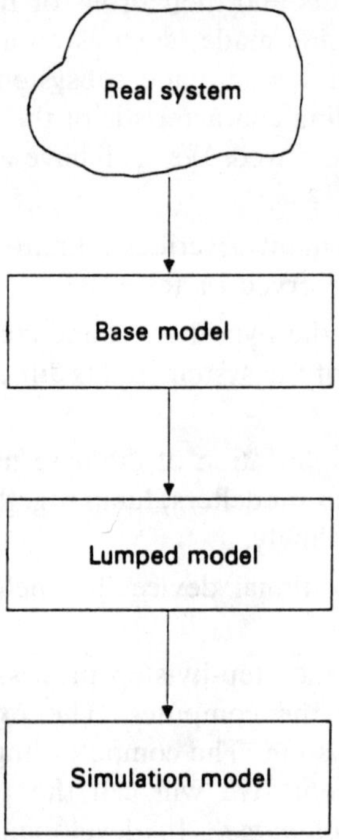

FIGURE 1.2 Ball Suspended from an Elastic Cord

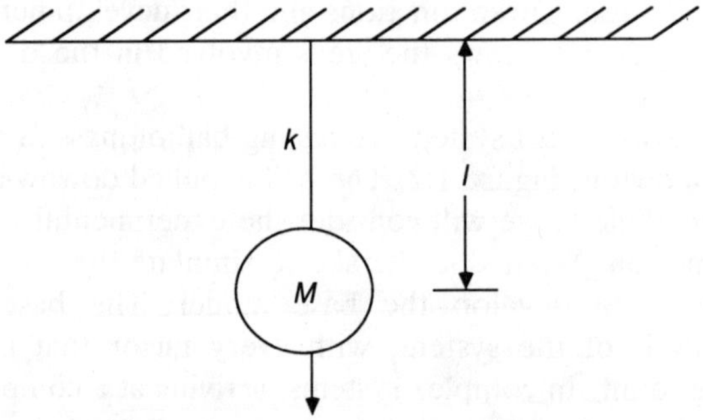

From the diagram of our system, we can determine the following elements of the base model:

A variable for the length of the elastic cord, l
A variable for the elasticity of the cord, k
A factor for the friction within the cord
A factor for the friction of the ball with the air
A factor for the initial displacement of the ball

It is in the next step—the transforming of the base model into a lumped model—that the art of modelling becomes critical. For a simulation to be successful, the model upon which it is based must be valid. What, we ask, are valid simplifications of our base model? What simpler system is reasonably analogous to ours? "Reasonably analogous" means that the behavior of the lumped model and that of the base model are sufficiently alike to provide the desired understanding of the base model.

Perhaps the first assumption we can make is that air friction is insignificant. By doing this, we eliminate a struggle with complex air-resistance equations. Second, we will assume that the elastic cord is linear, that is, the force of the stretched cord on the ball is directly proportional to how much the cord is stretched. Finally, we will assume that the mass of the cord is negligible compared to that of the ball.

Our simplifying assumptions give us our lumped model. The lumped model is what we will actually simulate as an image of the real system. Turning to elementary physics, we find that the behavior of our lumped model can be described as simple harmonic motion (SHM). The equations that describe SHM are well known and easily programmed on a computer. With our computer program, we can specify different values for the various components of our model. We can instruct the computer to print out the position of the ball, say every hundredth of a second, after specifying a length and elasticity of the cord, along with the ball's mass and initial displacement. To validate our model, we would compare the results obtained from the computer with results measured in the laboratory. We would find that, for a steel ball hanging by an ordinary rubber band, our simulation would work well, because it corresponds closely with our lumped model. However, this simulation would work poorly if we were to use a table tennis ball instead of a steel ball, because we could no longer assume that air friction is insignificant and that the mass of the cord is negligible compared to the mass of the ball. Given a set of initial conditions, the computer could accurately predict the ball's vertical position at a future time. Thus, one of simulation's primary capabilities is clear: when we program the mathematical equations for a validated model into a computer, simulation allows us to predict the future state of a system. By varying the inputs to the simulation program, we can predict the results of different actions on the system.

1.3 USES AND PURPOSES OF SIMULATION

Simulation can be used in three main areas:

1. Situations in which mathematical and statistical techniques using analytic methods fail because the problems are too complex and other techniques must be derived to predict the behavior of the system; for example, the prediction of heart rate and blood pressure in the cardiovascular system, and the average waiting time of a message in a large-scale computer communication network. Here, one must be careful not to regard simulation as merely a second-best alternative, as a substitute to be used when analytic methods fail. Simulation of complex systems can yield valuable insight into which variables in the system are more important than others and how these more important variables interact.

2. Situations in which it is impossible to gain practical experience because the phenomena being considered have not yet occurred; for example, the design of new aircraft, supercomputers, and medical equipment.

3. Situations that develop over a long period of time in which an adequate understanding of the phenomena would take too long; for example, the growth of a forest over several centuries. The simulation method speeds up the real-life situation through the use of time scaling.

Simulation has the following purposes:

1. To predict the consequences of changes in policy, conditions, or methods; for example, grade point average of a university's graduates after the admission policy changes from considering high school records alone to considering high school records and standardized test scores; stopping distance of a car travelling 40 mph when the pavement changes from dry to wet; and information dissemination when the source changes from exclusively newspapers to exclusively radio.

2. To learn about new systems in order to redesign or redefine them; for example, high-definition TV, space stations, and new weapons systems.

3. To educate people regarding how a real-life system or situation actually operates; for example, nuclear power plant operation, and air traffic control systems.

4. To verify or demonstrate a new proposal dealing with the system; for example, new data structure for an existing program.

5. To project the future behavior of the system; for example, the effect of higher interest rates on housing starts.

6. To generate less-"biased" data when the data are obtained by survey methods in which the subjects become aware of the objectives of the survey and answer accordingly; for example, public opinion polls and psychological testing.

1.4 PITFALLS OF SIMULATION

Although simulation can have many useful results, there are pitfalls to the process. Simulation is both an art and a science. To simulate any complex system, judgment, keen attention to detail, and experience are all important in constructing the base model, simplifying it to arrive at the lumped model, and implementing the mathematical model with a computer program. In our simple system, we assumed that air friction was an insignificant factor. If our ball were not made of steel, but was instead an inflated balloon, our assumption would probably invalidate the lumped model. No matter how carefully we programmed the computer, the bad link in the previous step would have poisoned the resulting simulation.

In simple systems, the techniques of simulation are relatively straightforward, and room for serious judgment errors is relatively limited. But, in complex systems, models can become enormously intricate, and what seem to be trivial simplifications can markedly affect the validity of a model. These hazards are particularly great when the internal workings of a system are not fully understood. For example, in simulating the atmosphere to predict weather, numerous volumes of data are candidates for inclusion. As modellers, how do we treat wind, air, and sea temperatures, sunspot activity, pollution, and other factors? As we shall see in the following chapters, using simulation effectively in the social sciences often involves facing equally challenging modelling subtleties.

Recently, natural scientists and engineers have been trying to aid social scientists and behavioral scientists in designing and understanding complex social systems by using the techniques developed for understanding natural and physical phenomena. It is important, however, to point out that the physical scientists have been modelling natural phenomena, whereas the engineers and social scientists have been modelling and simulating man-made systems. The engineers have been more successful than the social scientists in the use of simulation because of the way the simulation tools have been used and because of the emphasis on the final objectives of the simulation studies. The engineer uses the simulation model to design a new or better system, and builds a physical or iconic model for testing. A distinguishing feature of a physical model is that it "looks like" the entity being simulated. The physical model can be scaled-up, scaled-down, or full-scale, and consequently can be used for demonstration. The social scientist often uses simulation to try to explain an existing system, and attempts to use controlled experiments and games as "physical" models of complex systems.

The difficulties of using simulation can be classified into two general categories:[23]

1. *Inherent limitations of simulation.* One inherent limitation is that general relationships among entities are obtained from a limited set of experimental data by induction. The effects of this limitation can be reduced through the use of statistical testing, sensitivity analysis, and, when available, larger data samples. A second inherent limitation is that simulation methods alone cannot solve fuzzy and/or complex systems. Studies where the analysis takes into account more of the real world than the program is equipped to handle usually fail to determine the important relationships among entities. It is interesting to note that the power of computer simulation methods fosters the development of large scale models that fail to simplify reality, and therefore, defeat the objective of model construction. A third limitation is that simulation might fail to give an answer or may provide a single answer when more than one answer exists.

2. *Human misuses and misapplications of simulation.* Most of these abuses are committed by mathematicians and people in the "exact" sciences who tend to view the problems facing the social scientist as similar to the problems encountered in their own fields, which can be solved using "known" and "proven" theories. The blind transfer of techniques which have proven effective in the exact sciences is based on naïveté. Suppose the government of Chile imposes a tariff on American-made radios. How will the U.S. government react? A physical scientist or engineer, familiar with Newton's third law of motion (for every action there is an equal and opposite reaction), might incorrectly conclude that the United States will impose an equal tariff on the import of Chilean radios. The human misuse here is failure to take into account that Chile does not export radios!

Such misapplication causes many aberrations in physical science, engineering, and social science research. If the same mathematical equation appears in two different fields, it is neither enough to prove a relationship between them nor to explain one in terms of the other. It is ridiculous to propose that a relationship exists between the two different fields if the basic mechanisms are entirely different. As Einstein stated, "science is fully justified in finding some relations between fields only after the equality of mathematical methods has been reduced to an equality of the real nature of concepts."[11] However, there are many systems whose basic mechanisms appear to be entirely different but which are described by the same mathematical relationship. Examples are diffusion and the spread of disease through a population, and heart attack and the self-destruction of a bridge as it oscillates and collapses.

Some researchers apply computer packages indiscriminately, using crude analytical techniques that are full of meaningless details without bothering to learn the use of more exact, powerful, and appropriate techniques. This tendency has been encouraged by the availability of simulation packages, which give computer "outputs" that may produce meaningless results when employed by scientists who understand neither the mechanics nor the limitations of these packages. The impressive-looking "printout" and animations often make the fallacious results even more deceptive. Nevertheless, such printouts and graphics animations are commonly used by policy makers and designers to reach important decisions. In addition, researchers often use a technique of first identifying their conclusion and then developing computerized confirmation of the conclusion arrived at by other means. The validity of this approach is questionable at best.

Another criticism of researchers is that their simulations have often been validated superficially, if at all. As C. Hermann[46] puts it, "insofar as their validation is concerned, it is premature to reject or accept the value of most simulations and games in the behavioral sciences." The models are usually not validated because of the following:

1. The monetary and psychological resources of the researchers are often so exhausted in programming the system model that validation is simply not possible. With the advent of general-purpose simulation languages, however, much less time is required for programming, leaving the researcher more time and money to address validation. This excuse should therefore no longer be accepted.

2. Appropriate data do not exist. Again, this excuse should not be accepted, because researchers in most fields have gathered massive amounts of data. The data might not be perfectly adequate for testing the simulation model, but they should be adequate for most conditions. And if data are not available, a strategy to gather them must be a part of the simulation modelling process.

3. Knowledge about various mathematical and statistical techniques in evaluating simulations is inadequate.

At this point we would like to emphasize that the validation of the simulation model depends on knowledge of the system to be simulated. There is no substitute for understanding the fundamental relationships that govern the behavior of the system under study. We often return to the theories of the basic physical sciences and social sciences. This caution is not intended to prevent you from using simulation for experimentation but rather to remind you to evaluate the results of your simulations critically before recommending policy or design.

1.5 CLASSIFICATION OF MODELS

Models of real systems are classified according to their variables of interest and can be continuous, discrete, or combined, and deterministic or stochastic. The classification is based on the behavior of the dependent variables of the model. We usually consider time to be the independent variable.

If the dependent variable can take on any value within a specified range, the model is continuous. For example, our suspended ball model is continuous because the position of the ball (dependent variable) varies continuously over time.

If the dependent variable can take on only specific values, then the model is discrete. Consider a model of a library where the dependent variable is the number of books checked out every day. Since a book cannot be divided into parts and must be considered as a whole, the number of books will be expressed as an integer only. Therefore, this model is discrete.

Models of systems can include both discrete and continuous dependent variables. For example, in computing the balance in a checking account with continuous compounding of interest, we must also provide for discrete jumps due to deposits and withdrawals.

In a deterministic model, the dependent variables are completely determined by the initial conditions, states of the model, and input(s) to the model. For example, when you purchase a 91-day certificate of deposit at your local bank, you know the amount of principal, the interest rate, the time period, and the amount of money you will receive at the end of 91 days.

In a stochastic model, the dependent variables cannot be evaluated with certainty based on knowledge of the initial conditions, states of the model, and inputs to the model. For example, the average number of times you must dial your phone to reach out and touch Mom on the telephone company's favorite holiday, Mother's Day, cannot be predicted with certainty. The average number of times you dial is a random variable. The model is, therefore, stochastic. However, when all telephone systems offer an automatic redial feature, this example will be out-dated.

Just in case, let us consider a small ice cream stand at the beach with only one server. The first customer to arrive will get immediate service. Subsequent customers may have to wait in line for service, depending on the status of the server. Because the time it takes to serve a customer is a random variable, the number of customers waiting in line at any moment in time is also a random variable. This is the single-server, single-queue model.

It is important to understand that although a real system may be continuous, the model developed need not be continuous. Similarly, if the real system is discrete, then the model need not be discrete; and so on. We, as modellers, can choose which classification best suits our objectives.

1.6 SIMULATION LANGUAGES

Digital computer simulation techniques for simulating models can be classified as continuous-system simulation languages (CSSLs), discrete-event simulation languages, or combined simulation languages. The goals of these languages are to reduce the programming effort and, sometimes, to provide a framework for developing models. The choice for students is often based on availability of the simulation language (processor or package).

CSSLs are used when the independent variable, time, is continuous. The most popular CSSLs are ACSL (A Continuous Simulation Language), DYNAMO (DYNAmic MOdels), NDTRAN (Notre Dame TRANslator), and CSMP (Continuous System Modelling Program). Discrete-event simulation languages (processors) are used to simulate systems in which the dependent variables change values only at specified moments in time. Some of the most popular languages are SLAM (Simulation Language for Alternative Modelling), SIMSCRIPT, GPSS (General Purpose System Simulation), SIMULA, and SIMAN (SImulation Modeling and ANalysis). Despite the availability of simulation languages, most simulations use programming languages. These include FORTRAN, Pascal, PL/1, BASIC, and C.

Throughout the text, examples are presented to help you gain an understanding of the methods of simulation defined as model construction and model use. You are encouraged to perform the simulations and exercise the models presented. We believe the best way to learn is to actually do the simulation and interpret the results. As in physics, chemistry, or engineering, there is no substitute for hands-on experimentation.

1.7 SUMMARY

Simulation is a very common and very powerful tool, used in almost every field to provide systematic approaches to solving problems. However, its effectiveness in supplying accurate and dependable information depends on the careful and deliberate development of simulation models. We, as modellers, must learn as much as possible about the real system and consider many different variables and interrelationships to make our model reflect the real system as accurately as possible.

Simulation languages and programming languages have been developed so that computers can exercise complex models with greater efficiency, completing computations in a fraction of the time previously required. Through the structure of these languages, we are given an organized approach to problems.

With all the advantages simulation affords, we must be careful to avoid the pitfalls of misuse and misapplication. It is the responsibility of the modeller to be aware of these dangers and act as a critic.

Clearly, simulation is not only a science, based on specific rules, hard facts, and procedure, but also an art, where the modeller's insight into and sensitivity to simulation objectives play a crucial role in its success. The following chapters will expand on these concepts and methods.

1.8 REVIEW PROBLEMS

1. What are the elements of simulation? Discuss each element.
2. List examples of social system models. Find examples of existing models in your field of interest.
3. List examples of physical science and engineering system models. Find examples in your field of interest.
4. Discuss the uses and purposes of simulation. What use or purpose is most applicable to your field of study?
5. What are the two basic categories of simulation pitfalls?
6. Discuss the advantages and disadvantages of simulation.
7. Using Newton's laws, develop a mathematical model for the system of Figure 1.2.

1.9 REFERENCES FOR FURTHER READING

1. Abelson, R.P., "Simulation of Social Behavior," in *The Handbook of Social Psychology*, 2nd ed., edited by G. Lindzey and E. Aronson, vol. 1 (Reading, MA: Addison-Wesley, 1968), pp. 274–356.
2. Aburdene, M.F., and Kochenburger, R.J., "On Computer Simulation of Social Systems," in *Modeling and Simulation*, edited by William G. Vogt and Marlin H. Mickle, vol. 4 (Pittsburgh: Instrument Society of America, 1973), p. 419.
3. Alvarez, Patria G.; Gilliom, Greg L.; Obermeyer, John R.; Rogers, Paul L.; and Woodward, Malcolm E., "Simulation Ensures Working First-Pass VLSI Computer System," *Hewlett-Packard Journal* 36, no. 9 (September 1985): 13–16.
4. Armstrong, J.R.; Tront, J.G.; and Li, K.W., "Modelling and Simulation of the Effects of Internal Transient Upsets on Microprocessors," *Transactions of the Society for Computer Simulation* 2, no. 1 (May 1985): 73–93.
5. Banta, Robert H., Jr.; Dorward, Peter H.; and Scampini, Steven A., "New Cardiograph Family with ECG Analysis Capability," *Hewlett-Packard Journal* 36, no. 9 (September 1985): 23–28.
6. Bekey, G.A., "Simulability of Dynamical Systems," *Transactions of the Society for Computer Simulation* 2, no. 1 (May 1985): 57–72.
7. Beshers, J.M., *Computer Methods in the Analysis of Large-Scale Social Systems* (Cambridge, MA: MIT Press, 1968).
8. *Bibliography of Selected RAND Corporation Publications*: *Computer Simulation*, The RAND Corporation, 513–1042, August 1972.
9. Birtwistle, G.M., *Discrete Event Modelling on SIMULA* (London: Macmillan, 1979).
10. Birtwistle, G.; Lomow, G.; Unger, B.; and Luker, P., "Process Style Packages for Discrete Event Modelling: Experience from the Transaction, Activity, and Event Approaches," *Transactions of the Society for Computer Simulation* 2, no. 1 (May 1985): 27–56.
11. Boss, Alan P., "The Origin of the Moon," *Science*, January 24, 1986, pp. 341–45.
12. Bottom, Joseph; Bernard, Alan; and Anderson, Kevin, "The Art of Modelling," *Datamation* 31, no. 22 (November 1985): 140–46.
13. Boyle, Graeme R.; Cohn, Barry M.; Pederson, Donald O.; and Solomon, James E., "Macromodeling of Integrated Circuit Operational Amplifiers," *IEEE Journal of Solid-State Circuits*, vol. SC-9, no. 6 (December 1974): 353–63.
14. Caldwell, D.K., and Dawson, B., eds., *Proceedings of the Fourth Annual Simulation Symposium* (New York: Gordon & Breach, 1971).
15. Campbell, B. Dell, "Grasping the Concepts behind Optimization Methods for Control," *Control Engineering* 26, no. 11 (November 1979): 59–62.
16. Caplin, W.D., *Simulation in the Study of Politics* (Chicago: Markham, 1968).
17. Chase, P.E., "Control Theory and the Nuclear Arms Race," *Bendix Technical Journal* 1, no. 3 (Autumn 1968): 43–54.
18. Chorafas, D.N., *Systems and Simulation* (New York: Academic Press, 1965).
19. Chow, We-Min; MacNair, Edward; and Saure, Charles H., "Analysis of Manufacturing Systems by the Research Queueing Package," *IBM Journal of Research and Development* 29, no. 4 (July 1985): 330–42.

20. Churchman, C.W., "An Analysis of the Concept of Simulation," in *Symposium on Simulation Models: Methodology and Applications to the Behavioral Sciences*, edited by A.C. Hoggatt and F.E. Balderston (Cincinnati: South-Western, 1963), pp. 1–12.

21. Davisson, William I., and Uhran, John J., Jr., *A Primer for NDTRAN: A Continuous System Interpreter* (Notre Dame, IN: University of Notre Dame, 1979).

22. Doue, Jone C., and Vallance, Anthony G., "Computer-aided ECG Analysis," *Hewlett-Packard Journal* 36, no. 9 (September 1985): 29–34.

23. Dov, A.G. Beged, Guest Editorial, *Simulation* 14, no. 1 (January 1970): 5–6.

24. Druckman, D., "Understanding the Operation of Complex Social Systems: Some Uses of Simulation Design," *Simulation and Games* 2, no. 2 (June 1971): 173–95.

25. Dym, C.L., and Ivery, E.S., *Principles of Mathematical Modeling* (New York: Academic Press, 1980).

26. Engelke, H.; Grotrian, J.; Scheuing, C.; Schmackpfeffer, A.; Schwarz, W.; Solf, B.; and Tomann, J., "Integrated Manufacturing Modeling System," *IBM Journal of Research and Development* 29, no. 4 (July 1985): 343–55.

27. Evans, G.W.; Wallace, G.F.; and Sutherland, G.L., *Simulation Using Digital Computers* (Englewood Cliffs, NJ: Prentice-Hall, 1967), Chapter 1.

28. Feigenbaum, E.A., and Feldman, J., eds., *Computers and Thought* (New York: McGraw-Hill, 1963), pp. 375–86.

29. Forrester, J.W., *Industrial Dynamics* (Cambridge, MA: MIT Press, 1961).

30. Forrester, J.W., *Principles of Systems* (Cambridge, MA: Wright-Allen, 1968).

31. Forrester, J.W., *Urban Dynamics* (Cambridge, MA: MIT Press, 1969).

32. Forrester, J.W., *World Dynamics* (Cambridge, MA: Wright-Allen, 1971).

33. Gaur, S.P.; Habitx, P.A.; Park, Y.J.; Cook, R.K.; Hauang, U.S.; and Wagner L.F., "Two-dimensional Device Simulation Program: 2DP," *IBM Journal of Research and Development* 29, no. 3 (May 1985): 242–51.

34. Gershwin, Stanley B.; Akell, Ramakrishna; and Choong, Yong F., "Short-Term Production Scheduling of an Automated Manufacturing Facility," *IBM Journal of Research and Development* 29, no. 4 (July 1985): 392–400.

35. Gleick, James, "Moon's Creation Now Attributed to Giant Crash," *New York Times Science Times*, June 3, 1986, p. C1.

36. Gordon, G., *The Application of GPSS V to Discrete Event Simulation* (Englewood Cliffs, NJ: Prentice-Hall, 1975).

37. Greenblat, C.S., "Compiled Bibliography: Gaming and Simulation in the Social Sciences: A Guide to the Literature," *Simulation and Games* 3, no. 4 (December 1972): 477–91.

38. Griffiths, David H., "Physics Teaching: Does It Hinder Intellectual Development?" *American Journal of Physics* 44, no. 1 (January 1976): 81–85.

39. Guetzkow, Harold S., ed., *Simulation in the Social Sciences: Readings* (Englewood Cliffs, NJ: Prentice-Hall, 1962).

40. Guetzkow, Harold S., ed., *Simulation in International Relations: Developments for Research and Teaching* (Englewood Cliffs, NJ: Prentice-Hall, 1963).

41. Guetzkow, Harold S., ed., *Simulation in Social and Administrative Science* (Englewood Cliffs, NJ: Prentice-Hall, 1972).

42. Haines, C.L., "An Algorithm for Carrier Routing in a Flexible Material-handling System," *IBM Journal of Research and Development* 29, no. 4 (July 1985): 356–62.

43. Hall, Ronald T., "Deep Submergence Simulation—A Case History," *Sperry Engineering Review* 20, no. 2 (1967): 2–7.

44. Hamann, Jerry C., and Jacquot, Raymond G., "Analog Simulation on the Personal Computer," *COED Journal* 5, no. 2 (April–June 1985): 12–15.

45. Harris, Henry E., Jr., and Smith, Robert E., "Correlation of Computer Simulation and Full-Scale Data," *Sperry Engineering Review* 20, no. 2 (1967): 31–37.

46. Hermann, C., "Validation Problems in Games and Simulation," *Behavioral Science* 12, no. 3 (May 1967): 216–30.

47. Hillier, F.S., and Lieberman, G.J., *Introduction to Operations Research* (San Francisco: Holden-Day, 1968).

48. Hockney, R.W., and Eastwood, J.W., *Computer Simulation Using Particles* (New York: McGraw-Hill, 1981).

49. Hoover-Siegel, Laurette L., "GMR's Driving Simulator: Motionless Driving," *Search* 14, no. 6 (November–December 1979).

50. Inbar, M., and Stall, C.S., *Simulation and Gaming in the Social Sciences* (New York: The Free Press, 1972).

51. Jeswiet, J., and Mulvanna, C.A., "The Simulation of Plane Kinematic Mechanisms on a Graphic Computer: Student Project," *COED Journal* 5, no. 3 (July–September 1985): 25–28.

52. Knepper, R.W.; Gaur, S.P.; Chang, F.Y.; and Srinivasan, G.R., "Advanced Bipolar Transistor Modelling: Process and Device Simulation Tools for Today's Technology," *IBM Journal of Research and Development* 29, no. 3 (May 1985): 218–28.

53. Knight, D.E.; Curtis, H.W.; and Fogel, L.F., eds., *Cybernetics, Simulation, and Conflict Resolution* (London: Spartan Books, 1971).

54. Laux, Steven E., and Byrnes, Robert G., "Semiconductor Device Simulation Using Generalized Mobility Models," *IBM Journal of Research and Development* 29, no. 3 (May 1985): 289–301.

55. Lave, C.A., and March, J.G., *An Introduction to Models in the Social Sciences* (New York: Harper & Row, 1975).

56. Law, Averill M., and Kelton, W. David, *Simulation Modeling and Analysis* (New York: McGraw-Hill, 1982).

57. Leigh, I.R., *Modelling and Simulation* (London: Peter Peregrinus, 1983).

58. Martin, F.F., *Computer Modeling and Simulation* (New York: Wiley, 1968).

59. McLeod, J., ed., *Simulation: The Dynamic Modeling of Ideas and Systems with Computers* (New York: McGraw-Hill, 1968).

60. Miller, D.K., "Validation of Computer Simulation in the Social Sciences," in *Modeling and Simulation*, edited by William G. Vogt and Marlin H. Mickle, vol. 6 (Pittsburgh: Instrument Society of America, 1975), pp. 743–46.

61. Mitchell, E.L., and Gauthier, J.S., *ACSL User Guide/Reference Manual*, Concord, Mass.: Mitchell and Gauthier Associates, 1981.

62. Mitchiner, R.G., and Leonard, R.G., "Simulation of Dynamic Mechanical Systems at Virginia Polytechnic Institute and State University," *COED Journal* 3, no. 3 (May–June 1983): 2–7.

63. Nygard, Kendall E., and Iyengar, Ashok K., "An Interactive System for Analyzing and Generating Input Process in Simulation," *COED Journal* 3, no. 3 (May–June 1983): 8–12.

64. O'Brien, R.R.; Hsieh, C.M.; Moore, J.S.; Lever, R.F.; Brannon, K.W.; Srinivasan, G.R., Knepper, R.W.; and Murley, P.C., "Two-dimensional Process Modelling: A Description of the SAFEPRO Program," *IBM Journal of Research and Development* 29, no. 3 (May 1985): 229–41.

65. Orcutt, G.H., "Views on Simulation and Models of Social Systems," in *Symposium on Simulation Models: Methodology and Application to the Behavioral Sciences*, edited by A.C. Hoggatt and F.E. Balderston (Cincinnati: South-Western, 1963), pp. 221–36.

66. Payne, James A., *Introduction to Simulation: Programming Techniques and Methods of Analysis* (New York: McGraw-Hill, 1982).

67. Pegden, C. Dennis, *Introduction to SIMAN* (State College, PA: Systems Modeling Corporation, 1986).

68. Pfister, Gregory F., "The IBM Yorktown Simulation Engine," *Proceedings of the IEEE* 74, no. 6 (June 1986): 850–60.

69. Pritsker, A. Alan B., *Introduction to Simulation and SLAM* II, 3rd ed. (New York: Wiley, Halstead Press, 1986).

70. Pugh, Alexander L., III, *Dynamo* II *User's Manual* (Cambridge, MA: MIT Press, 1976).

71. Quinn P.J.; Salmon, J.K.; and Zurek, W.H., "Primordial Density Fluctuations and the Structure of Galactic Haloes," *Nature*, July 24, 1986, pp. 329–34.

72. Raser, J.R., *Simulation and Society* (Boston: Allyn & Bacon, 1971).

73. Ringwood, A.E., "Terrestrial Origin of the Moon," *Nature*, July 24, 1986, pp. 323–28.

74. Ripley, Brian D., *Stochastic Simulation* (New York: Wiley, 1987).

75. Salmas, Sandra, "A Bank Teller's Delicate Balance," *New York Times Business Day*, December 9, 1985, pp. D1–D6.

76. Schmidt, J.W., and Taylor, R.E., *Simulation and Analysis of Industrial Systems* (Homewood, IL: Richard D. Irwin, 1970).

77. Shannon, Robert E., *Systems Simulation: The Art and Science* (Englewood Cliffs, NJ: Prentice-Hall, 1975).

78. Shecter, M., "On the Uses of Computer Simulation for Research," *Simulation and Games* 2, no. 1 (March 1971): 73–89.

79. Soler, Francisco DeP., and Schuster, Richard E., "Compound Growth and Related Situations: A Problem-solving Approach," *Mathematics Teacher* 75, no. 8 (November 1982): 640–44.

80. Spinoza, Baruch, *The Road to Inner Freedom: The Ethics*, edited by Dagober D. Runes (New York: Philosophical Library, 1957), p. 24.

81. Werner, R., and Werner, J., *Bibliography of Simulation: Social Systems and Education, 2000 Items* (La Jolla, CA: Western Behavioral Science Institute, 1969).

82. Whitman, David L., and Terry, Ronald E., "A Graphical Simulation of Vapor-Liquid Equilibrium for Use as an Undergraduate Laboratory Experiment and to Demonstrate the Concept of Mathematical Modeling," *COED Journal* 5, no. 2 (April–June 1985): 8–11.

83. Wittrock, Robert J., "Scheduling Algorithms for Flexible Flow Lines," *IBM Journal of Research and Development* 29, no. 4 (July 1985): 401–12.

84. Zeigler, B.P., "An Introduction to the Theory of Modeling and Simulation," in *Proceedings of the Sixth Annual Conference on Modeling and Simulation*, vol. 6, part 2, (Pittsburgh: Instrument Society of America, 1975), pp. 731–38.

85. Zeigler, B.P., *Theory of Modelling and Simulation* (New York: Wiley, 1984).

2 Simulation Examples

2.1 INTRODUCTION

This chapter introduces some applications of simulations. The examples that are developed represent simplified systems whose behavior can be simulated by hand using pencil and paper. The models for these systems also have analytic solutions to predict system behavior. By comparing the simulated solution with the analytic solution, we should begin to feel confident in our use of the concepts and methods of simulation. These concepts and methods will later help us to simulate more complex systems.

The first step in simulating a real system is to determine what questions we wish to answer; that is, we must define the goals for the simulation project. The second step is to collect and quantify data characterizing the system. The third step is to analyze these data to help identify the laws that govern the system's behavior. In the fourth step, we construct a mathematical model for the system based on these laws. This step is equivalent to the lumped model discussed in Chapter 1. In the fifth step, we "exercise" the model to simulate the behavior of the system. In the sixth step, we compare the original data (if they exist) to the simulated data to determine whether our model adequately represents the behavior of the real system. If it does not, we must modify the model or construct another model by repeating the above sequence of steps beginning with either the second, third, fourth, or fifth step. The sixth step introduces us to the topics of verification and validation which will be discussed in detail in Chapter 13. For present purposes, we define verification as the correct translation of a mathematical model into a computer program for simulation. Validation, on the other hand, examines the degree of agreement between the real system's behavior and the behavior predicted by the simulation model. Figure 2.1 shows the six simulation steps graphically.

2.2 SIMULATION OF STOCHASTIC SYSTEMS: THE SALESPERSON PROBLEM

The behavior of some systems cannot be predicted with certainty but rather follows the laws of probability. As noted in Chapter 1, these systems are called

FIGURE 2.1 Steps in a Simulation

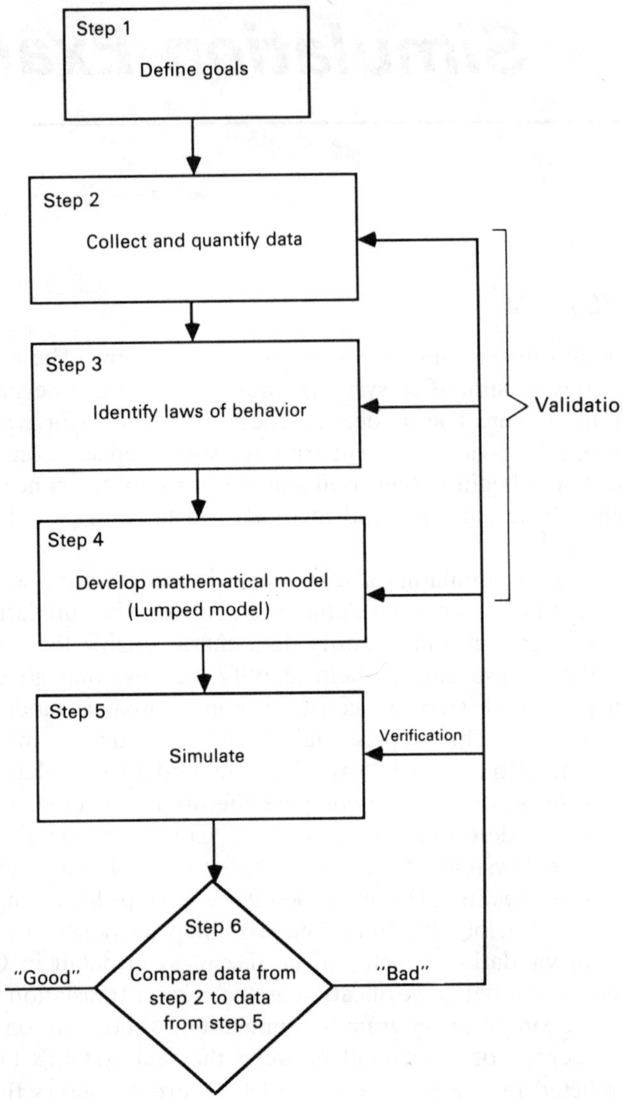

stochastic systems. The simulation of stochastic systems is often referred to as Monte Carlo simulation because one constructs a "game of chance," using random numbers, to analyze the behavior of the system. Random numbers can be obtained from a table of random digits, from the toss of a die, or with a roulette wheel or a computer program. Consider the following example.

The Basic Example

A salesperson calls on a customer to sell his company's product. The probability of a sale during each sales call is a constant, P. What is the total probability p_N for a sale in up to N sales calls?[2]

The probability of no sale on each sales call is

$$1 - P \tag{2.1}$$

It follows that the probability of failing to sell the product in N calls is

$$F_N = (1 - P)^N \tag{2.2}$$

and the probability of a sale is

$$p_N = 1 - F_N \tag{2.3}$$

or

$$p_N = 1 - (1 - P)^N \tag{2.4}$$

The observable data in our system are the number of sales and the number of calls the salesperson makes. The experimental frame limits the values of P to between zero and one inclusive. N must be an integer greater than or equal to one. The base model and the lumped model are the same in our simple example, and are described by equation (2.4). The simple computations in this and the following examples can be solved with pencil and paper.

Hand Simulation

Let $P = 0.5$ and $N = 2$. We can generate the random number (RN) by the toss of a coin. We will assume that a head represents a random number such that $0 \leq \text{RN} \leq 0.5$, and that a tail represents a random number such that $0.5 < \text{RN} \leq 1.0$.

For purposes of our simulation, the salesperson is given two chances to sell the product. If the first coin comes up heads, we will consider it a sale and the salesperson moves on to another customer. If the first coin toss comes up tails, there is no sale and the salesperson makes a return visit. If the second toss is heads, the total sales will increase by one. If the second toss is tails, we assume that we are dealing with a tough customer, the total sales remain unaffected, and our salesperson moves on to the next customer. Table 2.1 summarizes the results of 10 salesperson visits.

Try simulating this experiment on your own. How do your results compare with the results obtained here?

TABLE 2.1 Summary of Salesperson Results

Customer Number (MCount)	*Results of First Visit (NCount = 1)*	*Results of Second Visit (NCount = 2)*	*Total Number of Sales*
1	H		1
2	H		2
3	T	T	2
4	T	H	3
5	H		4
6	T	H	5
7	H		6
8	T	T	6
9	T	H	7
10	H		8

Note: Total probability = $\frac{8}{10} = 0.8$.

For a small number of customers, we expect that the total probability of a sale will vary widely from one set of customers to another. A single customer among a small number of customers represents a large fraction of the total, with the decision to buy or not buy causing large variation in the total probability of a sale. Conversely, a single customer among a large number of customers represents a much smaller fraction of the total, with the buy/no buy decision accounting for a smaller change in the total probability of a sale.

Analytic Solution

The analytic solution of equation (2.4) is given by

$$p_N = 1 - (1 - P)^N = 1 - (1 - 0.5)^N = 1 - 0.25 = 0.75$$

Computer Simulation

When the number of customers, M, is small, the salesperson problem can be simulated with pencil and paper. For a larger number of customers, one way to compute the total probability of a sale in N visits is to write a computer simulation program. A program in Pascal is presented in Algorithm 2.1. We can use the program to simulate the salesperson problem with various other combinations of values for P, N, and M. Thus, even in this simple situation, the value of the computer as a computational tool is clear. Sample programs in other computer languages appear in Section 2.7.

ALGORITHM 2.1 Salesperson Algorithm Presented in Pascal

```
PROGRAM Sales(INPUT,OUTPUT);
VAR
   N,MCount,NCount,NOfSales,M:INTEGER;
   TotalProb,P:REAL;
BEGIN
   WRITELN('Input the probability of a sale during any one visit');
   WRITELN('and the number of visits you wish to try.');
   READLN(P,N);
   WHILE ((P>1) OR (P<0)) DO
       BEGIN
           WRITELN('The probability must be between 1 and 0.');
           READLN(P);
       END;
   WRITELN('Now input the number of times you wish this experiment');
   WRITELN('to be repeated.');
   READLN(M);
   NOfSales:=0; {Initialize the number of sales}
   MCount:=0;   {Initialize the number of experiments}
   WHILE (MCount<M) DO
       BEGIN
           NCount:=0; {Initialize the number of visits}
           WHILE (NCount < N) DO
               BEGIN
                   IF (RANDOM<=P) THEN  {Random is a computer generated}
                       BEGIN            {random number between 0 and 1.}
                                        {Check with your computer center}
                                        {for the function your computer has.}
                           NOfSales:=NOfSales+1;
                           NCount:=N
                       END
                   ELSE
                       NCount:=NCount+1;
               END;
           MCount:=MCount+1;
       END;
   TotalProb:=NOfSales/M;
   WRITELN('If the probability of a sale for an individual visit is ',P:5:4);
   WRITELN('the probability of a sale after up to ',N,' visits is');
   WRITELN('approximately ',TotalProb:5:4);
END.
```

Comparison of Methods

Table 2.2 shows the results of all three problem-solving methods: hand simulation, computer simulation, and analytic solution. The estimate of the total probability of a sale from any simulation will usually differ from the analytic result when the number of customers is small. However, variations in results should fall within an interval. This issue of sample size will be discussed more fully in Chapter 5.

TABLE 2.2 Comparison of Salesperson Simulation Results

	Hand Simulation	*Computer Simulation*	*Analytic Solution*
Total probability of sale	0.8	0.75	0.75

2.3 SIMULATION OF DYNAMIC CONTINUOUS SYSTEMS: THE INFORMATION DIFFUSION PROBLEM

Systems whose behavior changes with time are called dynamic systems. When the changes in behavior develop over time and can occur at every instant of time t, the system is called a dynamic continuous system. When the changes in behavior can occur only at specific moments of time, the system is called a dynamic discrete system. Here, we will develop a mathematical model for a dynamic continuous system relating to the diffusion of information through a population.

The diffusion of information through a group is similar to the spread of a disease through a population. At first, a small fraction of the population has the information. As this group comes in contact with the rest of the population, the information spreads from one person to another until the entire population learns the information. In general, the diffusion process can be described as follows. There exists a population of N units, where units may be people or groups of people. Information is transmitted to the population from a source either at an initial point or continuously. The population will consist of h hearers and i ignorants. Assume that we are interested in the number of hearers at time $t = T$. Let us develop the model and obtain a solution by simulation.

The Basic Example

The system under consideration is a population, N, which can be considered to be a single group or formed from separate groups. The hearers can become spreaders of the information or just converts. The observable data are the number of hearers, ignorants, spreaders, and converts. The experimental frame for the present application is that the population is a single group. Other simplifying assumptions are that all hearers are spreaders, and that the population is completely intermixed. However, only a fraction of meetings between hearers and ignorants results in the spreading of information, changing ignorants to hearers.

To develop the lumped model, assume that the contacts between hearers, $h(t)$, and ignorants, $i(t)$, are proportional to the product $h(t)i(t)$. In reality, this assumption is never satisfied exactly in human relations, but it is essential to the development of the law that governs the behavior of the system.

The change in the number of hearers between time t and time $t + \Delta t$ is given by

$$\Delta h(\Delta t) = kh(t)i(t)\Delta t \qquad \textbf{(2.5)}$$

where k is a proportionality constant. At any given time, the number of hearers is

new number of hearers = old number of hearers + change in number of hearers

Mathematically, this can be expressed as

$$h(t+\Delta t) = h(t) + \Delta h(\Delta t) \tag{2.6}$$

and $$h(t+\Delta t) = h(t) + kh(t)i(t)\Delta t \tag{2.7}$$

or $$h(t+\Delta t) = h(t) + kh(t)(N - h(t))\Delta t \tag{2.8}$$

where N = the size of the population

$i(t) = N - h(t)$

Hand Simulation

If we know the size of the population, N, the conversion coefficient, k, and the initial value of $h(t)$, then, by using hand simulation, we can solve either equation (2.7) or (2.8). Notice that during computation each $h(t)$ first occurs on the left side of the equations, then *recurs* on the right side. This type of equation is called recursive and is useful for computer simulation because the same equation is used to generate new values of $h(t)$ from the old values of $h(t)$ and is easily programmed. An algorithm presented in Pascal for obtaining the number of hearers and ignorants with respect to time is shown in Algorithm 2.2.

ALGORITHM 2.2 Information Diffusion Algorithm Presented in Pascal

```
PROGRAM Rumor (INPUT,OUTPUT);
VAR

    Pop,Hearers,PerH,PerI,TotTime,ClockT,ConvCoef,TPT,DelT,DeltPrint:REAL;

BEGIN
    ClockT:=0;
    WRITELN('Please input the total population and initial number of hearers.');
    READLN(Pop,Hearers);
    WHILE (Hearers >= Pop) DO
        BEGIN
            WRITELN('The total population must be larger than');
            WRITELN('the initial number of hearers.');
            WRITELN('Please try again.');
            READLN(Pop,Hearers);
        END;
    WRITELN('Now input the length of time you want this experiment to go on,');
    WRITELN('the coefficient of conversion, DelT and DeltPrint.');
    READLN(TotTime,ConvCoef,DelT,DeltPrint);
    WHILE(ConvCoef < 0) DO
        BEGIN
            WRITELN('The coefficient must be greater than 0.');
            WRITELN('Please try again.');
            READLN(ConvCoef);
        END;
    WRITELN('Time            %Hearers          %Ignorants');
    TPT:=-DelT/2;      {Initialize Time To Print}
    REPEAT
        PerH:=Hearers/Pop*100;
        PerI:=100-PerH;
        IF (ClockT>=TPT) THEN
```

continues

ALGORITHM 2.2 Information Diffusion Algorithm Presented in Pascal (Continued)

```
                BEGIN
                    WRITELN(ClockT:4:2,'                 ',PerH:4:2,'                  ',PerI
:4:2);
                    TPT:=TPT+DeltPrint;
                END;
            Hearers:=Hearers+ConvCoef*Hearers*(Pop-Hearers)*DelT;
            ClockT:=ClockT+DelT;
        UNTIL (ClockT > TotTime);
END.
```

Computer Simulation

Algorithm 2.2 can be executed with a computer program. The results of a computer simulation run are shown in Table 2.3. It is evident from the table that the results obtained after time $t = 0.2$ are erroneous. The number of hearers has begun to decrease, and this violates our assumptions. Where is the error in our calculation? In our model, in our algorithm, or in our program? Before answering this, let us investigate the analytic solution. Then we will return to this wayward simulation.

Analytic Solution

To obtain the analytic solution to this problem, we must perform some algebraic manipulation and elementary calculus. We can rearrange equation (2.8) into the following form:

$$\frac{h(t+\Delta t) - h(t)}{\Delta t} = kh(t)(N - h(t)) \qquad \textbf{(2.9)}$$

Taking the limit as Δt approaches zero, we obtain

$$\frac{dh(t)}{dt} = kh(t)(N - h(t)) \qquad \textbf{(2.10)}$$

or

$$\frac{dh(t)}{dt} = kNh(t) - kh^2(t) \qquad \textbf{(2.11)}$$

TABLE 2.3 Computer Simulation of Information Diffusion System

t	h	i
0.0	1.0	99.0
0.1	11.86	88.14
0.2	96.89	3.11
0.3	92.88	7.12

Notes: $\Delta t = 0.05$, $N = 100$, $k = 0.5$.

FIGURE 2.2 Analytic Solution to the Information Diffusion System

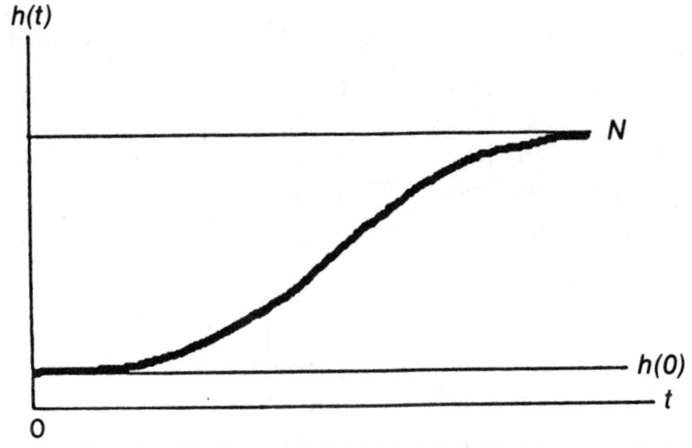

which has the solution*

$$h(t) = \frac{Nh(0)}{h(0) + (N - h(0))e^{-kNt}} \tag{2.12}$$

Finally, let us graph the analytic solution. We can see from the graph in Figure 2.2 that the analytic solution of our model appears to satisfy our intuition.

Now we can return to our computer simulation. Carefully reviewing our reasoning, we find that the error in the computer simulation lies in the assumption that $h(t)$ and $i(t)$ in equation (2.7) are constant over the interval Δt, when, in fact, they are variable. This assumption—the change in the number of hearers in Δt time units—can be considered a good assumption when Δt is "small." Figure 2.3 illustrates the approximation $\Delta h(\Delta t)/k\Delta t$ for a rather large value of Δt.

FIGURE 2.3 Error in the Information Diffusion Computation

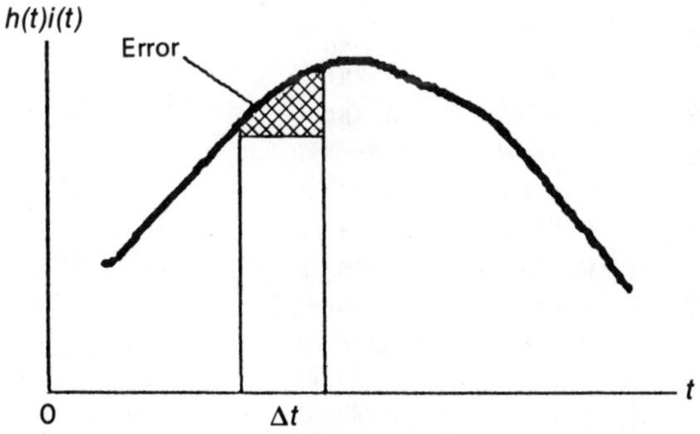

* This equation is a special case of Bernoulli's differential equation called Verhalst's equation. Bernoulli's differential equation is a special type of nonlinear equation that can be reduced to a linear equation by proper transformation.

TABLE 2.4 Simulation Results of Information Diffusion System

Time	*h*	*i*	$\Delta h(\Delta t)$
0.00000	1.00000	99.00000	0.00000
0.01000	1.49500	98.50500	0.49500
0.02000	2.23132	97.76868	0.73632
0.03000	3.32209	96.67791	1.09077
0.04000	4.92796	95.07204	1.60587
0.05000	7.27051	92.72948	2.34256
0.06000	10.64147	89.35853	3.37096
0.07000	15.39600	84.60400	4.75453
0.08000	21.90882	78.09118	6.51282
0.09000	30.46325	69.53675	8.55443
0.10000	41.05482	58.94518	10.59158
0.11000	53.15474	46.84526	12.09992
0.12000	65.60498	34.39502	12.45024
0.13000	76.88741	23.11259	11.28242
0.14000	85.77274	14.22726	8.88534
0.15000	91.87430	8.12570	6.10155
0.16000	95.60702	4.39298	3.73272
0.17000	97.70702	2.29298	2.10000
0.18000	98.82722	1.17278	1.12020
0.19000	99.40673	0.59327	0.57951
0.20000	99.70161	0.29839	0.29487
0.21000	99.85036	0.14964	0.14875
0.22000	99.92506	0.07494	0.07471
0.23000	99.96250	0.03750	0.03744
0.24000	99.98125	0.01875	0.01874
0.25000	99.99062	0.00938	0.00937
0.26000	99.99531	0.00469	0.00469
0.27000	99.99765	0.00235	0.00235
0.28000	99.99883	0.00117	0.00117
0.29000	99.99941	0.00059	0.00059
0.30000	99.99970	0.00030	0.00029
0.31000	99.99985	0.00015	0.00015
0.32000	99.99992	0.00008	0.00008
0.33000	99.99996	0.00004	0.00004
0.34000	99.99998	0.00002	0.00002
0.35000	99.99998	0.00002	0.00001
0.36000	99.99999	0.00001	0.00001
0.37000	99.99999	0.00001	0.00000
0.38000	99.99999	0.00001	0.00000
0.39000	99.99999	0.00001	0.00000
0.40000	99.99999	0.00001	0.00000

Notes: $N = 100,\ h(0) = 1,\ k = 0.5,\ \Delta t = 0.01.$

FIGURE 2.4 Plot of $h(t)$ and $i(t)$ vs. Time for the Information Diffusion Example

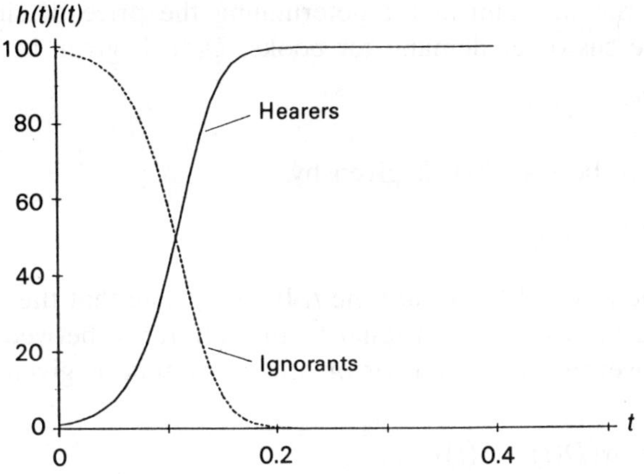

The accumulation of errors in the approximation of a sequence of $\Delta h(\Delta t)$'s caused the incorrect results. It should be clear that the computer cannot calculate $h(t)$ exactly because of the use of a finite Δt. Computer roundoff and truncation errors also make it difficult to compute $h(t)$ exactly. In this example, the true value of $h(t)$ can be approximated using *rectangular integration*, which is a method for determining the area under a curve by breaking up the area into rectangles and adding together the individual rectangular areas. The accuracy of this method depends on the size of Δt, which, in this example, corresponds to the width of the rectangle. Table 2.4 shows the computer simulation results from Algorithm 2.2 with $\Delta t = 0.01$ unit. Figure 2.4 is a plot of these data and concurs with our original intuition. Computer programs for this problem in other languages appear in Section 2.8.

2.4 EQUILIBRIUM OF DYNAMIC SYSTEMS: THE SUPPLY AND DEMAND PROBLEM

The behavior of real systems exhibits one or more of the following characteristics:

1. *Growth.* The observable data continue to increase in value over time.
2. *Decay.* The observable data decrease to a minimum value over time.
3. *Equilibrium.* The observable data increase to an equilibrium (stable) level if the initial value is below the equilibrium level; the observable data decrease to an equilibrium level if the initial value is above the equilibrium level; and the value of the data remains the same if the initial value is at the equilibrium level.
4. *Oscillations.* The observable data attain values above and below some level for a period of time.

The Basic Example

Let us look at the mechanism for determining the price of simulation books. Assume that the customer demand for books, $D(t)$, is given by

$$D(t) = 50 - p(t) \tag{2.13}$$

and the supply of books, $S(t)$, is given by

$$S(t) = 10 + p(t) \tag{2.14}$$

where $p(t)$ is the price of books at time t. If we assume that the rate of change of the price of the books is proportional to the difference between $D(t)$ and $S(t)$, then the law governing the behavior of price over time is given by

$$\frac{\Delta p(t)}{\Delta t} = \alpha(D(t) - S(t)) \tag{2.15}$$

where α is the proportionality constant. The simulation model for the price of the book at any time t is

$$p(t + \Delta t) = p(t) + \Delta p(\Delta t) \tag{2.16}$$

or

$$p(t + \Delta t) = p(t) + \alpha(40 - 2p(t))\Delta t \tag{2.17}$$

Now let us observe how $p(t)$ changes over time.

Case 1 Let us assume that $p(t = 0) = p(0) = 20$. From equation (2.17), we have

$$p(0 + \Delta t) = p(0) + \alpha(40 - 2p(0))\Delta t$$
$$p(\Delta t) = p(0) = 20$$
$$p(2\Delta t) = p(\Delta t) = p(0) = 20$$

and, in general,

$$p(k\Delta t) = p(0)$$

This result indicates that the system has reached equilibrium at the price of $20 per book.

Looking a little more carefully, we see that equilibrium is reached when

$$\frac{\Delta p(t)}{\Delta t} = 0$$

Equilibrium occurs when $40 - 2p(t) = 0$ and $p(t) = 20$ is the equilibrium price. Knowing the equilibrium values of our observable data can often help us to determine that the simulation process is correct. If we do not obtain the equilibrium values, then we should doubt our simulation results. However, obtaining the equilibrium values does not necessarily indicate that our simulation process is correct.

Case 2 If we assume that $p(0) < 20$, then $\Delta p(t)$ is positive, indicating that the new price should be higher. The price continues to increase until the equilibrium price of $20 per book is reached.

Case 3 If we assume that $p(0) > 20$, then $\Delta p(t)$ will be negative, and we expect the price to decrease until the equilibrium price of $20 per book is obtained.

These three cases indicate that our supply and demand system will always reach the equilibrium price of $20. This type of system is said to have a *stable equilibrium*. The system is stable due to the negative slope of the demand curve and the positive slope of the supply curve. There will be excess demand when $p(t)$ is below the equilibrium price, and excess supply when $p(t)$ is above the equilibrium price. The rate at which the equilibrium price is reached also is determined by α and is often called the *delay factor* or, more correctly in this example, the *first-order delay factor*.

Analytic Solution

We obtain analytic solution of equation (2.17) by solving

$$\frac{dp}{dt} = \alpha(40 - 2p(t)) \tag{2.18}$$

to yield

$$p(t) = 20 + (p(0) - 20)e^{-2\alpha t} \tag{2.19}$$

Figure 2.5 is a plot of $p(t)$, where the equilibrium price turns out to be $20. This

FIGURE 2.5 Plot of Price vs. Time for the Supply and Demand System, $\Delta t = 0.1$

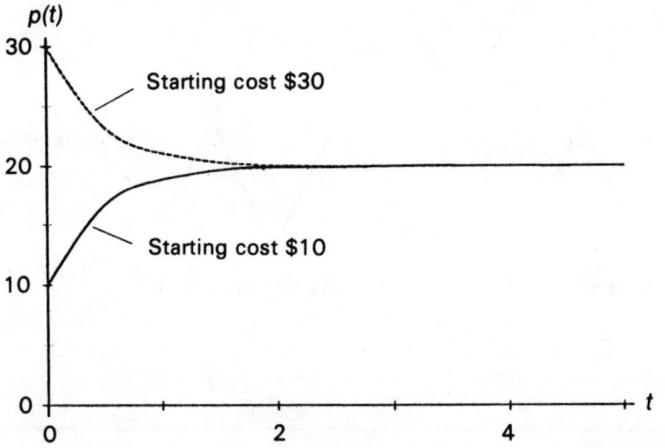

FIGURE 2.6 Supply and Demand Curves

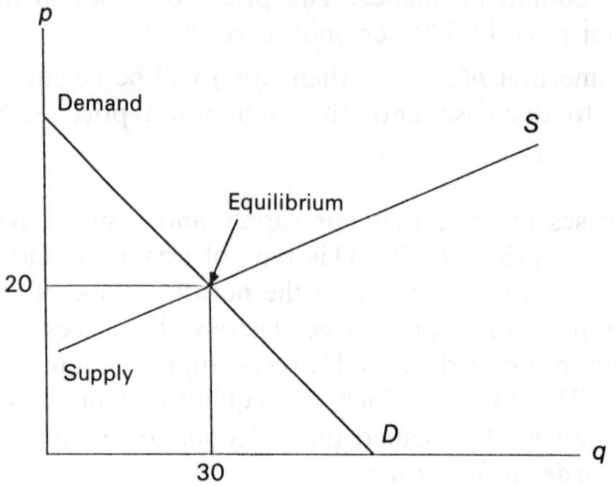

equilibrium price is consistent with the solution obtained from the intersection of the supply and demand curves shown in Figure 2.6. The quantity of books is q.

Simulation Results

For $\Delta t = 1.0$ and $\alpha = 1$, the simulation results exhibit the behavior shown in Figure 2.7. This behavior is inconsistent with our analytic solution. Again, the culprits are the integration method and the step size Δt. For $\Delta t = 0.1$, the result is closer to the analytic solution.

FIGURE 2.7 Plot of $p(t)$ vs. Time for the Supply and Demand System, $\Delta t = 1.0$

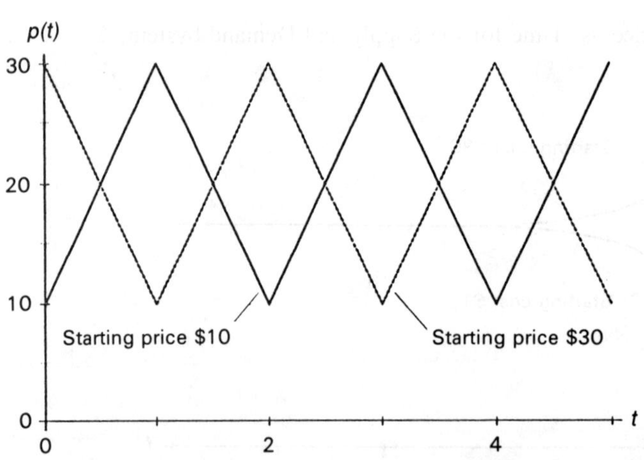

2.5 SUMMARY

The basic concepts of simulation were introduced using examples to help you develop a sense of confidence in the simulation method and a knowledge of some of its pitfalls. The simulations addressed systems whose outcomes cannot be predicted with certainty over time. Simulation of such stochastic systems requires the ability to generate random numbers. You might think about how you would simulate our salesperson system by hand if the probability of a sale on each visit was $\frac{1}{6}$ or, say, $\frac{1}{10}$. In our second example, the information diffusion equation simulated a continuous dynamic system and introduced us to the rectangular integration and problems involved in simulating these systems. Other, more accurate methods of integration include the trapezoidal and the Runge-Kutta. The third example introduced us to equilibrium. This concept is helpful when we examine the results of simulation and decide on the quality of our simulation model in simulating the real system.

2.6 REVIEW PROBLEMS

1. Assume that we want to use the computer simulation model of the salesperson problem developed in this chapter.
   - **a.** What characteristic should the product have to make our model an appropriate one?
   - **b.** What kind of customer is assumed in the model so that the conditions of constant P will hold true?
   - **c.** What factors influence P?
   - **d.** What factors influence N?
   - **e.** In what other situations can we use this model?
   - **f.** What is the $\left.\frac{P}{p_N}\frac{\Delta p_N}{\Delta P}\right|_N$, where $\left(\frac{P}{p_N}\frac{\Delta p_N}{\Delta P}\right)$ is the sensitivity function of the total probability p_N with respect to the probability of a sale P, keeping N constant? You can also do this by simulation as well as analytically.
   - **g.** What is the $\left.\frac{N}{p_N}\frac{\Delta p_N}{\Delta N}\right|_P$? You can do this by simulation as well as analytically.
2. The information diffusion model developed in this chapter assumed that the number of contacts between hearers and ignorants is proportional to the product of hearers and ignorants.
   - **a.** What does this assumption imply in terms of interaction within the population? Will this assumption lead to an overestimate or an underestimate in the number of hearers with respect to time?
   - **b.** What is the significance of k, the conversion coefficient?
   - **c.** Examine the behavior of $h(t)$ around $t = 0$.
3. Consider a population with a large number of individuals (P) where the nourishment is bad, the food supply and individual space are limited, and the population is isolated. Develop a simulation model for the population growth if the birth rate (BR) is constant and the total death rate is given by DR + kP, where k represents a coefficient of resistance to an increase in population. Let DR be constant.

4. Use the information diffusion model or the results of problem 3 to explain how a new process, such as hybrid corn in the 1930s, or a new product, such as television in the 1950s, spreads throughout the population. Explain the eventual slowdown in the rate of the spread (psychological and mathematical reasons). Repeat the simulation of the information diffusion system and compare the results.

5. Develop a stochastic model for our information diffusion process. Let the probability of conversion be k. Simulate the system and compare the results with our model.

6. Using the rectangular integration method, find the area under the curve $y(x) = 2 - x$ over the interval $x = [0, 2]$. Compare the results with the analytic solution. What is the relationship between the error and Δx? Define the error as the analytic value minus the computed value. Let the relative error equal the error divided by the analytic value. What is the relative error for $\Delta x = 0.1$? A computer solution is suggested.

7. A Democratic candidate lives in a Republican district. The Republicans outnumber the Democrats 2 to 1. Using simulation, determine if it is worthwhile for the Democrat to campaign in the district if the following conditions hold:
   - **a.** If the candidate encounters a Republican, he increases the probability of voting for either candidate from 0.5 to 0.6, and the probability of voting Democrat from 0.2 to 0.4.
   - **b.** If the candidate encounters a Democrat, he increases the Democrat's probability of voting for either candidate from 0.4 to 0.7, and the probability of voting Democrat from 0.8 to 0.9.

8. Andy challenges Mark to play the following game. A coin is tossed until either the number of heads is four greater than the number of tails tossed or the number of tails exceeds the number of heads tossed by four. Mark must pay Andy $1 for each toss of the coin, but at the end of the game Andy will pay Mark $9. The game is played until Mark's or Andy's money runs out. Assume Andy has $50 and Mark has $100 initially. By simulation determine who has the better odds of winning. After how many tosses on the average will the money run out for either player?

9. Let p be the probability of an error occurring during the transmission of a character from your terminal to your computer network.
   - **a.** Using simulation, compute the probability that a message of length L characters is transmitted with some error. Let $L = 80$ and $p = 1 \times 10^{-10}$.
   - **b.** Using an analytical method, compute the probability of some error in the message.
   - **c.** Let $L = 40$ and $p = 1 \times 10^{-6}$. Repeat parts **a** and **b**.
   - **d.** How would you determine p?

10. In an ethernet local network, a 10-megabit-per-second data link is used to transmit packets from one node to another. Tests show that the probability of a packet requiring retransmission due to errors is $p = 10^{-6}$. Let us define the effective data rate as the number of data bits transmitted per second. Using simulation, determine the effective data rate of the link when the packet is as shown in Figure 2.8.

FIGURE 2.8 Problem 10

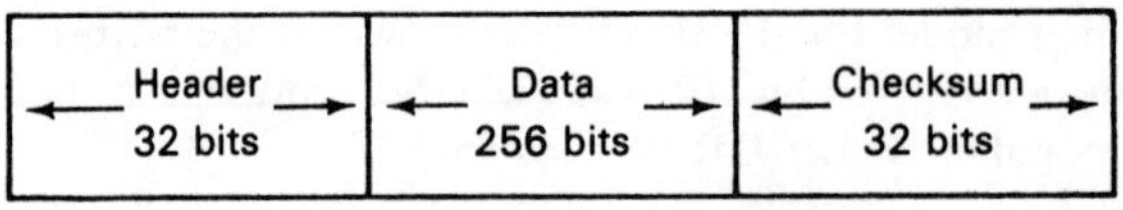

11. Consider the computer network shown in Figure 2.9. The probability of a failure of a link between any two nodes in a 1-hour period is 10^{-8}. A network is defined as disconnected if at least one node is unreachable from all other nodes. Determine the probability that the network will become disconnected within a year. Assume that it takes 24 hours to repair a failed link.

FIGURE 2.9 Problem 11

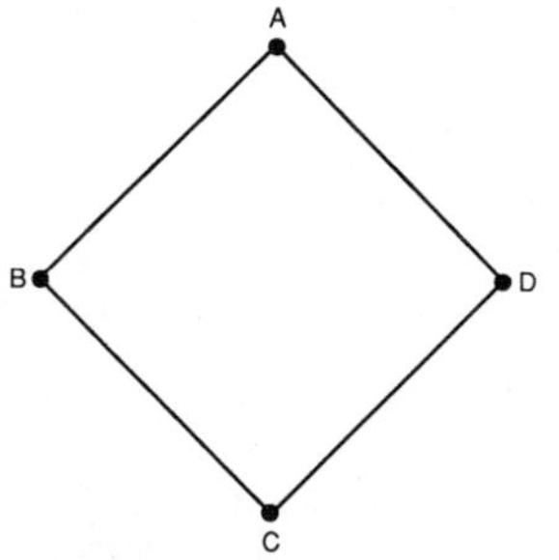

2.7 APPENDIX: SIMULATION OF THE SALESPERSON PROBLEM

APL Sample Program

```
    ∇SALES[□]∇
 ∇ TPR← N SALES P
[1]D1←?(M)ρ1000
[2]D2←?(M)ρ1000
[3]L1←(1000×P)≤D1
[4]TS1←+/L1
[5]TS2←+/(~L1)/(1000×P)≤D2
[6]TOTSALES←TS1+TS2
[7]TPR←TOTSALES÷M
 ∇
    M←1000
    N←2
    P←.5


      N SALES P
0.759
      N SALES P
0.744
      N SALES P
0.748
      N SALES P
0.746
```

continues

APL Sample Program (Continued)

```
      N SALES P
0.735
      N SALES P
0.751
      N←3
      N SALES P
0.765
      N SALES P
0.752
      N SALES P
0.752
```

BASIC Sample Program

```
10 REM *********************
20 REM PROGRAM TO SIMULATE THE PROBABILITY OF AND TOTAL
30 REM NUMBER OF SALES BY A SALESMAN.
40 REM IF NO SALE IS MADE ON FIRST VISIT, A SECOND VISIT IS MADE
50 INPUT: "ENTER NUMBER OF CUSTOMERS   ",N
60 DIM P1(100),P2(100)
70 PRINT " "
80 PRINT "CUSTOMER";TAB(16);"1ST VISIT";TAB(30);"2ND VISIT";TAB(45);"SALES"
90 PRINT"-----------------------------------------------------"
100 TSALE=0
110 REM LOOP FOR DETERMINING INDIVIDUAL AND TOTAL SALES TO CUSTOMERS
120 FOR I=1 TO N
130 P1(I)=RND
140 REM CASE IF P1(I) IS GREATER THAN OR  EQUAL TO 0.5
150 IF P1(I) >= 0.5 THEN
160   P1(I)=1
170   P2(I)=0
180   GOTO 350
190 REM  CASE FOR NO SALE ON FIRST VISIT
200 ELSE
210   P1(I)=0
220 REM USING DIFFERENT RANDOM NUMBER FOR SECOND VISIT
230   P2(I)=RND
240 REM CASE- NO SALE ON FIRST OR SECOND VISIT
250   IF P2(I) < 0.5 THEN
260      P2(I)=0
270      P1(I)=0
280      GOTO 340
290 REM CASE- NO SALE ON FIRST VISIT BUT SALE ON SECOND VISIT
300   ELSE
310      P2(I)=1
320      P1(I)=0
330      GOTO 340
340   ENDIF
350 ENDIF
360 REM FIND TOTAL SALES MADE
370     TSALE=TSALE + P1(I) + P2(I)
380 REM              OUTPUT
390 PRINT I;TAB(20);P1(I);TAB(32);P2(I);TAB(46);TSALE
400 NEXT I
410 PRINT " "
420 PRINT"-----------------------------------------------------"
430 PRINT " "
440 PRINT" "
450 PRINT "STOCHASTIC PROB. OF SALE=",TSALE/N
```

continues

BASIC Sample Program (Continued)

```
460 REM PAVE=PROBABILITY OF A SALE ON ANY VISIT
470 PAVE=.5
480 PRINT" "
490 PRINT" "
500 REM THEORETICAL # OF SALES
510 PTOT=1-(1-PAVE)*(1-PAVE)
520 PRINT "MODEL PROB. OF  SALE  WITH ",N,"PERSONS =",PTOT
530 PRINT" "
540 PRINT" "
550 STOP
560 END
```

FORTRAN Sample Program

```
C
C VARIABLES:
C      P           PROBABILITY OF A SALE IN ONE VISIT
C      N           NUMBER OF VISITS
C      M           NUMBER OF TRIALS
C      R           RANDOM NUMBER (BTWN 0.0 & 1.0)
C      NSALES      COUNT FOR NUMBER OF SALES
C      MCOUNT      COUNT FOR NUMBER OF TRIALS
C      NCOUNT      COUNT FOR NUMBER OF VISITS
C      TPROB       APPROXIMATED TOTAL PROBABILITY OF A SALE
C
      REAL P,N,M,R,NSALES,MCOUNT,NCOUNT,TPROB
C
C INITIALIZE VARIABLES:
      NSALES = 0.0
      MCOUNT = 0.0
C
      PRINT*,'ENTER THE PROBABILITY OF A SALE DURING ONE'
      PRINT*,'VISIT AND THE NUMBER OF TIMES YOU WISH TO'
      PRINT*,'APPROACH A POTENTIAL CUSTOMER.'
      READ*,P,N
      REPEAT 10, WHILE (P.GT.1.0)
           PRINT*,'THE PROBABILITY MUST BE BETWEEN 0 AND 1.0'
           READ*,P
10    CONTINUE
      PRINT*,'ENTER THE NUMBER OF TIMES YOU WISH THIS'
      PRINT*,'EXPERIMENT REPEATED.'
      READ*,M
      REPEAT 20,WHILE (MCOUNT.LT.M)
           NCOUNT = 0
           REPEAT 30,WHILE (NCOUNT.LT.N)
           CALL RANDOM(R)
                IF (R.LE.P) THEN
                     NSALES = NSALES + 1
                     NCOUNT = N
                   ELSE
                     NCOUNT = NCOUNT + 1
                ENDIF
30         CONTINUE
           MCOUNT = MCOUNT + 1
20    CONTINUE
```

continues

FORTRAN Sample Program (Continued)

```
      TPROB = NSALES/M
      PRINT*,'IF THE PROBABILITY OF A SALE PER VISIT IS ',P
      PRINT*,'THEN THE PROBABILITY OF A SALE AFTER ',N
      PRINT*,'VISITS IS APPROXIMATELY ',TPROB
      STOP
      END
```

```
ENTER THE PROBABILITY OF A SALE DURING ONE
VISIT AND THE NUMBER OF TIMES YOU WISH TO
APPROACH A POTENTIAL CUSTOMER.
?.5,2

ENTER THE NUMBER OF TIMES YOU WISH THIS
EXPERIMENT REPEATED.
?1000

IF THE PROBABILITY OF A SALE PER VISIT IS  .5000000
THEN THE PROBABILITY OF A SALE AFTER  2.000000
VISITS IS APPROXIMATELY  .7280000
 *STOP*

ENTER THE PROBABILITY OF A SALE DURING ONE
VISIT AND THE NUMBER OF TIMES YOU WISH TO
APPROACH A POTENTIAL CUSTOMER.
?.3,2

ENTER THE NUMBER OF TIMES YOU WISH THIS
EXPERIMENT REPEATED.
?1000

IF THE PROBABILITY OF A SALE PER VISIT IS  .3000000
THEN THE PROBABILITY OF A SALE AFTER  2.000000
VISITS IS APPROXIMATELY  .5050000
 *STOP*

ENTER THE PROBABILITY OF A SALE DURING ONE
VISIT AND THE NUMBER OF TIMES YOU WISH TO
APPROACH A POTENTIAL CUSTOMER.
?.75
?1

ENTER THE NUMBER OF TIMES YOU WISH THIS
EXPERIMENT REPEATED.
?1000

IF THE PROBABILITY OF A SALE PER VISIT IS  .7500000
THEN THE PROBABILITY OF A SALE AFTER  1.000000
VISITS IS APPROXIMATELY  .7380000
 *STOP*
```

SLAM Sample Program

```
LIMITS,2,2,500;
NETWORK;
      CREATE, 1,,, 1000, 1;   INITIATE A VISIT
      ACTIVITY,, 0.5, SALE;   VISIT #1-PROBABILITY OF SALE=0.5
      ACTIVITY,, 0.5, NSALE;           -PROBABILITY OF NO SALE=0.5
 SALE GOON,1;
      ACTIVITY/1,,,FINAL;     NUMBER OF ENTITIES THAT GO THROUGH
;                             ACTIVITY 1 IS THE TOTAL NUMBER OF SALES
NSALE GOON, 1;
      ACTIVITY,, 0.5, SALE;   VISIT #2 (IF VISIT #1 WAS A NO SALE)
;                                      -PROBABILITY OF A SALE=0.5
      ACTIVITY/2,, 0.5;                -PROBABILITY OF NO SALE=0.5
;                             NUMBER OF ENTITIES THAT GO THROUGH
;                             ACTIVITY 2 IS THE TOTAL NUMBER OF
;                             NO SALES
      TERMINATE;
;
FINAL TERMINATE;
      ENDNETWORK;
INITIALIZE, 0, 1000;
SIMULATE;
FIN;
```

LOTUS® 1 2 3 Sample Program

```
 MCOUNT VISIT #1             VISIT #2            # OF SALES
************************************************************
      1 @IF(@RAND<=0.5,0,1) @IF(B3=1," ",E3)    +F3       @IF(@RAND<=0.5,0,1) @IF(B3+E3>=1,1,0)
      2 @IF(@RAND<=0.5,0,1) @IF(B4=1," ",E4)    +D3+F4    @IF(@RAND<=0.5,0,1) @IF(B4+E4>=1,1,0)
      3 @IF(@RAND<=0.5,0,1) @IF(B5=1," ",E5)    +D4+F5    @IF(@RAND<=0.5,0,1) @IF(B5+E5>=1,1,0)
      4 @IF(@RAND<=0.5,0,1) @IF(B6=1," ",E6)    +D5+F6    @IF(@RAND<=0.5,0,1) @IF(B6+E6>=1,1,0)
      5 @IF(@RAND<=0.5,0,1) @IF(B7=1," ",E7)    +D6+F7    @IF(@RAND<=0.5,0,1) @IF(B7+E7>=1,1,0)
      6 @IF(@RAND<=0.5,0,1) @IF(B8=1," ",E8)    +D7+F8    @IF(@RAND<=0.5,0,1) @IF(B8+E8>=1,1,0)
      7 @IF(@RAND<=0.5,0,1) @IF(B9=1," ",E9)    +D8+F9    @IF(@RAND<=0.5,0,1) @IF(B9+E9>=1,1,0)
      8 @IF(@RAND<=0.5,0,1) @IF(B10=1," ",E10)  +D9+F10   @IF(@RAND<=0.5,0,1) @IF(B10+E10>=1,1,0)
      9 @IF(@RAND<=0.5,0,1) @IF(B11=1," ",E11)  +D10+F11  @IF(@RAND<=0.5,0,1) @IF(B11+E11>=1,1,0)
     10 @IF(@RAND<=0.5,0,1) @IF(B12=1," ",E12)  +D11+F12  @IF(@RAND<=0.5,0,1) @IF(B12+E12>=1,1,0)
------------------------------------------------------------
PROB. OF A SALE =           +D12/A12
------------------------------------------------------------
```

```
 MCOUNT VISIT #1 VISIT #2 # OF SALES
****************************************
      1        1                    1
      2        1                    2
      3        1                    3
      4        0        0           3
      5        0        1           4
      6        0        0           4
      7        0        1           5
      8        1                    6
      9        0        1           7
     10        1                    8
----------------------------------------
PROB. OF A SALE =      0.8
----------------------------------------
```

GPSS Sample Program

```
             GPSS SIMULATION FOR THE SALESPERSON PROBLEM
      SIMULATE
      GENERATE    1                      CREATE CUSTOMERS
      ADVANCE     1
      TRANSFER    .500,AGAIN,SALE
AGAIN TRANSFER    .500,NSALE,SALE
SALE  TERMINATE   1
NSALE TERMINATE   1
      START       1000                   SIMULATE FOR 1000 CUSTOMERS
      END
```

LISP Sample Program for the Connection Machine®

```
;;;;;;
;;;;;; Simulation of Salesperson Problem
;;;;;; Lisp program for The Connection Machine from
;;;;;;;   Thinking Machines Corporation
;;;;;; In this example 10000 simulations are done in parallel
;;;;;;
;;;;;;
(*defun sales-person ()
;;;  Check if a sale occurred  on the first visit and return a 1
;;;
  (if!! (<!! (random!! (!! 100)) (!! 50))
        (!! 1)
;;;;;; Check if a sale occurred on the second visit
        (if!! (<!! (random!! (!! 100)) (!! 50))
              (!! 1)
              (!! 0))))
;;;;;;
;;;;;;
;;;;;;
;;;;;; Compute the sum of sales
(defun compute-sum ()
  (cm:time
    (*all
      (*when (<!! (self-address!!)(!! 10000))
        (*sum (sales-person))))))
;;;;;;;;;;;
;;;;;;;;;;;
;;;;;;;;;;; Sample simulations
Command: (compute-sum)
Evaluation of (*ALL
                (*WHEN (<!! (SELF-ADDRESS!!) (!! 10000))
                  (*SUM (SALES-PERSON)))) took 0.051878 seconds of elapsed time,

during which the CM was active for 0.004859 seconds or 9.37% of the total elapse
d time.
7493
Command: (compute-sum)
Evaluation of (*ALL
                (*WHEN (<!! (SELF-ADDRESS!!) (!! 10000))
                  (*SUM (SALES-PERSON)))) took 0.036953 seconds of elapsed time,

during which the CM was active for 0.004843 seconds or 13.11% of the total elaps
ed time.
7498
Command: (compute-sum)
Evaluation of (*ALL
                (*WHEN (<!! (SELF-ADDRESS!!) (!! 10000))
                  (*SUM (SALES-PERSON)))) took 0.036679 seconds of elapsed time,

during which the CM was active for 0.004845 seconds or 13.21% of the total elaps
ed time.
7520
```

2.8 APPENDIX: SIMULATION OF THE INFORMATION DIFFUSION PROBLEM

ACSL Sample Program

ACSL file

```
PROGRAM SPREAD OF RUMOR SIMULATION
CONSTANT K=0.9 , TSTOP = 1.0 , HO = 1.0 , N = 100
CINTERVAL CINT = 0.01
NSTEPS NSTEP = 10
   DHDT = K * HT * (N-HT)
         HT = INTEG(DHDT,HO)
         IT = N-HT
TERMT(T.GE.TSTOP)
END
```

Data file

```
SET TITLE ="DIFFUSION PROCESS IN A POPULATION SAMPLE (K=.9,.5,.1)"
PREPAR T,HT
OUTPUT 'NCIOUT'=1,T,HT
START
RANGE 'ALL'
PLOT 'XAXIS'=T,HT,...
      'HI'=100.,'LO'=0.,'TAG'='HEARERS',...
      'XHI'=0.3,'XLO'=0.,'XTAG'='TIME'
'CHANGE VALUE OF K'
SET K=.5
START
RANGE 'ALL'
PLOT 'XAXIS'=T,HT,...
      'HI'=100.,'LO'=0.,'TAG'='HEARERS',...
      'XHI'=0.3,'XLO'=0.,'XTAG'='TIME'
'CHANGE VALUE OF K'
SET K=.1
START
RANGE 'ALL'
PLOT 'XAXIS'=T,HT,...
      'HI'=100.,'LO'=0.,'TAG'='HEARERS',...
      'XHI'=0.3,'XLO'=0.,'XTAG'='TIME'
STOP
```

LOTUS® 1 2 3 Sample Program

```
DIFFUSION OF INFORMATION IN A POPULATION SAMPLE
=====================================================
   N=100        H(0)=1        K=0.5     DELTA T=0.01
=====================================================
   Time          h              i        Del h(del t)
*****************************************************
0            1             99           0
+A7+0.01     +B7+D8        +C7-D8       +B7*C7*0.01*0.5
+A8+0.01     +B8+D9        +C8-D9       +B8*C8*0.01*0.5
+A9+0.01     +B9+D10       +C9-D10      +B9*C9*0.01*0.5
+A10+0.01    +B10+D11      +C10-D11     +B10*C10*0.01*0.5
```

continues

LOTUS 1 2 3 Program (Continued)

```
DIFFUSION OF INFORMATION IN A POPULATION SAMPLE
=====================================================
   N=100        H(0)=1        K=0.5      DELTA T=0.01
=====================================================
    Time          h             i        Del h(del t)
*****************************************************
+A11+0.01    +B11+D12     +C11-D12     +B11*C11*0.01*0.5
+A12+0.01    +B12+D13     +C12-D13     +B12*C12*0.01*0.5
+A13+0.01    +B13+D14     +C13-D14     +B13*C13*0.01*0.5
+A14+0.01    +B14+D15     +C14-D15     +B14*C14*0.01*0.5
+A15+0.01    +B15+D16     +C15-D16     +B15*C15*0.01*0.5
+A16+0.01    +B16+D17     +C16-D17     +B16*C16*0.01*0.5
+A17+0.01    +B17+D18     +C17-D18     +B17*C17*0.01*0.5
+A18+0.01    +B18+D19     +C18-D19     +B18*C18*0.01*0.5
+A19+0.01    +B19+D20     +C19-D20     +B19*C19*0.01*0.5
+A20+0.01    +B20+D21     +C20-D21     +B20*C20*0.01*0.5
+A21+0.01    +B21+D22     +C21-D22     +B21*C21*0.01*0.5
+A22+0.01    +B22+D23     +C22-D23     +B22*C22*0.01*0.5
+A23+0.01    +B23+D24     +C23-D24     +B23*C23*0.01*0.5
+A24+0.01    +B24+D25     +C24-D25     +B24*C24*0.01*0.5
+A25+0.01    +B25+D26     +C25-D26     +B25*C25*0.01*0.5
+A26+0.01    +B26+D27     +C26-D27     +B26*C26*0.01*0.5
+A27+0.01    +B27+D28     +C27-D28     +B27*C27*0.01*0.5
+A28+0.01    +B28+D29     +C28-D29     +B28*C28*0.01*0.5
+A29+0.01    +B29+D30     +C29-D30     +B29*C29*0.01*0.5
+A30+0.01    +B30+D31     +C30-D31     +B30*C30*0.01*0.5
+A31+0.01    +B31+D32     +C31-D32     +B31*C31*0.01*0.5
+A32+0.01    +B32+D33     +C32-D33     +B32*C32*0.01*0.5
+A33+0.01    +B33+D34     +C33-D34     +B33*C33*0.01*0.5
+A34+0.01    +B34+D35     +C34-D35     +B34*C34*0.01*0.5
+A35+0.01    +B35+D36     +C35-D36     +B35*C35*0.01*0.5
+A36+0.01    +B36+D37     +C36-D37     +B36*C36*0.01*0.5
+A37+0.01    +B37+D38     +C37-D38     +B37*C37*0.01*0.5
+A38+0.01    +B38+D39     +C38-D39     +B38*C38*0.01*0.5
+A39+0.01    +B39+D40     +C39-D40     +B39*C39*0.01*0.5
+A40+0.01    +B40+D41     +C40-D41     +B40*C40*0.01*0.5
+A41+0.01    +B41+D42     +C41-D42     +B41*C41*0.01*0.5
+A42+0.01    +B42+D43     +C42-D43     +B42*C42*0.01*0.5
+A43+0.01    +B43+D44     +C43-D44     +B43*C43*0.01*0.5
+A44+0.01    +B44+D45     +C44-D45     +B44*C44*0.01*0.5
+A45+0.01    +B45+D46     +C45-D46     +B45*C45*0.01*0.5
+A46+0.01    +B46+D47     +C46-D47     +B46*C46*0.01*0.5
*****************************************************
```

```
DIFFUSION OF INFORMATION IN A POPULATION SAMPLE
================================================
   N=100       H(0)=1        K=0.5    DELTA T=0.01
================================================
    Time         h             i       Del h(del t)
************************************************
   0.00000     1.00000     99.00000      0.00000
   0.01000     1.49500     98.50500      0.49500
```

continues

LOTUS 1 2 3 Program (Continued)

DIFFUSION OF INFORMATION IN A POPULATION SAMPLE

N=100 H(0)=1 K=0.5 DELTA T=0.01

Time	h	i	Del h(del t)
0.02000	2.23132	97.76868	0.73632
0.03000	3.32209	96.67791	1.09077
0.04000	4.92796	95.07204	1.60587
0.05000	7.27051	92.72949	2.34256
0.06000	10.64147	89.35853	3.37095
0.07000	15.39600	84.60400	4.75453
0.08000	21.90881	78.09119	6.51282
0.09000	30.46324	69.53676	8.55443
0.10000	41.05482	58.94518	10.59158
0.11000	53.15473	46.84527	12.09992
0.12000	65.60497	34.39503	12.45024
0.13000	76.88740	23.11260	11.28242
0.14000	85.77274	14.22726	8.88534
0.15000	91.87429	8.12571	6.10156
0.16000	95.60701	4.39299	3.73272
0.17000	97.70701	2.29299	2.10000
0.18000	98.82722	1.17278	1.12020
0.19000	99.40673	0.59327	0.57951
0.20000	99.70161	0.29839	0.29487
0.21000	99.85036	0.14964	0.14875
0.22000	99.92507	0.07493	0.07471
0.23000	99.96251	0.03749	0.03744
0.24000	99.98125	0.01875	0.01874
0.25000	99.99062	0.00938	0.00938
0.26000	99.99531	0.00469	0.00469
0.27000	99.99765	0.00235	0.00234
0.28000	99.99883	0.00117	0.00117
0.29000	99.99941	0.00059	0.00059
0.30000	99.99971	0.00029	0.00029
0.31000	99.99985	0.00015	0.00015
0.32000	99.99993	0.00007	0.00007
0.33000	99.99996	0.00004	0.00004
0.34000	99.99998	0.00002	0.00002
0.35000	99.99999	0.00001	0.00001
0.36000	100.00000	0.00000	0.00000
0.37000	100.00000	0.00000	0.00000
0.38000	100.00000	0.00000	0.00000
0.39000	100.00000	0.00000	0.00000
0.40000	100.00000	0.00000	0.00000

continues

LOTUS 1 2 3 Program (Continued)

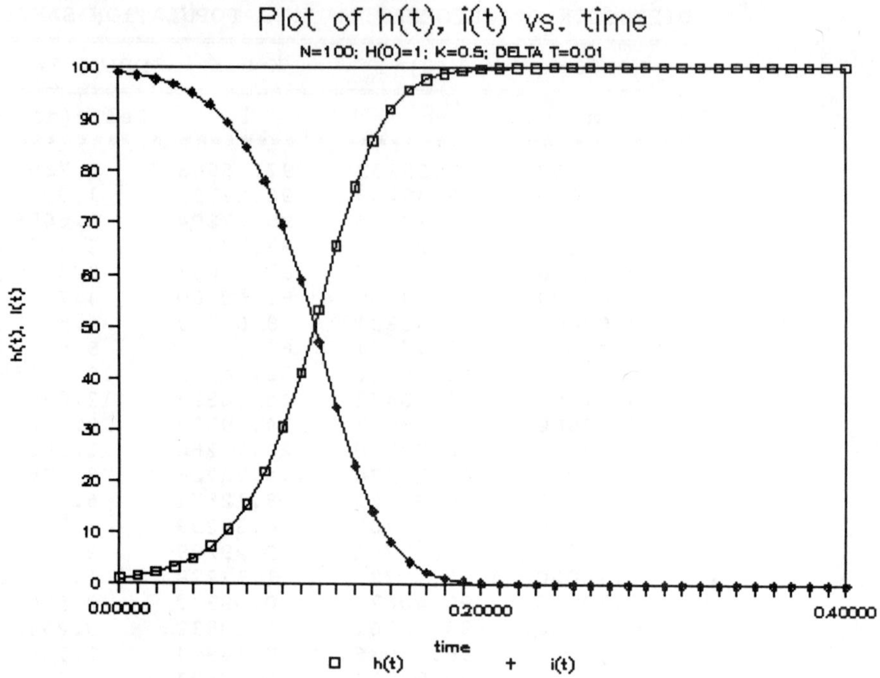

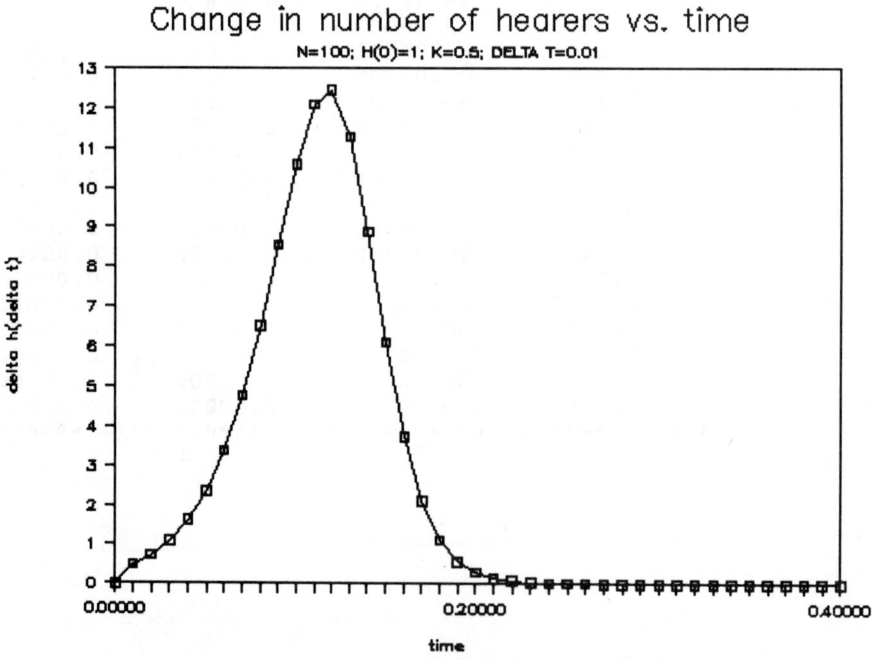

Pascal Sample Program

```
PROGRAM Rumor (INPUT,OUTPUT);
VAR
    Pop,Hearers,PerH,PerI,TotTime,ClockT,ConvCoef,TPT,DelT,DeltPrint:REAL;
BEGIN
    ClockT:=0;
    WRITELN('Please input the total population and initial number of hearers.');
    READLN(Pop,Hearers);
    WHILE (Hearers >= Pop) DO
        BEGIN
            WRITELN('The total population must be larger than');
            WRITELN('the initial number of hearers.');
            WRITELN('Please try again.');
            READLN(Pop,Hearers);
        END;
    WRITELN('Now input the length of time you want this experiment to go on,');
    WRITELN('the coefficient of conversion, DelT and DeltPrint.');
    READLN(TotTime,ConvCoef,DelT,DeltPrint);
    WHILE(ConvCoef < 0) DO
        BEGIN
            WRITELN('The coefficient must be greater than 0.');
            WRITELN('Please try again.');
            READLN(ConvCoef);
        END;
    WRITELN('Time            %Hearers           %Ignorants');
    TPT:=-DelT/2;       {Initialize Time To Print}
    REPEAT
        PerH:=Hearers/Pop*100;
        PerI:=100-PerH;
        IF (ClockT>=TPT) THEN
            BEGIN
                WRITELN(ClockT:4:2,'                 ',PerH:4:2,'                    ',PerI
:4:2);
                TPT:=TPT+DeltPrint;
            END;
        Hearers:=Hearers+ConvCoef*Hearers*(Pop-Hearers)*DelT;
        ClockT:=ClockT+DelT;
    UNTIL (ClockT > TotTime);
END.
```

2.9 APPENDIX: SIMULATION OF THE SUPPLY AND DEMAND SYSTEM PROBLEM

CSMP Sample Program

```
TITLE SUPPLY AND DEMAND SIMULATION
* DPDT IS THE RATE OF CHANGE OF THE PRICE
INCON  IC=0
DPDT=A*(40.-2,*P)
P=INTGRL(IC.DPDT)
S=10.+P
D=50.-P
PARAMETER A=.75
TIMER   FINTIM=10.,OUTDEL=0.1,PRDEL=0.1,DELT=0.1
LABEL    SUPPLY AND DEMAND SIMULATION
PRTPLOT         D,S,P
END
STOP
```

DYNAMO Sample Program

```
*       SUPPLY AND DEMAND SIMULATION
N       TIME=0
N       P=0    INITIAL VALUE OF PRICE
L       P.K=P.J+DT*RP.JK
R       RP.KL=A*(40.-2*P.K)
S       D.K=50.-P.K
S       S.K=10.+P.K
C       A=0.5
PLOT    D=D,S=S,P=P
PRINT   D,S,P
SPEC    DT=.01,LENGTH=10/PRTPER=0.1/PLTPER=0.1
RUN     1
```

2.10 REFERENCES

1. Aburdene, Maurice, "A Diffusion of Rumor Example Demonstrating the Characteristics of Computer Simulation," *COED Journal* 2, *no*. 2 (March–April 1983): 3–6.

2. Vaswani, Ram, "An Appreciation of Computer Simulation," in *Winter Simulation Conference*, vol. 2 (Jan. 14–16, 1974): 593–99.

3 System Description

3.1 INTRODUCTION

This chapter introduces the basic vocabulary and notation used to quantitatively and qualitatively describe systems. Several criteria used to classify systems are also considered. The description and classification of systems will help us in the model construction stage of the simulation process.

We saw in Chapter 1 that a system can be defined as a set of objects and the relationships among them. A system can be either real or conceptual. Real systems are natural systems in which the objects perceived or inferred from observations exist independently of the observer. Examples are the solar system, the weather, and forest growth. Conceptual systems are symbolic constructs (logical or mathematical) corresponding to reality. Examples are the salesperson, information diffusion, and supply and demand systems described in Chapter 2.

A system can be defined by a set of m inputs $[u_1, u_2, \ldots, u_m]$; n internal states, levels, or conditions $[x_1, x_2, \ldots, x_n]$; and l outputs $[y_1, y_2, \ldots, y_l]$. With some systems, it is difficult or impossible to know anything about the internal states. Systems of this type are discussed later in this chapter.

The system's boundary must first be defined to determine whether an entity should be classified as a system input, an internal state, or a system output. Sometimes a system's boundary is obvious and needs no elaboration. For example, when we simulate the heating system of a house, the boundaries are the house's interior walls and ceiling. For other systems, it can be a tricky and challenging task to define the system's boundary, and then to determine what should be considered inputs, internal states, or outputs. An example is the formation of the galaxies and their origins. These issues constitute the "art" aspect of the simulation process. The block diagram in Figure 3.1 illustrates the concept of a system.

3.2 SYSTEM DESCRIPTIONS

There are two principal ways to describe systems mathematically: by external descriptions and by internal descriptions. An *external description* of a system is basically a functional representation where the system itself is considered to be a

FIGURE 3.1 Block Diagram of a System

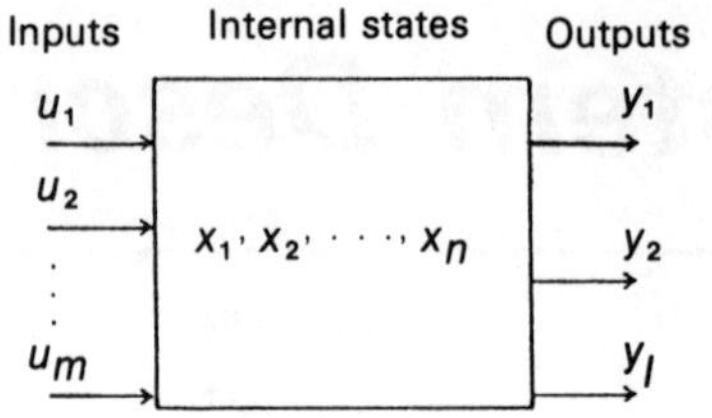

"black box" and the system description is given in terms of a causal relationship between inputs $(u_1, u_2, \ldots, u_m)$ and outputs $(y_1, y_2, \ldots, y_l)$. External descriptions are often used in economics and engineering to relate outputs to inputs. The mathematical form of this representation for continuous systems is

$$\begin{aligned} y_1 &= f_1(u_1, u_2, \ldots, u_m, t) \\ y_2 &= f_2(u_1, u_2, \ldots, u_m, t) \\ &\vdots \\ y_l &= f_l(u_1, u_2, \ldots, u_m, t) \end{aligned} \tag{3.1}$$

where t denotes time. A similar representation can be developed for discrete systems. Now let us look at some examples.

EXAMPLE 3.1 **RC Circuit** Consider the resistor-capacitor (RC) circuit in Figure 3.2. The external description for this continuous system is given by

$$y_1(t) = v_o(t) = f(V_i, t) = V_i(1 - e^{-t/RC}) \tag{3.2}$$

where V_i = the input voltage to the system
v_o = the output voltage of the system

This applies only when $v_o(0) = 0$ and V_i is a constant.

FIGURE 3.2 RC Circuit

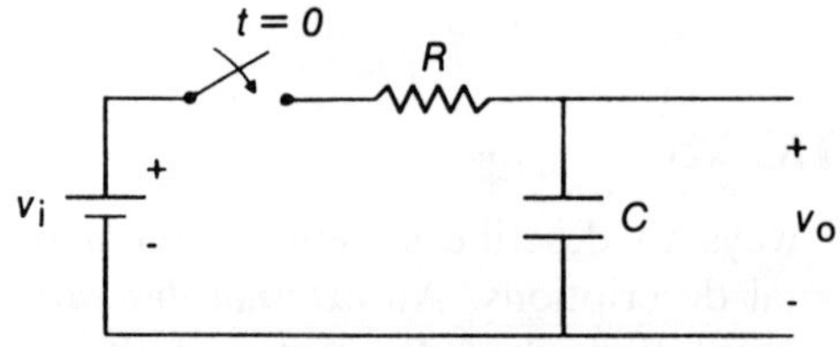

EXAMPLE 3.2 **Securities Market** This example represents an external description of a discrete system. A securities market is described by

$$R_{it} = \alpha_i + \beta_i R_{mt} + e_{it} \qquad (t = 1, 2, \ldots, T) \tag{3.3}$$

where R_{it} = the return on security i for period t

R_{mt} = the return on the market for period t

α_i and β_i are usually estimated for security i

e_{it} is the error and is usually random

An *internal description* is a structural representation that defines the relationship among the inputs to a system, the system's internal states, and outputs from the system. It is usually a good way to describe continuous-time and discrete-time systems. We define a continuous-time system as one whose internal state(s) can change values at any instant in time, and a discrete-time system as one whose internal states can change only at specific instants of time.

For a continuous-time system, the internal description can often be given by the following differential equations:

$$\begin{aligned}
\dot{x}_1(t) &= f_1(x_1, x_2, \ldots, x_n, u_1, u_2, \ldots, u_m, t) \\
\dot{x}_2(t) &= f_2(x_1, x_2, \ldots, x_n, u_1, u_2, \ldots, u_m, t) \\
&\vdots \\
\dot{x}_n(t) &= f_n(x_1, x_2, \ldots, x_n, u_1, u_2, \ldots, u_m, t) \\
y_1(t) &= g_1(x_1, x_2, \ldots, x_n, u_1, u_2, \ldots, u_m, t) \\
y_2(t) &= g_2(x_1, x_2, \ldots, x_n, u_1, u_2, \ldots, u_m, t) \\
&\vdots \\
y_l(t) &= g_l(x_1, x_2, \ldots, x_n, u_1, u_2, \ldots, u_m, t)
\end{aligned} \tag{3.4}$$

Together, equations (3.4) constitute a *state variable description* of a system.

For the first set of equations (3.4), note that the current value of any internal state can be a function of all states, inputs, and time. Likewise, the outputs of the system are also functions of the states of the system, inputs, and time.

The set of variables $(\dot{x}_1(t), \dot{x}_2(t), \ldots, \dot{x}_n(t))$ on the left side of the equations represents the time rate of change of states (conditions or levels) of the system. The notation used is called dot notation and is often abbreviated as follows:

$$\begin{aligned}
\dot{x}_1(t) &= \frac{dx_1(t)}{dt} \\
\dot{x}_2(t) &= \frac{dx_2(t)}{dt} \\
&\vdots \\
\dot{x}_n(t) &= \frac{dx_n(t)}{dt}
\end{aligned} \tag{3.5}$$

The right side of equations (3.5) corresponds to the relationships $(f_1, f_2, \ldots, f_n)$ of all states of the system $(x_1, x_2, \ldots, x_n)$, of all inputs $(u_1, u_2, \ldots, u_m)$, and of time. Remember that the above discussion applies to systems whose behavior is continuous in time. That is, the state of the system can change at any instant of time.

For example, in Chapter 2 the internal description of the information diffusion system was given as

$$\frac{dh(t)}{dt} = kNh(t) - kh^2(t) \tag{3.6}$$

If we let the internal state $x_1(t) = h(t)$, then

$$\frac{dx_1(t)}{dt} = \dot{x}_1 = \frac{dh(t)}{dt} \tag{3.7}$$

and equation (3.7), the state variable description, becomes

$$\frac{dx_1(t)}{dt} = k(Nx_1(t) - x_1^2(t)) \tag{3.8}$$

The relationship f_1 is given by

$$f_1(x_1, t) = k(Nx_1(t) - x_1^2(t)) \tag{3.9}$$

If we consider the output as the number of hearers at any time t, then

$$y_1(t) = x_1(t) \tag{3.10}$$

and $$g_1(x_1(t), t) = x_1(t) \tag{3.11}$$

Notice that this system does not have any external inputs. The dynamics of the system depend only on the internal state $x_1(t)$ and its initial value $x_1(0)$.

Now we will consider other examples of state variable system descriptions. These are lumped models of systems and are intended to help you develop skills for the model construction stage of the simulation process.

EXAMPLE 3.3 **Modelling of a Greenhouse** Greenhouses can increase crop yield by a factor of 10 compared with field-grown plants. To obtain this increase, temperature and other factors must be controlled. Controlling the temperature of the greenhouse using traditional heating systems becomes economically prohibitive. However, heat can be supplied to the greenhouse by allowing heated water from a nearby power plant to flow through pipes in a concrete floor, which will decrease heating costs. The rate of heat flow, Q, is controlled by opening and closing valves at the entrance to the greenhouse.

Let us develop a simulation model to help predict the temperature of the greenhouse, T_g. If we assume that temperature T_g is uniform in the greenhouse (accomplished by proper installation of water pipes and air-circulating fans) and that the temperature outside (ambient temperature) is also uniform, at T_a, then we can greatly simplify our modelling process. For this real system, we will consider Q as the input variable that can be controlled and T_g as the output variable to be controlled. The objective of the simulation is to determine T_g when Q and T_a are known.

The important fundamental physical relationship here is that the total rate of change of heat inside the greenhouse equals the rate of heat supplied by the hot water pipes, Q, minus the rate at which heat is transferred to the outside. The specific heat of an object is given by

$$C = \frac{\Delta H}{\Delta T} \tag{3.12}$$

where ΔH is the amount of heat that must be supplied to change the temperature by ΔT. The heat balance relationship leads to

$$\frac{d}{dt}(\Delta H) = Q - K(T_g - T_a) \tag{3.13}$$

where $d/dt\,(\Delta H)$ is the total rate of change of heat inside the greenhouse. Remember that Q is the rate of heat supplied by the water pipes and that $K(T_g - T_a)$ is the rate of heat transferred to the outside from the greenhouse due to the temperature difference. K is a constant that depends on the construction and contents of the greenhouse. For example, a better insulated greenhouse has a lower value of K than a less well insulated structure.

From equations (3.12) and (3.13), we obtain

$$C\frac{d}{dt}(T_g - T_a) = -K(T_g - T_a) + Q \tag{3.14}$$

which leads to

$$\frac{dT_g}{dt} = -\frac{K}{C}(T_g - T_a) + \frac{1}{C}Q \tag{3.15}$$

Let $a = K/C$ and $b = 1/C$. Then

$$\frac{dT_g}{dt} = -a(T_g - T_a) + bQ \tag{3.16}$$

If we let $T = T_g - T_a$ and if T_a is constant, then we obtain

$$\frac{dT}{dt} = -aT + bQ \tag{3.17}$$

and

$$T_g = T + T_a \tag{3.18}$$

as our model. The input to the system is Q, the internal state is T, and T_g is the output.

This model can be simulated using the following equation:

$$T(t + \Delta t) = T(t) + \Delta T(t) \tag{3.19}$$

which leads to

$$T(t + \Delta t) = T(t) + (-aT(t) + bQ(t))\Delta t \tag{3.20}$$

$$T_g(t) = T(t) + T_a \tag{3.21}$$

If Q is constant, heat is supplied to the greenhouse at a constant rate Q_0.

If we let $K = 0.005$ and $C = 0.01$, then $a = 0.5$ and $b = 10.0$. Using $\Delta t = 0.1$, $Q = 1$, and $T(0) = -5.0$, we obtain

$$T(t + \Delta t) = T(t) + (-0.5T(t) + 10.0)\Delta t \tag{3.22}$$

The results of this simulation are shown in Table 3.1, and plotted in Figure 3.3.

The analytic solution for this model is given by

$$T_g(t) = T_a + (T_g(0) - T_a)e^{-at} + Q_0 \frac{b}{a}(1 - e^{-at}) \tag{3.23}$$

or

$$T_g(t) = T_a + (T_g(0) - T_a)e^{-(K/C)t} + \frac{Q_0}{K}(1 - e^{-(K/C)t}) \tag{3.24}$$

As t approaches infinity,

$$T_g = T_a + Q_0 \frac{b}{a} = T_a + \frac{Q_0}{K} \tag{3.25}$$

That is, the final temperature or equilibrium temperature of the greenhouse exceeds the ambient temperature by Q_0/K.

Remember, the model presented is simplified and many of our assumptions are subject to criticism. Which of our assumptions would you criticize as being unrealistic or inappropriate?

TABLE 3.1 Temperatures and Times for Greenhouse Simulation

t	$T(t)$	T_g
0	−5°	55°
0.1	−3.75	56.25
0.2	−2.57	57.43
0.3	−1.43	58.57
0.4	−0.36	59.64
0.5	0.66	60.66
0.6	1.62	61.62
0.7	2.54	62.54
0.8	3.41	63.41
0.9	4.24	64.24
1.0	5.03	65.03
1.5	8.42	68.42
2.0	11.04	71.04
10.0	19.85	79.85

FIGURE 3.3 Plot of Temperature vs. Time for the Greenhouse Simulation, $T_a = 60°$

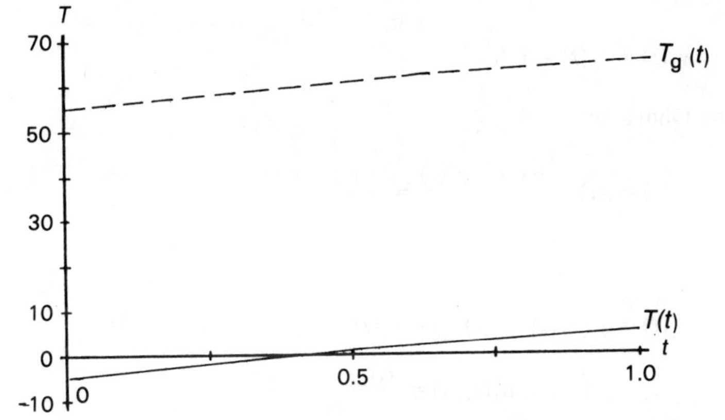

EXAMPLE 3.4 **Electrical Circuit** Let us develop a state variable description of the electrical circuit shown in Figure 3.4. Assume that the state is the voltage v_o.

Applying Kirchhoff's current law at node A, we obtain the equation

$$C\frac{dv_o(t)}{dt} + \frac{v_o(t) - v_i(t)}{R} = 0 \tag{3.26}$$

Rearranging, we get

$$\frac{dv_o(t)}{dt} + \frac{1}{RC}v_o(t) = \frac{1}{RC}V_i(t) \tag{3.27}$$

If we let $x_1(t) = v_o(t)$, $\alpha = 1/RC$, and $u_1(t) = v_i(t)$, then we have

$$\frac{dx_1(t)}{dt} = -\alpha x_1(t) + \alpha u_1(t) \tag{3.28}$$

and this yields

$$\dot{x}_1(t) = f(x(t), u_1(t), t) = -\alpha x_1(t) + \alpha u_1(t) \tag{3.29}$$

FIGURE 3.4

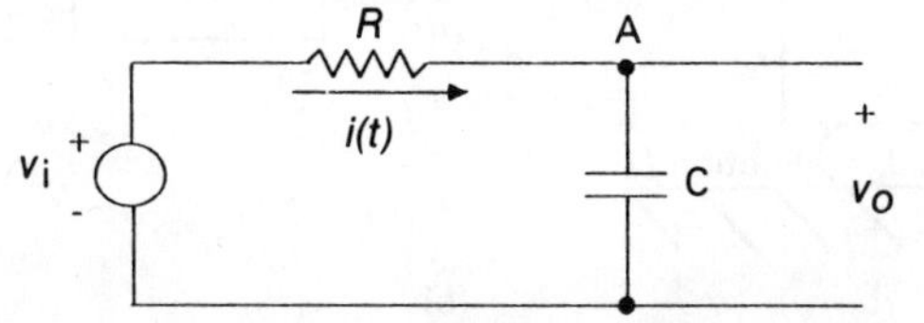

If we consider voltage $v_o(t)$ and current $i(t)$ to be the variables of interest and also the outputs, then

$$y_1(t) = v_o(t) = x_1(t)$$

And, by Ohm's law,

$$y_2(t) = i(t) = \frac{v_i(t) - v_o(t)}{R} = \frac{u_1(t) - x_1(t)}{R} \tag{3.30}$$

This results in

$$y_1(t) = g_1(x_1(t), u_1(t), t) = x_1(t) \tag{3.31}$$

and $$y_2(t) = g_2(x_1(t), u_1(t), t) = \frac{u_1(t) - x_1(t)}{R} \tag{3.32}$$

EXAMPLE 3.5 **Mechanical System** Let us develop a model for the mechanical system shown in Figure 3.5. Assume the output variables of interest to be position $x(t)$ and velocity $v(t)$ of mass M. Let K and D represent the spring and friction coefficient, respectively. Consider $u(t)$ as the input force applied to the system.

When we apply Newton's second law to the force diagram in Figure 3.5(b), we obtain

$$M\frac{dv(t)}{dt} + Dv(t) + Kx(t) = u(t) \tag{3.33}$$

But

$$v(t) = \frac{dx(t)}{dt}$$

and $$\frac{dv(t)}{dt} = \frac{d^2x(t)}{dt^2} \tag{3.34}$$

FIGURE 3.5 (a) Mechanical System (b) Free-Body Diagram

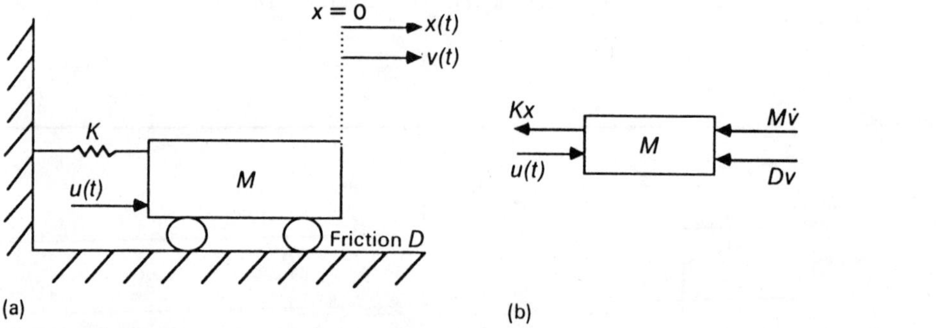

Equation (3.33) therefore becomes

$$M \frac{d^2x(t)}{dt^2} + D \frac{dx(t)}{dt} + Kx(t) = u(t) \tag{3.35}$$

Differential equation (3.35) is a description of the system and can be put in state variable form. Let $x_1(t) = x(t)$ and

$$x_2(t) = v(t) = \frac{dx(t)}{dt} = \frac{dx_1(t)}{dt} \tag{3.36}$$

From equation (3.36), we have $\dot{x}_1(t) = x_2(t)$ and

$$\dot{x}_2(t) = \frac{d^2x_1(t)}{dt^2} = \frac{d^2x(t)}{dt^2} \tag{3.37}$$

Solving equation (3.35) leads to

$$\dot{x}_2(t) = \frac{d^2x(t)}{dt^2} = -\frac{K}{M} x_1(t) - \frac{D}{M} x_2(t) + \frac{1}{M} u(t) \tag{3.38}$$

Since we are interested in $x(t)$ and $v(t)$ as our outputs, we note that

$$y_1(t) = x_1(t)$$
$$y_2(t) = x_2(t)$$

Collecting our results and letting $u_1(t) = u(t)$, we obtain

$$\begin{aligned} \dot{x}_1(t) &= f_1(x_1(t), x_2(t), u_1(t), t) = x_2(t) \\ \dot{x}_2(t) &= f_2(x_1(t), x_2(t), u_1(t), t) = -\frac{K}{M} x_1(t) - \frac{D}{M} x_2(t) + \frac{1}{M} u_1(t) \\ y_1(t) &= g_1(x_1(t), x_2(t), u_1(t), t) = x_1(t) \\ y_2(t) &= g_2(x_1(t), x_2(t), u_1(t), t) = x_2(t) \end{aligned} \tag{3.39}$$

Furthermore, equations (3.39) can be represented as

$$\begin{aligned} \begin{bmatrix} \dot{x}_1 \\ \dot{x}_2 \end{bmatrix} &= \begin{bmatrix} 0 & 1 \\ -K/M & -D/M \end{bmatrix} \begin{bmatrix} x_1(t) \\ x_2(t) \end{bmatrix} + \begin{bmatrix} 0 \\ 1/M \end{bmatrix} u_1(t) \\ \begin{bmatrix} y_1 \\ y_2 \end{bmatrix} &= \begin{bmatrix} 1 & 0 \\ 0 & 1 \end{bmatrix} \begin{bmatrix} x_1(t) \\ x_2(t) \end{bmatrix} \end{aligned} \tag{3.40}$$

The state variable description of systems is suitable for computer solution, as we will see in later chapters.

EXAMPLE 3.6 **Another Electrical Circuit** Refer to Figure 3.6. Let us assume our input to be voltage v_i and our outputs to be voltage v_o and current i_2. We can choose our internal states as the voltages at node A and node B. It should be clear that knowing the node voltages at A and B (the internal states) along with v_i provides all the information we need to describe the

FIGURE 3.6 Electrical Circuit

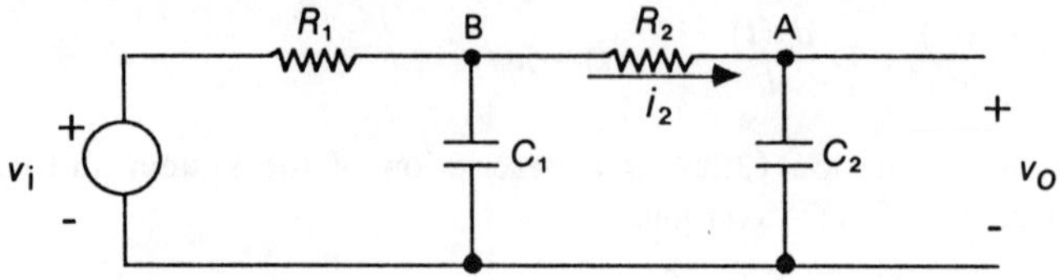

system's behavior. Knowledge of $v_i(t)$, $v_A(t)$, and $v_B(t)$ enables us to compute all currents, energy dissipation, and energy storage in the system. We can use Kirchhoff's current law to develop the state variable description.

Applying Kirchhoff's current law at node A, we obtain

$$C_2 \frac{dv_A}{dt} + \frac{v_A - v_B}{R_2} = 0 \tag{3.41}$$

Similarly, applying Kirchhoff's current law at node B, we get

$$C_1 \frac{dv_B}{dt} + \frac{v_B - v_A}{R_2} + \frac{v_B - v_i}{R_1} = 0 \tag{3.42}$$

Equations (3.41) and (3.42) lead to

$$\frac{dv_A}{dt} = -\frac{v_A}{R_2C_2} + \frac{v_B}{R_2C_2} \tag{3.43}$$

and

$$\frac{dv_B}{dt} = +\frac{v_A}{R_2C_1} - \frac{v_B}{R_2C_1} - \frac{v_B}{R_1C_1} + \frac{v_i}{R_1C_1}$$

$$= \frac{v_A}{R_2C_1} - v_B\left(\frac{1}{R_2C_2} + \frac{1}{R_1C_1}\right) + \frac{v_i}{R_1C_1} \tag{3.44}$$

If we let $x_1(t) = v_A(t)$, $x_2(t) = v_B(t)$, and $u(t) = v_i(t)$, we have

$$\begin{aligned} \dot{x}_1(t) &= -\frac{1}{R_2C_2} x_1(t) + \frac{1}{R_2C_2} x_2(t) \\ \text{and} \quad \dot{x}_2(t) &= \frac{1}{R_2C_1} x_1(t) - \left(\frac{1}{R_2C_2} + \frac{1}{R_1C_1}\right) x_2(t) + \frac{1}{R_1C_1} u(t) \end{aligned} \tag{3.45}$$

Outputs $v_o(t)$ and $i_2(t)$ can be expressed in terms of the internal states as

$$\begin{aligned} y_1(t) &= v_o(t) = v_A(t) = x_1(t) \\ y_2(t) &= i_2(t) = \frac{v_B(t) - v_A(t)}{R_2} = \frac{x_2(t)}{R_2} - \frac{x_1(t)}{R_2} \end{aligned} \tag{3.46}$$

The matrix representation of equations (3.45) and (3.46) is

$$\dot{\mathbf{x}}(t) = \begin{bmatrix} \dot{x}_1 \\ \dot{x}_2 \end{bmatrix} = \begin{bmatrix} \dfrac{1}{R_2C_2} & \dfrac{1}{R_2C_2} \\ \dfrac{1}{R_2}C_1 & -\left(\dfrac{R_1C_1 + R_2C_2}{R_1C_1\ R_2C_2}\right) \end{bmatrix} \begin{bmatrix} x_1(t) \\ x_2(t) \end{bmatrix} + \begin{bmatrix} 0 \\ \dfrac{1}{R_1C_1} \end{bmatrix} u(t)$$

$$\mathbf{y} = \begin{bmatrix} y_1(t) \\ y_2(t) \end{bmatrix} = \begin{bmatrix} 1 & 0 \\ \dfrac{1}{R_2} & -\dfrac{1}{R_2} \end{bmatrix} \begin{bmatrix} x_1(t) \\ x_2(t) \end{bmatrix} \tag{3.47}$$

or

$$\begin{aligned} \dot{\mathbf{x}}(t) &= [A]\mathbf{x}(t) + \mathbf{b}u(t) \\ \mathbf{y}(t) &= [C]\mathbf{x}(t) \end{aligned} \tag{3.48}$$

An analogous set of equations, called *difference equations*, can be used to describe discrete systems as conceptualized in Figure 3.1. In this case, the state of the system can change values only at specific instants of time. Digital systems, such as computers, are discrete systems. Difference equations take the following form:

$$\begin{aligned} x_1(k+1) &= f_1(x_1(k), x_2(k), \ldots, x_n(k), u_1(k), u_2(k), \ldots, u_m(k), k) \\ x_2(k+1) &= f_2(x_1(k), x_2(k), \ldots, x_n(k), u_1(k), u_2(k), \ldots, u_m(k), k) \\ &\vdots \\ x_n(k+1) &= f_n(x_1(k), x_2(k), \ldots, x_n(k), u_1(k), u_2(k), \ldots, u_m(k), k) \\ y_1(k) &= g_1(x_1(k), x_2(k), \ldots, x_n(k), u_1(k), u_2(k), \ldots, u_m(k), k) \\ y_2(k) &= g_2(x_1(k), x_2(k), \ldots, x_n(k), u_1(k), u_2(k), \ldots, u_m(k), k) \\ &\vdots \\ y_l(k) &= g_l(x_1(k), x_2(k), \ldots, x_n(k), u_1(k), u_2(k), \ldots, u_m(k), k) \end{aligned} \tag{3.49}$$

where $k = 0, 1, 2, \ldots,$ with $t = k\Delta t$ and $t + \Delta t = (k+1)\Delta t$.

Given $x_1(k), x_2(k), \ldots, x_n(k), u_1(k), u_2(k), \ldots, u_m(k)$, calculation of $x_1(k+1), x_2(k+1), \ldots, x_n(k+1)$ is straightforward. In fact, digital simulation of equation (3.4) involves conversion to equation (3.49) for solution. We used a similar approach in Chapter 2 when we obtained the computer solution to the information diffusion example. Equations (3.4) or (3.49) can be used to describe most systems. The diffusion model simulated in Chapter 2 has the form of equations (3.4). Systems whose models have these forms can be simulated using numerical integration techniques, such as rectangular integration.

EXAMPLE 3.7 **Savings Plan** Let us consider a savings plan offered by a local bank. The interest rate for any time period depends on the prime interest rate. Let the interest rate for the kth period be $r(k)$. Assume that a customer makes a deposit $d(k)$ during the kth period and let $w(k)$ be the withdrawal during the kth period. Let $b(k)$ be the balance in the account at the kth period.

The balance at the $(k + 1)$st period is the balance at the kth period plus the change in the account. Change can occur due to a deposit in the kth period, $d(k)$, a withdrawal in the kth period, $w(k)$, and the addition of interest. The discrete equation describing the balance of the account is

$$b(k+1) = b(k) + d(k) - w(k) + r(k) \times (b(k) + d(k) - w(k)) \tag{3.50}$$

We can consider the internal state to be $x_1(k) = b(k)$, the output to be $b(k)$, and the inputs to be $u_1(k) = w(k)$ and $u_2(k) = d(k)$. The discrete set of equations becomes

$$\begin{aligned} x_1(k+1) &= (1 + r(k))x_1(k) - (1 + r(k))u_1(k) + (1 + r(k))u_2(k) \\ y_1(k) &= x_1(k) \end{aligned} \tag{3.51}$$

EXAMPLE 3.8 **Population Growth** Let us formulate a difference equation model for population growth of wild horses on an island. Assume that certain age groups have predictable birth and death rates. We will use the following notation:[2]

a. Let $x_j(k)$ denote the size of the population with members j years old at the beginning of the kth year.
b. Let b_j denote the number of births to a horse who is j years old.
c. Let d_j denote the fraction of the population of age j years who typically die during the year.
d. Let $y(k)$ denote the size of the population at the beginning of the kth year.
e. Assume that the oldest age of a horse in the population is N years.

Assume the states to be $x_1(k), x_2(k), \ldots, x_n(k)$ and the output to be $y(k)$. Then

$$\begin{aligned} x_1(k+1) &= b_1x_1(k) + b_2x_2(k) + \cdots + b_nx_n(k) \\ x_2(k+1) &= (1 - d_1)x_1(k) \\ x_3(k+1) &= (1 - d_2)x_2(k) \\ &\vdots \\ x_{n-1}(k+1) &= (1 - d_{n-2})x_{n-2}(k) \\ x_n(k+1) &= (1 - d_{n-1})x_{n-1}(k) \\ y(k) &= x_1(k) + x_2(k) + \cdots + x_n(k) \end{aligned} \tag{3.52}$$

These equations can be written as

$$\begin{aligned} \mathbf{x}(k+1) &= \begin{bmatrix} b_1 & b_2 & \cdots & b_{n-1} & b_n \\ 1 - d_1 & 0 & \cdots & 0 & 0 \\ 0 & 1 - d_2 & \cdots & 0 & 0 \\ \vdots & & & & \\ 0 & 0 & \cdots & 1 - d_{n-1} & 0 \end{bmatrix} \mathbf{x}(k) \\ y(k) &= [1 \cdots 1]\mathbf{x}(k) \end{aligned} \tag{3.53}$$

Note the similarity between this discrete form of the equations and the continuous form of the state variable equations:

$$\begin{aligned} \mathbf{x}(k+1) &= [A]\mathbf{x}(k) \\ y(k) &= [C]\mathbf{x}(k) \end{aligned} \tag{3.54}$$

FIGURE 3.7 System Containing Three D-Type Flip-Flops

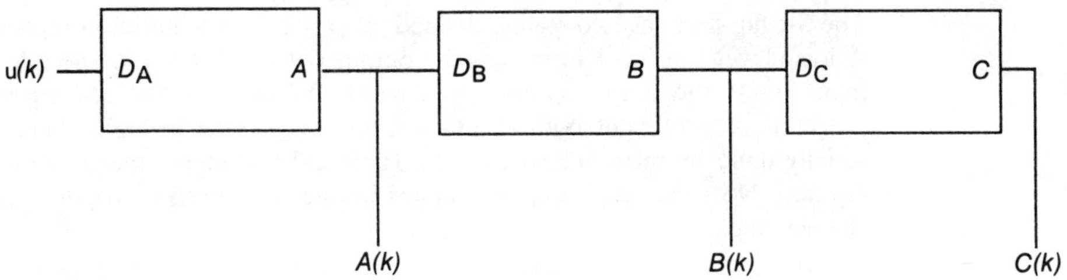

EXAMPLE 3.9 **Digital System (Logic System)** Consider the simple digital system shown in Figure 3.7. This represents a system with three D-type flip-flops. The next logic state of a D flip-flop is equal to its present input. The flip-flop behaves like a delay element. Thus, the basic equation governing the behavior of the D-type flip-flop shown in Figure 3.8 is given by

$$y(k+1) = D_x(k) = x(k) \tag{3.55}$$

Let us develop the state equations for the system shown in Figure 3.7. If we define the input as $u(k)$, the output as $C(k)$, and the states as $A(k)$, $B(k)$, and $C(k)$, the state variable description becomes

$$\begin{aligned} x_1(k+1) &= A(k+1) = u(k) \\ x_2(k+1) &= B(k+1) = A(k) = x_1(k) \\ x_3(k+1) &= C(k+1) = B(k) = x_2(k) \\ y(k) &= C(k) \end{aligned} \tag{3.56}$$

These equations can be written as

$$\begin{aligned} \mathbf{x}(k+1) &= \begin{bmatrix} 0 & 0 & 0 \\ 1 & 0 & 0 \\ 0 & 1 & 0 \end{bmatrix} \begin{bmatrix} x_1(k) \\ x_2(k) \\ x_3(k) \end{bmatrix} + \begin{bmatrix} 1 \\ 0 \\ 0 \end{bmatrix} u(k) \\ y(k) &= C(k) = [0 \quad 0 \quad 1] \begin{bmatrix} x_1(k) \\ x_2(k) \\ x_3(k) \end{bmatrix} \end{aligned} \tag{3.57}$$

FIGURE 3.8 D-Type Flip-Flop

EXAMPLE 3.10 **SR Flip-Flop System** The set-reset (SR) flip-flop has two inputs, S and R, and two outputs, Q and $\bar{Q}$. A block diagram representing the SR flip-flop is shown in Figure 3.9. The SR flip-flop has two states, defined by $Q = 1$, $\bar{Q} = 0$ and its complement $Q = 0$, $\bar{Q} = 1$. A logic level 1 at the S input sets the output Q to logic level 1, and a logic level 1 at the R input resets the output Q to logic level 0. When using an SR flip-flop, we must take precautions to prevent both R and S from being equal to logic 1 simultaneously. This is usually done by introducing a clock. Table 3.2 represents the transition table of the SR flip-flop. Note that the flip-flop remains in one state until a proper signal commands it to change state.

Now let us look at two SR flip-flops in series as shown in Figure 3.10. Let us assume the states of the system to be $Q_1(k)$ and $Q_2(k)$, the outputs to be $Q_1(k)$ and $Q_2(k)$, and the input to be $u(k)$. The logic equation describing the behavior of an SR flip-flop is given by

$$Q(k+1) = \bar{R}Q(k) + S(k) \tag{3.58}$$

Therefore, the discrete equation describing the behavior of the system is

$$\begin{aligned} Q_1(k+1) &= \overline{u(k)\bar{Q}_2(k)}Q_1(k) + \overline{u(k)}Q_2(k) \\ Q_2(k+1) &= (u_1(k) + \bar{Q}_1(k))Q_2(k) + u(k)\bar{Q}_1(k) \end{aligned} \tag{3.59}$$

Then the discrete state variable description becomes

$$\begin{aligned} Q_1(k+1) &= \overline{u(k)\bar{Q}_2(k)}Q_1(k) + \overline{u(k)}Q_2(k) \\ \bar{Q}_2(k+1) &= u(k)\bar{Q}_1(k) + (u_1(k) + \bar{Q}_1(k))Q_2(k) \\ y_1(k) &= Q_1(k) \\ y_2(k) &= Q_2(k) \end{aligned} \tag{3.60}$$

The logic system can be in one of four states or conditions:

$$\begin{aligned} (Q_1, Q_2) &= (0, 0) \\ (Q_1, Q_2) &= (1, 0) \\ (Q_1, Q_2) &= (0, 1) \\ (Q_1, Q_2) &= (1, 1) \end{aligned}$$

This can be represented graphically as shown in Figure 3.11. The transition from one state to another for the input sequence 0, 1, 1, 0, 1, 0, 0, 1 is shown in Figure 3.12. The initial state was $(Q_1, Q_2) = (0, 0)$. When a 0 input is applied to state 1, $(Q_1, Q_2) = (0, 0)$, the system remains in the same state; a 1 input into state 1 results in a transition to state 2; a 1 into state 2 keeps the system in state 2; a 0 into state 2 sends the system into state 3; and so on.

FIGURE 3.9 Block Diagram of an SR Flip-Flop

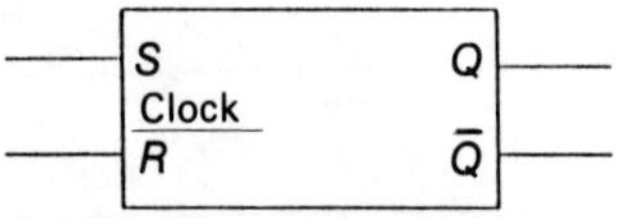

TABLE 3.2 Transition Table of SR Flip-Flop

S	*R*	*From Old State* $Q(k)$	*To New State* $Q(k+1)$
0	0	0	0
1	0	0	1
0	1	1	0
d	0	1	1

Note: d = do not care.

FIGURE 3.10 Flip-Flops in Series

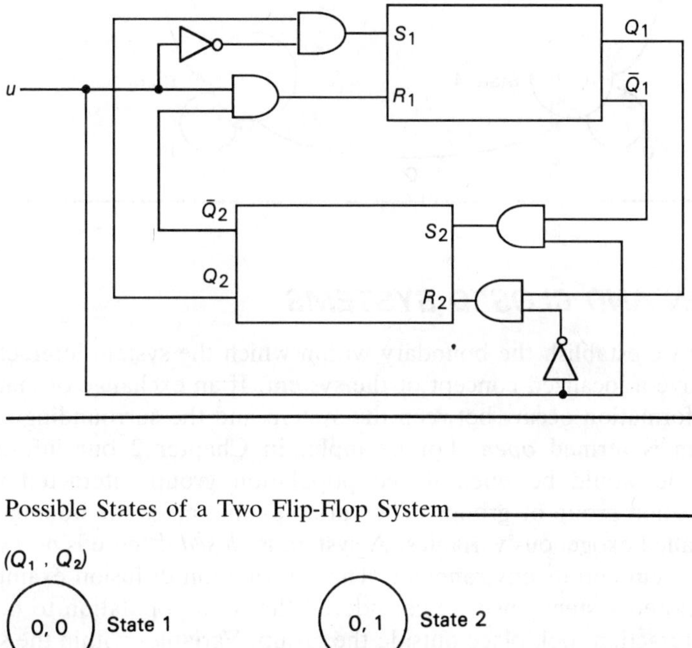

FIGURE 3.11 Possible States of a Two Flip-Flop System

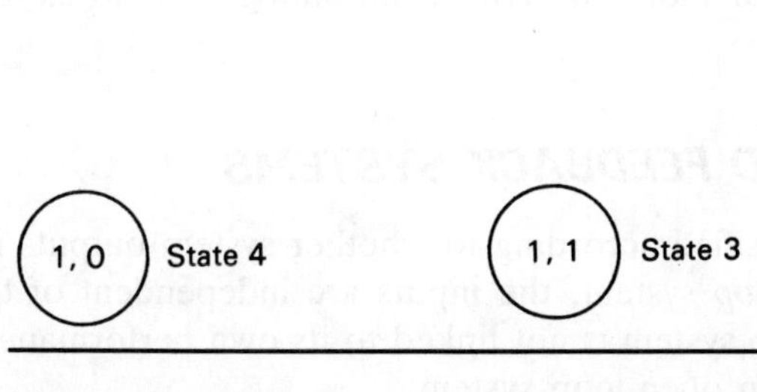

FIGURE 3.12 Transition Diagram

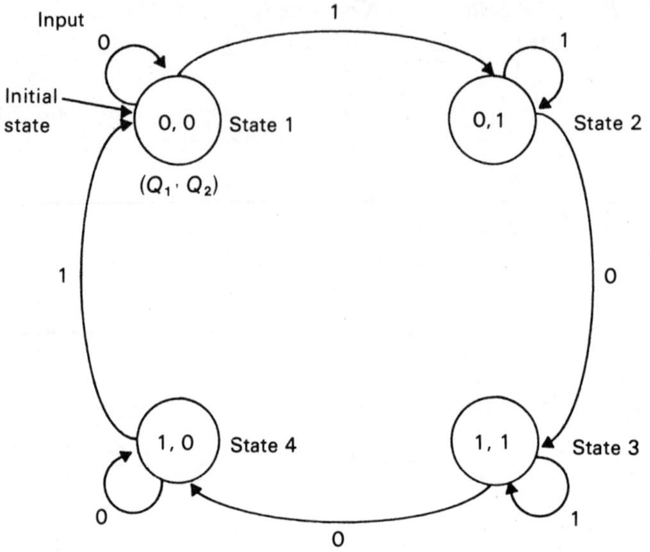

3.3 OPEN AND CLOSED SYSTEMS

Once we establish the boundary within which the system interaction takes place, we have a localized concept of the system. If an exchange of materials, energies, or information occurs between the system and the surrounding environment, the system is termed *open*. For example, in Chapter 2 our information diffusion example would be open if the population group interacted with an outside population group or groups. The variables that cross the boundary of the system are called exogenous variables. A system is *closed* if there is no exchange between the system and its environment. The information diffusion example of Chapter 2 is a closed system since we considered the total population to be one group and no interaction took place outside the group. Variables within the system boundary are called endogenous. An open system can become closed if interaction with the environment is cut off, or if we include the exogenous variables in the system. Likewise, closed systems can become open if we alter the boundary of the system so as to change one or more variables from endogenous to exogenous.

3.4 OPEN-LOOP AND FEEDBACK SYSTEMS

Systems are also classified according to whether system outputs influence system inputs. In an *open-loop* system, the inputs are independent of the outputs. The input to an open-loop system is not linked to its own performance. Figure 3.13 is a block diagram of an open-loop system.

FIGURE 3.13 Block Diagram of an Open-Loop System

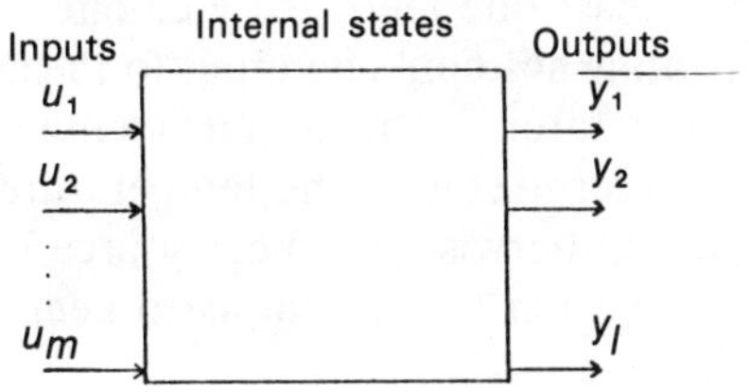

On the other hand, in a *feedback system* (also called a *closed-loop* feedback system), the outputs do influence the inputs. That is, the behavior of a feedback system is influenced by the system's past performance. Feedback systems are of two types: *negative feedback* or *positive feedback*. Feedback is generally used for control. Feedback is termed negative when the input acts in a manner that will reduce output variations, and positive when the input tends to reinforce the original output. Figure 3.14 depicts generalized block diagrams of closed-loop feedback systems.

FIGURE 3.14 (a) Negative Feedback Closed-Loop System (b) Positive Feedback Closed-Loop System

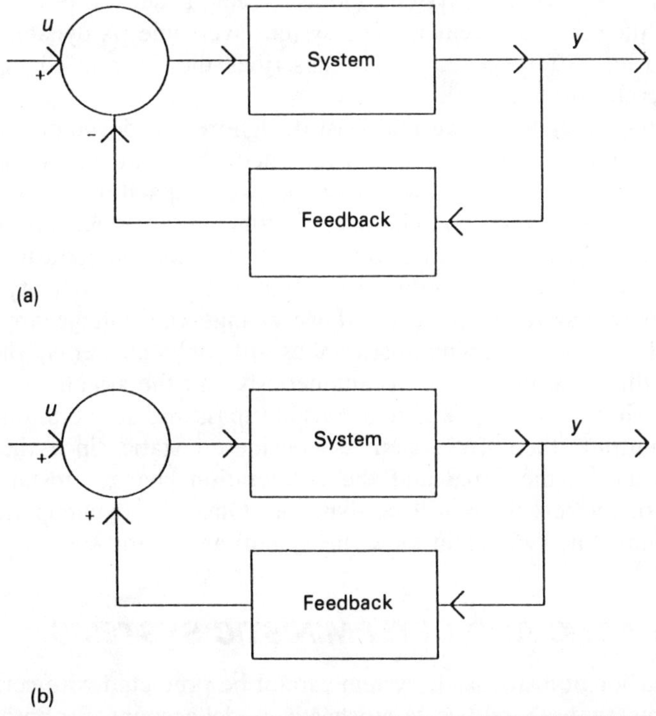

Complex systems can contain both negative and positive feedback elements. Most nontrivial systems include at least one feedback element. A familiar simple closed-loop system is the home heating system. The input to the system is the heat source, and the output is the temperature. The output (temperature) is sensed and compared to the setting of the thermostat. If the temperature is below the set point, the heat source is turned on. Otherwise, the heat source is turned off. This is a negative feedback closed-loop system that attempts to keep the temperature of the home at the desired setting.

An example of a positive feedback system is investing money in a savings account. The more money you deposit in the account, the larger your balance, and the more interest you receive, based on your increased balance. The interest is then added to your account, further increasing the balance. With every deposit or interest payment, the balance increases, which leads to even greater interest accumulation at each accounting period.

3.5 DYNAMIC AND STATIC SYSTEMS

A dynamic system is a system whose response is a function of time. Usually the response of a dynamic system to an input change is not instantaneous. Actually, in all physical systems the response to an input change is always delayed, even if by only picoseconds (trillionths of a second). Physical entities do not change state instantaneously. Systems are thus said to have memory of earlier states or conditions. A dynamic system model, using a set of differential equations, captures the continuous change in a system over time. A dynamic discrete system model, using difference equations, describes the discrete change in the system over discrete instants of time.

On the other hand, we may wish to ignore the development over time and assume that a system responds instantaneously to an input. Such systems are called static. For example, the voltage across a capacitor cannot change instantaneously, and so systems containing capacitors are dynamic. If a force is applied to a mass, the velocity of the mass does not change instantaneously. A system whose input is defined as force and whose output is defined as velocity is considered to be a dynamic system. However, if we change the voltage applied to an ideal electrical circuit containing resistors as the only elements, then the currents through the resistors change instantaneously and the system is considered static. Similarly, if a force is applied to a constant mass and acceleration is considered to be the output, then the system is considered static since the acceleration is proportional to the force and the acceleration changes instantaneously. Most systems of interest to us will be dynamic. Once the dynamic system reaches an equilibrium state, we might view the system as a static system.

3.6 STOCHASTIC AND DETERMINISTIC SYSTEMS

The behavior of a stochastic system cannot be predicted with certainty because it is governed by probabilities. A stochastic model accounts for both uncertainty and time variation. In such a system, the initial conditions, parameters, or inputs may be stochastic. The salesperson problem in Chapter 2 was a stochastic system because the sale of the product was not guaranteed on each sales visit to a customer. A system whose behavior can be predicted with certainty is called deterministic. If the response of a deterministic system will be the same regardless of when the same inputs are applied, the system is called a time-invariant system; otherwise, it is a time-varying system.

That is, the response of a time-invariant system depends only on the time that has elapsed since the application of the input. This behavior indicates that the system relationships do not change with time. If the system relationships do change with time, the system is time-varying. For example, in Chapter 2, if the value of k, the conversion coefficient in equation (2.5), does not vary with time, the system described by the equation is time-invariant. However, if k varies with time, the system becomes time-varying.

3.7 LINEAR AND NONLINEAR SYSTEMS

Linear system theory plays an important role in understanding the behavior of systems, because mathematically it is relatively easy to work with linear systems. It is therefore often convenient to approximate a system's behavior as linear, even if that assumption is known to be inaccurate, because this is the best we can do.

A system is linear when the effect of a sum of input changes is the same as the sum of the individual effects that would be caused by each change alone. We can express this characteristic mathematically as follows. Let $y = f(x_1, {}_2, \ldots, x_n, u, t)$ and let

$$y_1(t) = f(x_1, x_2, \ldots, x_n, u_1, t) \tag{3.61}$$

$$y_2(t) = f(x_1, x_2, \ldots, x_n, u_2, t) \tag{3.62}$$

$$y_3(t) = f(x_1, x_2, \ldots, x_n, u_1 + u_2, t) \tag{3.63}$$

where $y_1(t)$ = the system response to input u_1

$y_2(t)$ = the system response to input u_2

$y_3(t)$ = the system response to input $u_1 + u_2$

Then, if

$$y_1(t) + y_2(t) = y_3(t) \text{ for all inputs } u_1(t) \text{ and } u_2(t) \tag{3.64}$$

the system is linear. The greenhouse system is a linear system, whereas we have seen that information diffusion is an example of a nonlinear system.

Most physical and social systems are nonlinear. However, since linear system theory is well developed, we can often achieve a better understanding of a system by using a linear approximation of the system within a valid region of operation (linearization).

FIGURE 3.15 Causal Loop Diagram

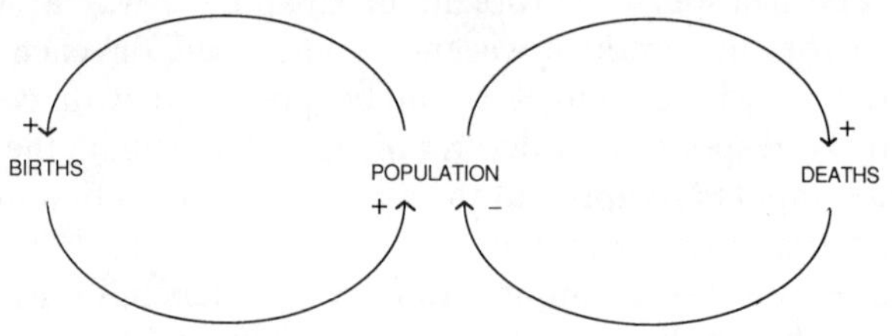

3.8 CAUSAL LOOP DIAGRAMS

Causal loop diagrams are a nonquantitative tool for showing the relationship between system variables. This technique helps us develop, understand, and document system models. Causal loop diagrams attempt to depict cause-and-effect relationships in a similar manner as the block diagram of Figure 3.13. For example, let us look at a large population of individuals; the population has a birth rate and a death rate. A causal loop diagram is shown in Figure 3.15.

The diagram indicates that as population increases, the number of deaths increases; and as the number of deaths increases, the population decreases. This is a negative feedback loop. The diagram also indicates that if the population increases, the number of births increases; and as the number of births increases, the population increases. This indicates a positive feedback loop.

The success of the simulation process depends heavily on proper identification of the system and proper development of a model for the system. The choice of system description can be viewed as a tool to use in the model construction phase of the simulation process. Again, remember that the model construction process depends on our knowledge of the system we are trying to simulate and on the objective of our simulation.

3.9 SUMMARY

The objective of this chapter was to help you develop lumped parameter models of systems. Our approach has been first to classify the system into one of several types and then to describe the system. The two principal methods of description, internal and external, were presented. The external description relates the inputs to the outputs of the system, and the internal description relates the states or conditions of the system to the inputs and to each other. The outputs are a combination of the internal states and inputs.

The state variable approach is appropriate for both continuous-time and discrete-time systems. As you have noticed from the examples, we always returned to the fundamental laws that govern the behavior of the system.

This chapter emphasized deterministic systems. The next few chapters will focus on stochastic or probabilistic systems. The student with computer programming background or simulation language experience may proceed with the simulation process by translating the lumped model description into a simulation program. What is interesting about the simulation process is that a student using the lumped model can learn more about the system than the model developer by studying the results of the simulation program.

3.10 REVIEW PROBLEMS

1. What is the difference between an internal and an external description of a system?
2. Why is computer simulation a natural method for the simulation of systems described by a set of difference equations?
3. Let $x(t)$ be an input to a system and $y(t)$ be the output of the system. Is the following system linear?

$$y(t) = a + x(t)$$

Assume that a is a constant.

4. List the different ways of classifying systems. Give an example of each.
5. Develop a state variable description of the electrical system shown in Figure 3.16. Let the input to the system be $v_i(t)$ and the output be $v_o(t)$.

FIGURE 3.16 Problem 5

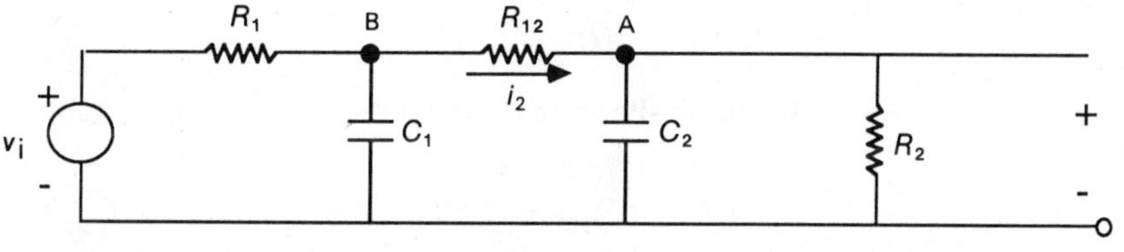

6. Develop a state variable model for the mechanical system shown in Figure 3.17. Define the input to the system to be $f(t)$ and the output of the system to be $\dot{x}_2 - \dot{x}_1$.

FIGURE 3.17 Problem 6

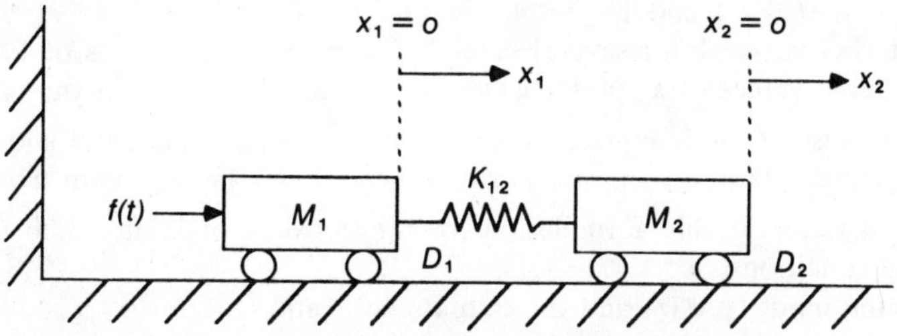

7. It can be demonstrated that the equations of motion for the system shown in Figure 3.18 are

$$(M + m)\ddot{x} + ml\ddot{\theta} = u$$
$$m\ddot{x} + ml\ddot{\theta} - mg\theta = 0$$

when the following approximations are used: cos $\theta = 1$, sin $\theta = 0$, $\dot{\theta} = 0$, and $\dot{x} = 0$. Develop a state variable model for the system. Let $m = 0.1$ kilogram, $M = 1.0$ kilogram, $l = 1$ meter, and $\theta = 0.5°$. (g = gravity = 9.8 meters per square second, m/s^2.)

FIGURE 3.18 Problem 7

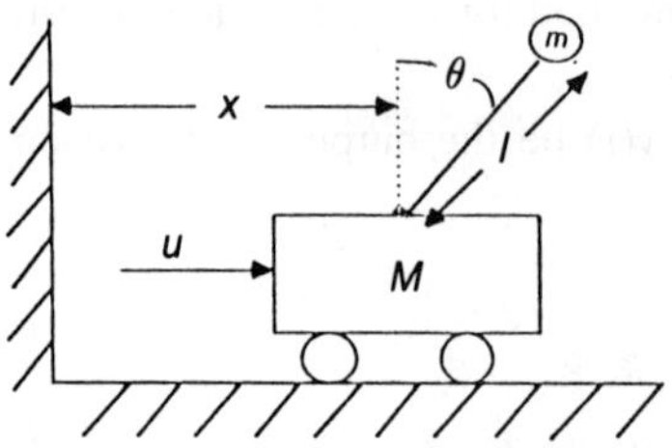

8. The purpose of this problem is to study the behavior of dynamic systems (population growth, information diffusion) described by the differential equation

$$\frac{dx(t)}{dt} = \lambda x(t)(1 - x(t))$$

Let $x(0) = x_o$. It can be shown that the solution is given by

$$x(t) = \frac{x_o e^{\lambda t}}{1 - x_o(1 - e^{\lambda t})}$$

a. Plot $x(t)$ and dx/dt versus t for several values of λ. Let $0 < x_o < 1$ and $1 < \lambda < 4$.
b. What are the equilibrium values of $x(t)$?

9. We wish to study the behavior of a dynamic system described by the following difference equation:

$$x(k+1) = \lambda x(k)(1 - x(k))$$

Assume that $x(0) < 1$ and $1 < \lambda < 4$.
a. Plot $x(k)$ versus k for several values of x_o and several values of λ.
b. For large values of k, plot $x(k)$ versus λ. See reference 8 at the end of the chapter.

10. Derive the discrete state equations for the module-8 binary counter shown in Figure 3.19. Assume initially that y_1, y_2, and y_3 are logic 0. Let x be the input and y be the output.

11. Develop a state variable formulation for the network of Figure 3.20 under each of the following conditions:
a. Let the input be $v_i(t)$ and the outputs be i_1 and i_2.
b. Let the input be $v_i(t)$ and the outputs be $i_1(t)$ and di_1/dt.

FIGURE 3.19 Problem 10

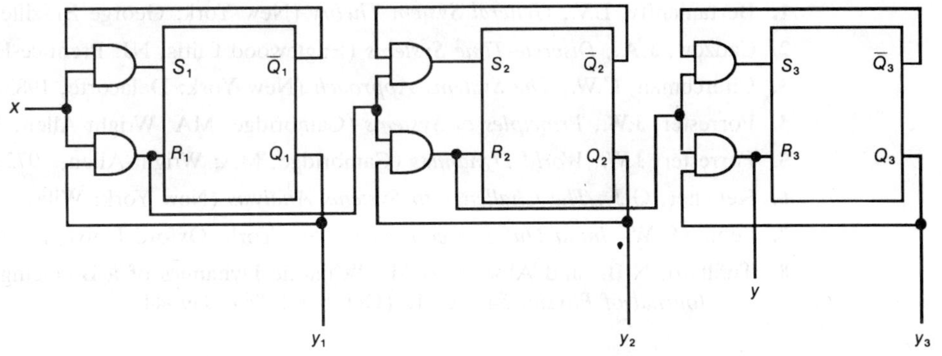

FIGURE 3.20 Problem 11

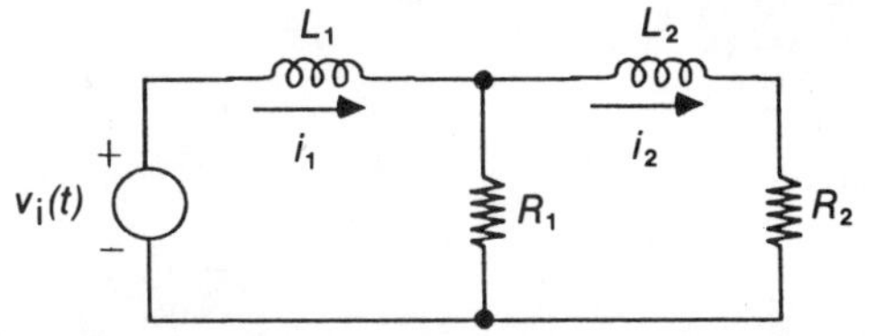

12. Assume that you invest P dollars at a fixed yearly interest rate r. Let P_k be the value of your investment after k years.
    - **a.** Develop a discrete time difference equation to determine the value of your investment after k years. Let $k > 0$.
    - **b.** Suppose that in year 0 you put P dollars into a tax-deferred investment. Assume that your tax rate is t at withdrawal and that the interest rate is fixed at r. What is your investment worth when it is withdrawn after k years? Assume no withdrawal penalty. Develop the discrete equation(s) to compute your answer.
    - **c.** Suppose that you put your money into a savings account and your tax rate is t and the interest rate is fixed at r. What is your investment worth after k years?

13. Consider a predator-prey model of wolves and rabbits. Assume that the following difference equations represent the system:

$$R(k) = aR(k-1) - bW(k-1)$$
$$W(k) = cW(k-1)$$

where $R(k)$ = the number of rabbits in year k
$W(k)$ = the number of wolves in year k

- **a.** Obtain a solution for $R(k)$ and $W(k)$ if $W(0) = 10.0$ and $R(0) = 5.0$, $a = 0.5$, $b = 0.2$, and $c = 0.9$.
- **b.** What conditions prevail when $c = 1$? What are $W(k)$ and $R(k)$? For what values of a and b will the population increase? Decrease? Remain the same?

3.11 REFERENCES

1. Bertanlanffy, L.V., *General System Theory* (New York: George Braziller, 1973).
2. Cadzow, J.A., *Discrete-Time Systems* (Englewood Cliffs, NJ: Prentice-Hall, 1973).
3. Churchman, C.W., *The Systems Approach* (New York: Delacorte, 1983).
4. Forrester, J.W., *Principles of Systems* (Cambridge, MA: Wright-Allen, 1968).
5. Forrester, J.W., *World Dynamics* (Cambridge, MA: Wright-Allen, 1972).
6. Kelleher, G.J., *The Challenge to Systems Analysis* (New York: Wiley, 1970).
7. Leontief, W., *Input Output Economics* (New York: Oxford University Press, 1986).
8. Tufillaro, N.B., and Albano, A.M., "Chaotic Dynamics of a Bouncing Ball," *American Journal of Physics* 54, no. 10 (October 1986): 939–44.

4 Basic Concepts of Probability Theory

4.1 INTRODUCTION

Many phenomena of interest to us cannot be expressed as explicit functions of time. Examples are the outside temperature over the course of a day, the Dow Jones average, the outcome of the Super Bowl, noise in communication systems, crop yield, the demand for a new cereal, a country's export and import rates, the unemployment rate, wars, and natural disasters. These phenomena are called random or stochastic, implying uncertainty about their exact nature. Probability theory is the science of studying and understanding stochastic processes. It serves as a tool for describing or predicting the descriptive statistics of a process, such as the mean or average of specified parameters. The brief exposition of the theory presented here includes introductory regression and correlation analysis. The statistical methods covered are widely used in simulation.

4.2 PROBABILITY

The probability $p(a)$ of an event a occurring in an experiment with a set of possible outcomes A is defined in the following manner:

$$p(a) = \lim_{n \to \infty} \frac{n_a}{n} \tag{4.1}$$

where n = the number of times the experiment is repeated

n_a = the number of times event a occurred

Intuitively, the greater the value of n, the more confidence we have that the probability of event a occurring is n_a/n.

Suppose we wish to compute a player's probability of making a foul shot in a basketball game. We would count the number of foul shots the player makes and divide this number by the total number of foul shot attempts. The greater the number of foul shot attempts, the more confident we feel about our computation of the probability that the player will make a foul shot. This definition of the probability of event a occurring is called the relative frequency of a and leads to three basic axioms of probability theory for discrete systems:

1. The probability of an event a occurring can never be negative; that is, $p(a) \geq 0$. Probability $p(a) = 0$ if and only if a is an impossible event.
2. An event whose probability of occurrence is one is called a certain event. Using our notation, if $p(a_o) = 1$, then a_o is a certain event.
3. If events a_1 and a_2 are mutually exclusive, then

$$p(a_1 + a_2) = p(a_1) + p(a_2) \tag{4.2}$$

where $p(a_1 + a_2)$ is read as the probability of a_1 or a_2, but not both. Note that it is not necessary that one of the events occurs.

In a single toss of a coin, the outcomes are either a head (H) or a tail (T). The set of all possible outcomes is

$$A = \{\mathrm{H}, \mathrm{T}\}$$

If a fair coin is tossed repeatedly and often, then

$$p(\mathrm{H}) = \frac{1}{2} \quad \text{and} \quad p(\mathrm{T}) = \frac{1}{2}$$

The probability that the event is either a head or a tail is the certain event a_o, and therefore $p(a_o) = 1$. It should be clear that the events head and tail are mutually exclusive and that $p(\mathrm{H + T}) = p(\mathrm{H}) + p(\mathrm{T}) = \frac{1}{2} + \frac{1}{2} = 1$.

4.3 THE CONCEPT OF A RANDOM VARIABLE

From elementary algebra, we know that, if to each element d in some set D there corresponds one and only one number $g(d) = r$, then g is called a function. This function is a mapping or transformation from the set of elements D, known as the *domain* of the function, to another set of elements, R, called the *range*. Each element of $\{D\}$ is mapped into one and only one element of $\{R\}$. It is possible, however, that more than one element of $\{D\}$ can be mapped into a single element in $\{R\}$, as shown in Figure 4.1.

FIGURE 4.1 Functional Relationship

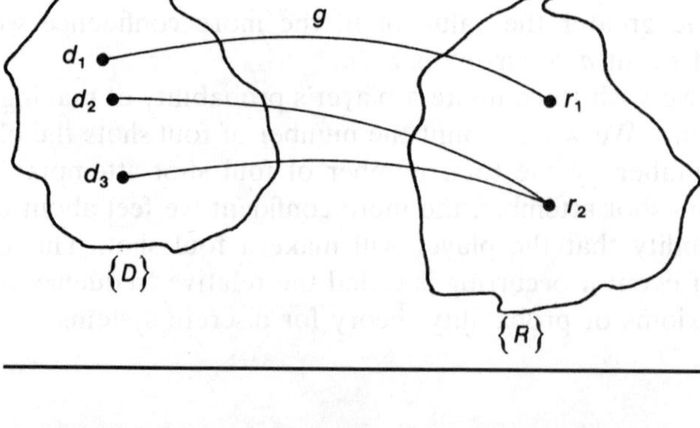

TABLE 4.1

d	$g(d)$
1	10
2	20
3	30
4	40
5	50
6	60

When we write $g(d) = r$, quantity r is called the value of function g at d. Elements of $\{D\}$ are called image points. Sometimes a table specifies a function, with the domain and range clearly indicated by the two columns as shown in Table 4.1.

Definition 4.1: A real random variable X is a function that assigns a number $g(d)$ to every outcome d of the experiment A, satisfying the following conditions:

1. The set $X \leq x$ is an event for any real number x.
2. The probability of events $X = +\infty$ and $X = -\infty$ equals zero. That is, $p(X = +\infty) = p(X = -\infty) = 0$.

Note: X is some random variable $g(d)$, and x is some specified value that random variable X might take.

EXAMPLE 4.1 Consider the tossing of a coin experiment. The event set A is defined as $\{H, T\}$, and $p(H) = \frac{1}{2}$, $p(T) = \frac{1}{2}$. An arbitrary definition of random variable X is

$$g(H) = 1 \quad \text{and} \quad g(T) = 0$$

Therefore,

$$\{D\} = \{T, H\}$$
$$\{R\} = \{0, 1\}$$

Let us expand on the meaning of $X \leq x$. Assume that random variable X of Example 4.1 is specified by Table 4.2. If an arbitrary but fixed number x is chosen and one finds all the right-column entries $g(d)$ that are less than or equal to x, then the corresponding d's at the left form a set. This set is denoted by $X \leq x$.

TABLE 4.2

d (*Outcome of Experiment*)	$g(d)$ (*Value of Random Variable* X)
T	0
H	1

EXAMPLE 4.2 In Example 4.1, if $x = 3$, then the set $X \leq 3$ is denoted by {H, T}. If $x = 0.5$, then the set $X \leq 0.5$ is the set {T}.

EXAMPLE 4.3 Let us consider the final grades for English Literature 101. Each student has a grade, represented by the random variable X. Many schools use the following definition of the random variable X for the grade point:

$$g(F) = 0 \qquad g(D) = 1 \qquad g(C) = 2 \qquad g(B) = 3 \qquad g(A) = 4$$

Therefore, the domain is the set {F, D, C, B, A}, and the range is the set {0, 1, 2, 3, 4}, as shown in the following table:

d (*Outcome of Final Exam*)	*g*(*d*) (*Value of Random Variable X*)
F	0
D	1
C	2
B	3
A	4

If x is chosen to be 3.5, then the set denoted by $X \leq 3.5$ is the set {F, D, C, B}. This set includes all students who received an F, D, C, or B for a final grade.

Therefore, we can view a random variable as a number representing the number of students in a class, the thermal voltage across a resistor, the number of computer users in the computer center, or any numeric representation of interest in the performance of an experiment.

4.4 DISCRETE AND CONTINUOUS RANDOM VARIABLES

In probability theory, we speak of two kinds of random variables. With discrete random variables, the difference between any two values that the variable may assume is finite and nonzero. Examples of discrete random variables are the number of people who can program a computer, the number of professors at a university, and the number of people entering their first year of schooling. Continuous random variables, on the other hand, are characterized by the fact that, within a defined range, the random variable may assume any value. The time a person waits in a queue, the temperature in a room between the hours of 1:00 p.m. and 5:00 p.m., and the time for conversion of a person to a political philosophy are all examples of continuous random variables. Characteristically, variables that are *counted* are discrete, and those that are *measured* are continuous.

4.5 THE DISTRIBUTION FUNCTION

Another important probability concept is the distribution function. A distribution function is the probability that the random variable $g(d)$ will take on a value less than the number x. We may define it formally as follows. Given a real number x and the set $X \leq x$, which is the outcome of d such that $g(d) \leq x$ is an event, the probability $p(X \leq x)$ is a function of x and is denoted by

$$F_X(x) \quad \text{or} \quad F(x)$$

$F(x)$ is called the distribution function of the random variable X, and

$$F(x) = p(X \leq x) \tag{4.3}$$

where $-\infty \leq x \leq +\infty$.

EXAMPLE 4.4 Determine the distribution function for Example 4.1.

Solution It is easy to show that the distribution function is a nondecreasing function with

$$F_X(-\infty) = 0 \quad \text{and} \quad F_X(+\infty) = 1$$

First, let us choose $x > 1$.

a. If $x > 1$, then the set $X \leq x$ corresponds to all the entries in the right column of Table 4.2, which means that the outcome of the experiment includes T or H. This is a certain event because

$$g(\text{H}) = 1 \leq x \quad \text{and} \quad g(\text{T}) = 0 \leq x$$

Therefore,

$$F(x) = p(X \leq x) = p(\text{H} + \text{T}) = 1$$

b. Similarly, if $0 < x < 1$, then the set $X \leq x = \text{T}$ because

$$g(\text{H}) = 1 > x \quad \text{and} \quad g(\text{T}) = 0 \leq x$$

Therefore,

$$F(x) = p(X \leq x) = p(\text{T}) = \frac{1}{2}$$

c. If $x < 0$, then the set $X \leq x$ is the impossible event, and

$$F(x) = 0$$

Therefore,

$$F(x) = \begin{cases} 1 & \text{when } x \geq 1 \\ 1/2 & \text{when } 0 \leq x < 1 \\ 0 & \text{when } x < 0 \end{cases}$$

This distribution function is shown in Figure 4.2.

FIGURE 4.2 Distribution Function of Example 4.4

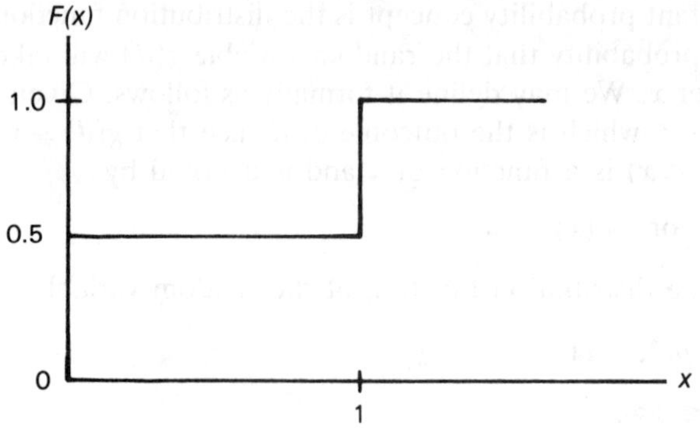

It should be clear by now that all distribution functions have the following properties:

1. $F(-\infty) = 0$, $F(+\infty) = 1$.
2. F is a nondecreasing function of x (i.e., $F(x_1) \leq F(x_2)$ for $x_1 \leq x_2$). Distribution functions are important in simulation of stochastic systems because they allow us to generate any type of random variable, as we will see in Chapter 5.

4.6 THE PROBABILITY DENSITY FUNCTION

A function related to the probability distribution function is the probability density function. The probability density function is a function $f(x)$ that gives the probability that random variable X occurs. The choice of $f(x)$ depends on the characteristics of the random variable.

The probability density function can be represented in three ways. The first type, the discrete probability density function, is depicted in Figure 4.3. In a discrete distribution, random variable X may assume i distinct values between a and b. If there is a population of n objects, then

$$f(x_i) = p(x_i) = \frac{n_{x_i}}{n} \tag{4.4}$$

where n_{x_i} is the number of objects that take on the value x_i. All values of $f(x_i)$ must sum to one. That is,

$$\sum_{i=1}^{n} f(x_i) = 1$$

FIGURE 4.3 Discrete Probability Density Function

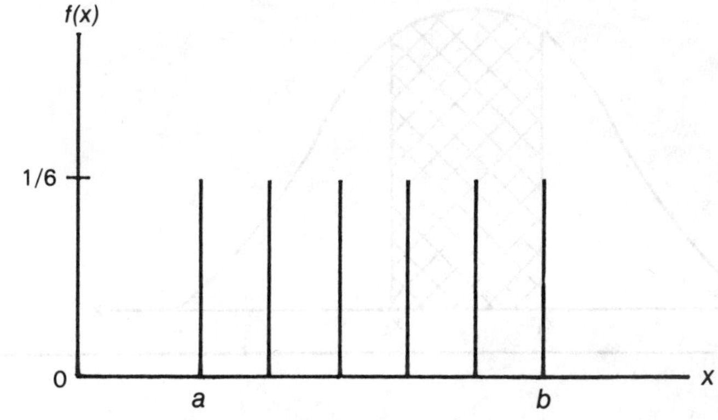

The second type of probability density function is represented by the histogram of Figure 4.4, where random variable X has many values which are classified into m distinct groups of width w. If the population consists of n objects and m intervals, then

$$f(w_i < x_i < w_{i+1}) = \frac{n_i}{n}$$

where n_i = the number of objects in the ith interval

$$\sum_{j=1}^{m} f(w_j < x_j < w_{j+1}) = 1$$

FIGURE 4.4 Histogram of a Probability Density Function

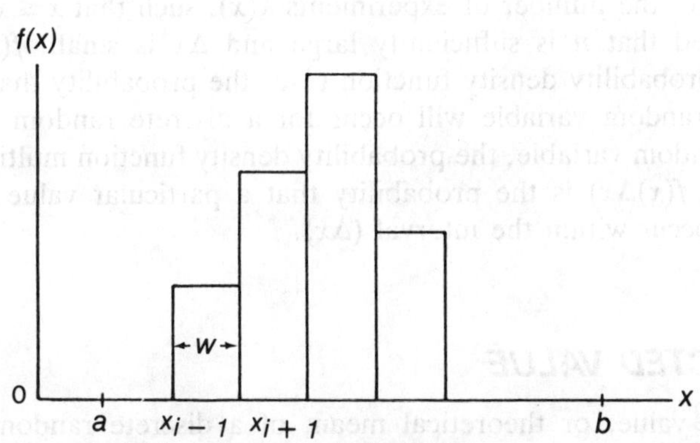

FIGURE 4.5 Continuous Probability Density Function

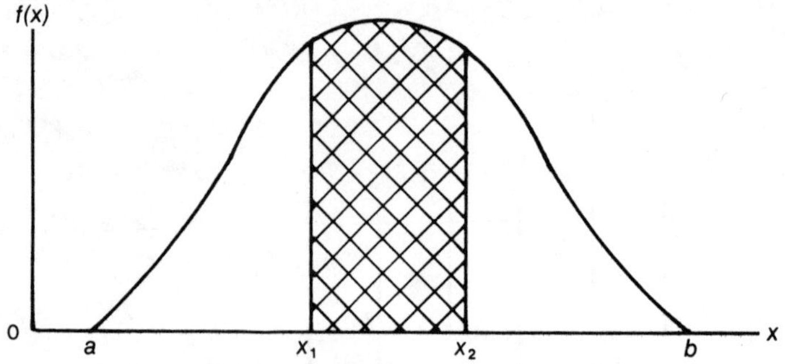

The third type is the continuous probability density function, which results when the width w of the histogram is made very small. Random variable X can take on any value between a and b if measured with sufficient accuracy. Therefore, the probability that the random variable takes on an exact value between a and b is zero. It is more meaningful to talk about the probability that random variable X takes on values between x_1 and x_2. This probability equals the shaded area in Figure 4.5.

The area under the entire curve of Figure 4.5 must be equal to one. This characteristic can be represented mathematically by

$$\int_a^b f(x)\,dx = 1 \tag{4.5}$$

Thus, when we move from discrete to continuous random variables, distribution functions involve integrals instead of finite sums. You can understand this phenomenon conceptually by recalling, from calculus, that an integral is the limit of a sum.

To determine $f(x)$ for a random variable X, we perform the experiment n times and count the number of experiments $k(x)$, such that $x \leq x(d) \leq (x + \Delta x)$. Then, provided that n is sufficiently large and Δx is small, $f(x) \cong k(x)/n$. In essence, the probability density function gives the probability that any particular value of the random variable will occur for a discrete random variable. For a continuous random variable, the probability density function multiplied by a small interval (e.g., $f(x)\Delta x$) is the probability that a particular value of the random variable will occur within the interval (Δx).

4.7 THE EXPECTED VALUE

The expected value, or theoretical mean, of a discrete random variable X is denoted by $E[\mathrm{X}]$ or μ. To obtain $E[\mathrm{X}]$, the experiment is repeated n times, where n is the size of the population and the outcomes $d_1, d_2, \ldots, d_m$ are observed

$n_1, n_2, \ldots, n_m$ times since random variable X has a value for each outcome. Then,

$$E[\mathrm{X}] = \mu = \frac{n_1 x(d_1) + n_2 x(d_2) + \cdots + n_m x(d_m)}{n} \quad \mathbf{(4.6)}$$

where $n = n_1 + n_2 + \cdots + n_m$. In mathematical notation, this can be written as

$$E[\mathrm{X}] = \frac{1}{n} \sum_{i=1}^{m} n_i x(d_i) = \sum_{i=1}^{m} x(d_i) p(d_i) \quad \mathbf{(4.7)}$$

For a continuous random variable, $E[\mathrm{X}]$ can be represented mathematically by

$$E[\mathrm{X}] = \lim_{\substack{a \to -\infty \\ b \to +\infty}} \frac{\int_a^b x f(x)\, dx}{\int_a^b f(x)\, dx} = \int_{-\infty}^{+\infty} x f(x)\, dx \quad \mathbf{(4.8)}$$

$E[\mathrm{X}]$ is often called the first moment of X.

4.8 THE VARIANCE

The variance, $\sigma^2[\mathrm{X}] = E[(\mathrm{X} - \mu)^2]$, is the spread about the mean, $E(\mathrm{X})$. The square root of the variance is called the standard deviation, σ. The second moment, $E[\mathrm{X}^2]$, is the expected value of X^2, which can be obtained numerically using

$$\frac{1}{n} \sum_{i=1}^{m} n_i x^2(d_i) \quad \mathbf{(4.9)}$$

for a discrete random variable. For a continuous random variable,

$$E[\mathrm{X}^2] = \int_{-\infty}^{\infty} x^2 f(x)\, dx \quad \mathbf{(4.10)}$$

The variance is thus given by

$$\sigma^2[\mathrm{X}] = E[\mathrm{X}^2] - E^2[\mathrm{X}] \geq 0 \quad \mathbf{(4.11)}$$

(Note: μ and σ are the mean and standard deviation respectively for the entire population of the random variable X. μ and σ have fixed numerical values.)

4.9 PROBABILITY DISTRIBUTION FUNCTIONS

We will now use the preceding definitions to explain some concepts of probability that are central to computer simulation of stochastic systems: the uniform distribution, the exponential distribution, the Poisson distribution, and the normal distribution.

The Uniform Distribution

The uniform distribution is widely applied in computer simulation. This distribution has a flat probability density function, which is given by

$$f(x) = \begin{cases} \dfrac{1}{b-a} & \text{when } a \leq x \leq b \\ 0 & \text{otherwise} \end{cases} \tag{4.12}$$

The uniform distribution is shown in Figure 4.6 and specifies that random variable X has an equal likelihood of occurring anywhere in the range $a \leq x \leq b$. The mean and variance are given by

$$E[\mathrm{X}] = \mu = \frac{b+a}{2} \tag{4.13}$$

and $$\sigma^2[\mathrm{X}] = \sigma_{\mathrm{X}}^2 = \frac{b^2 - 2ab + a^2}{12} = \frac{(b-a)^2}{12} \tag{4.14}$$

respectively. Note that the shaded area in Figure 4.6 must equal one because we are certain that the value of the random variable X will fall within the range $a \leq x \leq b$. If we take $a = 0$ and $b = 1$, then we have a standard uniform distribution. We used the uniform distribution for the simulation of the salesperson example in Chapter 2. As we will see in Chapter 5, standard uniformly distributed random variables can be used to generate random variables of other distributions.

As an exercise, sketch the probability distribution function for the uniform distribution of Figure 4.6.

FIGURE 4.6 Probability Density Function of the Uniform Distribution

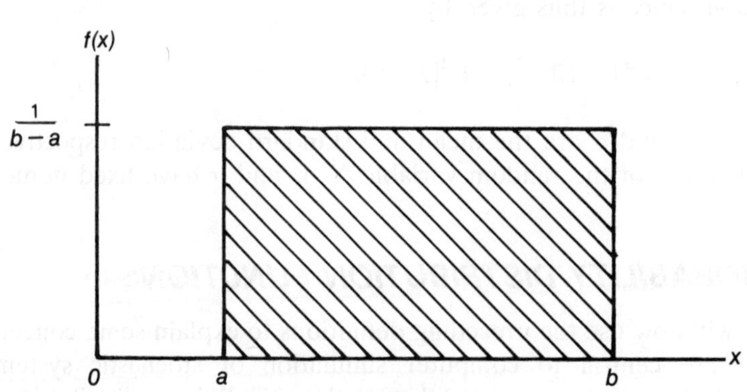

EXAMPLE 4.5 Show that $E[X]$ and $\sigma^2[X]$ for the uniform distribution are given by equations (4.13) and (4.14), respectively.

Solution

$$E[X] = \int_a^b xf(x)\,dx$$

$$= \int_a^b x \frac{1}{b-a}\,dx$$

$$= \frac{1}{b-a} \frac{x^2}{2} \bigg|_a^b$$

$$= \frac{1}{2} \frac{b^2 - a^2}{b-a}$$

$$= \frac{b+a}{2}$$

$$E[X^2] = \int_a^b x^2 f(x)\,dx$$

$$= \frac{1}{b-a} \frac{x^3}{3} \bigg|_a^b$$

$$= \frac{1}{3(b-a)} (b^3 - a^3)$$

$$= \frac{b^2 + ab + a^2}{3}$$

$$\sigma^2[X] = E[X^2] - E^2[X]$$

$$= \frac{b^2 + ab + a^2}{3} - \frac{b^2 + 2ab + a^2}{4}$$

$$= \frac{4(b^2 + ab + a^2) - 3(b^2 + 2ab + a^2)}{12}$$

$$= \frac{(b-a)^2}{12}$$

The Exponential Distribution

A random variable X is exponentially distributed with parameter λ if its probability density function has the following form:

$$f(x) = \lambda e^{-\lambda x} \qquad (0 \le x < \infty,\ e \cong 2.71828)$$
$$f(x) = 0 \qquad (-\infty < x < 0) \tag{4.15}$$

A sketch of this function is shown in Figure 4.7. From the definition presented

FIGURE 4.7 Probability Density Function of the Exponential Distribution

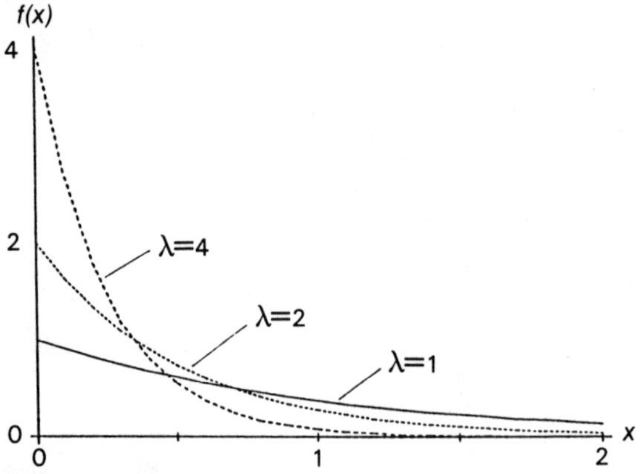

earlier, it can be shown that the distribution function is given by

$$F(x) = \int_0^{\infty} f(x)\, dx \qquad \textbf{(4.16)}$$

$$F(x) = 1 - e^{-\lambda x} \qquad (0 \leq x < \infty) \qquad \textbf{(4.17)}$$

Sketch the distribution function, $F(x)$.

The mean and variance of the random variable are given by

$$E[\mathrm{X}] = \frac{1}{\lambda} \qquad \textbf{(4.18)}$$

and

$$\sigma^2[\mathrm{X}] = \frac{1}{\lambda^2} \qquad \textbf{(4.19)}$$

respectively.

An exponential distribution can describe frequencies of births, deaths, and accidents; service times; message lengths; and even world conflicts. For example, consider a sequence of similar events, such as wars. Let $t_1, t_2, \ldots, t_n$ represent the starting time of the first, second, . . . , nth wars. Count the number of wars, n, that have occurred over a long period of time, T. If the average number of wars per unit time is λ, then the expected number of starting times in the time interval t is λt.

Let $f(t)$ be the probability density function of the intervals Δt between the starting time of wars. If a war occurred at time $t = 0$, then the probability that another war occurs within the interval $(0, t)$ is

$$\int_0^t f(t)\, dt \qquad \textbf{(4.20)}$$

The probability that no other war occurs in the time period $(0, t)$ is

$$1 - \int_0^t f(t)\, dt \tag{4.21}$$

The probability that the duration of the interval Δt between the start of two wars is between t and $t + dt$ is

$f(t)\, dt =$ probability that no war starts in time period $(0, t)$ times the probability that a war occurs in time period $(t, t + dt)$

This leads to

$$f(t)\, dt = \left(1 - \int_0^t f(t)\, dt\right) \lambda\, dt \tag{4.22}$$

and

$$f(t) = \left(1 - \int_0^t f(t)\, dt\right) \lambda \tag{4.23}$$

Taking the derivative of both sides yields

$$\frac{df(t)}{dt} = -\lambda f(t) \tag{4.24}$$

This leads to

$$\frac{df(t)}{f(t)} = -\lambda\, dt \tag{4.25}$$

and

$$f(t) = Ce^{-\lambda t} \tag{4.26}$$

where C is an integration constant.

It is also known that

$$\int_0^\infty f(t)\, dt = 1 \tag{4.27}$$

which leads to

$$f(t) = \lambda e^{-\lambda t} \tag{4.28}$$

Thus, the probability density function for the random variable for time between wars is exponentially distributed. The average time between wars is given by $1/\lambda$. Whether a stochastic process actually yields an exponential distribution is an empirical question. The answer to this question depends on the analyst's underlying assumptions.

An exponentially distributed random variable is characterized as "memoryless" or "ageless." To illustrate this property, let us assume that the waiting time of each shopper in line at a checkout counter is exponentially distributed. The probability of the customer waiting an additional 5 minutes after having waited 10 minutes is the same as the probability of waiting an additional 5 minutes after having waited only 2 minutes. In short, the waiting time from any point in time does not depend on how long the customer has already waited. As another example, let us suppose that the lifetime of highway bridges is exponentially distributed. This implies that a new bridge is just as likely as the Golden Gate Bridge (built 1937) to last another three years. Therefore, drive with confidence.

The Poisson Distribution

Unlike the uniform and exponential distributions, which are continuous, the Poisson distribution is discrete and relates to the number of occurrences of an event during an interval of time. The Poisson distribution usually describes a counting process, that is, the number of times an event might occur in t time units. The probability of exactly k events in time t is given by

$$p(k) = \frac{(\lambda t)^k}{k!} e^{-\lambda t} \tag{4.29}$$

where λ is the mean number of events per unit time. Figures 4.8 and 4.9 show two ways of viewing the Poisson distribution.

We can most effectively use the Poisson process to model systems when the following assumptions are satisfied. The systems to be modelled must

Deal with the number of events or elements
Occur continuously over time
Have a constant rate of events or elements occurring over a unit of time
Have a negligible probability of two or more events occurring simultaneously

FIGURE 4.8 Probability of Events vs. t for a Poisson Distribution with $\lambda = 1$

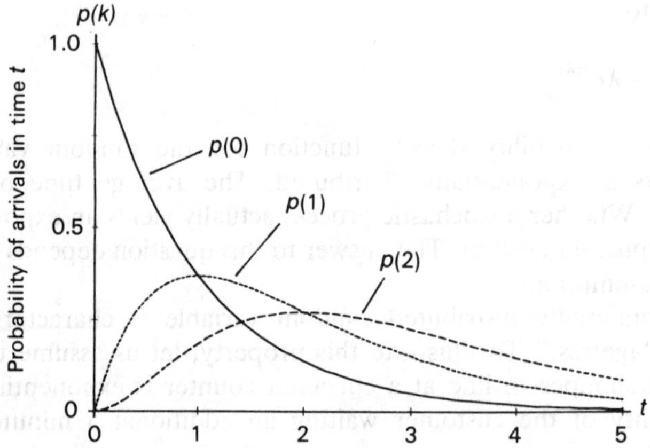

FIGURE 4.9 Discrete Probability Density Function for a Poisson Distribution with $\lambda = 1$

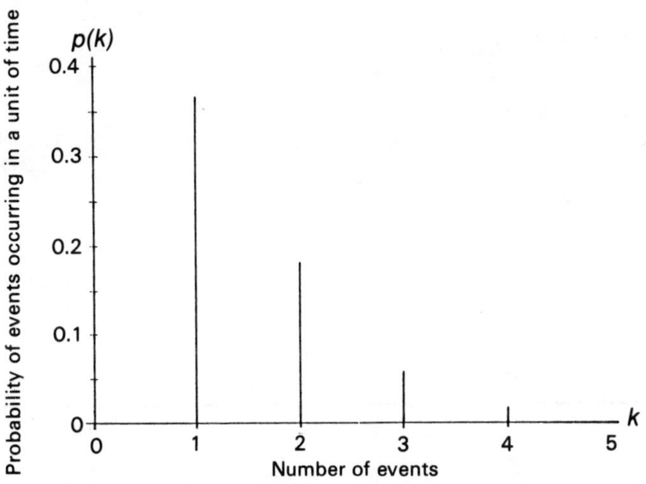

A Poisson distribution describes many phenomena in the real world. For example, the number of cars entering a tunnel during a 24-hour period can be very large. Yet, the probability of a car arriving at the tunnel within a small period of time is small.

L. Bortkiewicz described another Poisson distribution example.[3] He recorded the number of deaths due to kicks by a horse for 10 corps of the Prussian army over a period of 20 years. This example fits a Poisson distribution quite well, suggesting that the assumptions underlying Poisson distributions were well met for horse kicks in the Prussian army. Other examples of Poisson processes include lightning strikes within a specified area, occurrence of suicides, and occurrence of accidents.

The Poisson process is especially useful in simulating social systems and queueing systems because it represents a model whose assumptions mirror actual phenomena. Other examples include modelling the number of programming errors per line in a computer program, the arrival of messages entering a computer network, and the number of computer crashes per day.

The mean and variance of number of events per unit time for the Poisson distribution are equal and can be computed to be

$$\mu = \lambda \qquad \text{and} \qquad \sigma = \lambda$$

respectively.

The Normal Distribution

The normal distribution, often called the Gaussian distribution, is the most widely used distribution for continuous random variables. The probability density func-

FIGURE 4.10 Probability Density Function of the Normal Distribution

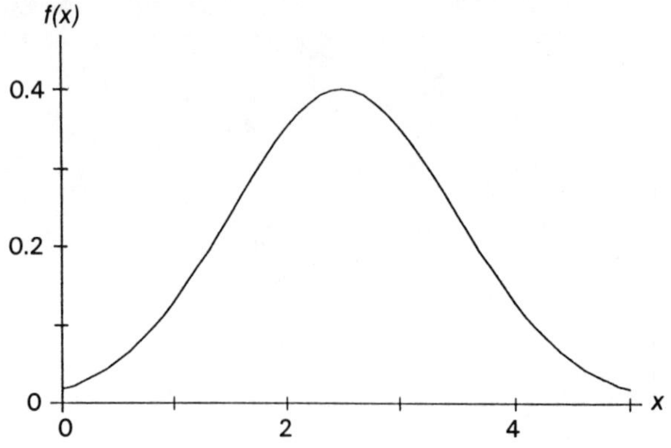

tion is bell-shaped, as shown in Figure 4.10, and is given by

$$f(x) = \frac{1}{\sqrt{2\pi\sigma^2}} \exp\left[-\frac{1}{2}\left(\frac{x-\mu}{\sigma}\right)^2\right] \tag{4.30}$$

where $-\infty < x < \infty$
μ = the mean
σ = the standard deviation

It should be clear that the normal distribution is completely specified when we identify values for the mean and standard deviation. The distribution function is shown in Figure 4.11.

FIGURE 4.11 Probability Distribution Function of the Normal Distribution

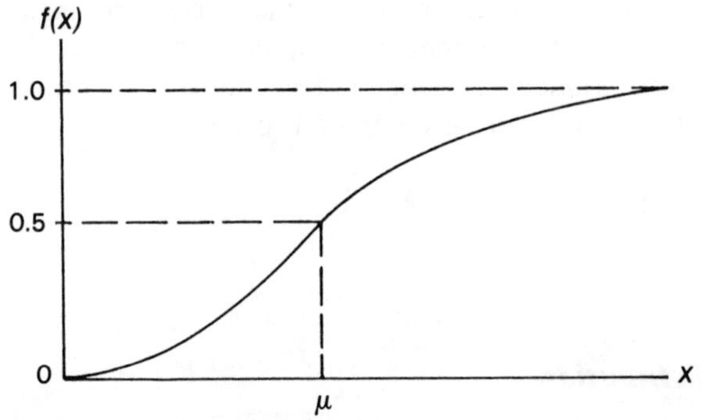

The standard normal distribution has $E[X] = \mu = 0$ and standard deviation $\sigma_X = 1$. However, any random variable Y, with expected value $E[Y]$ and variance $\sigma^2[Y]$, that is normally distributed can be converted to a standard normal distribution by the formula

$$X = \frac{Y - E[Y]}{\sigma[Y]} \quad \textbf{(4.31)}$$

The subtraction in the numerator "slides" the normal so that its mean is at zero. The division (denominator) "dilates" or "compresses" the normal so that its standard deviation is one.

EXAMPLE 4.6 Assume that Y is a normally distributed random variable with $E[Y] = 5$ and $\sigma^2[Y] = 9$. What is the probability that a randomly selected Y_i will lie between -4 and 6?

Solution See Figure 4.12. Conversion to the standard normal distribution leads to

$$x_1 = \frac{-4 - 5}{3} = -3 \quad \textbf{(4.32)}$$

and $$x_2 = \frac{6 - 5}{3} = 0.33 \quad \textbf{(4.33)}$$

The probability

$$p(-4 < Y < 6) = p(-3 < X < 0.33) = 0.4987 + 0.1293 = 0.6280 \quad \textbf{(4.34)}$$

can be obtained from the table on p. 126. The probability that a random number lies within one standard deviation is 0.6826; within two standard deviations, 0.9544; and within three standard deviations, 0.9974.

FIGURE 4.12 Conversion to a Standard Normal Distribution

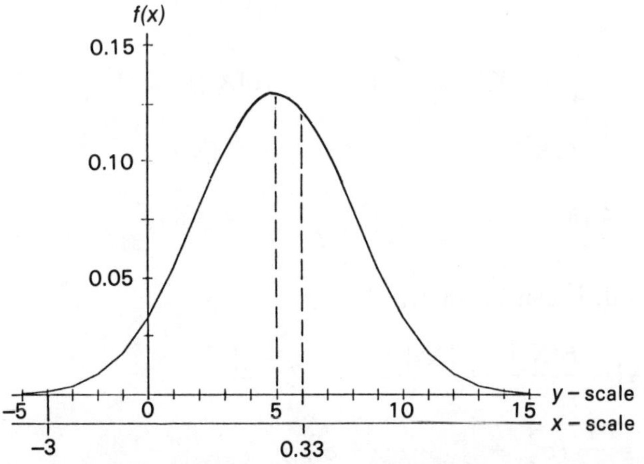

4.10 SAMPLING

Up to this point, we have been computing the mean and standard deviation of a population. That is, we have sampled every element and computed the population parameters μ and σ. Because this process can be expensive, we often sample a subset of the population, called the sample, and estimate parameters μ and σ by the sample mean and the sample standard deviation, respectively.

Let $X_1, X_2, \ldots, X_n$ be n random samples of a population, where n is the sample size. The observed sample values X_i are referred to as random variables. The sample mean is given by

$$\bar{X} = \frac{X_1 + X_2 + \cdots + X_n}{n} \tag{4.35}$$

EXAMPLE 4.7 Let X_1 and X_2 be random variables. Find $E[\bar{X}]$ and $\sigma^2_{\bar{X}}$. Assume that $E[X_1X_2] = E[X_1]E[X_2]$.

Solution

$$\bar{X} = \frac{X_1 + X_2}{2}$$

$$E(\bar{X}) = E\left[\frac{X_1 + X_2}{2}\right] = \frac{1}{2}(E[X_1] + E[X_2])$$

$$\sigma^2_{\bar{X}} = E[\bar{X}^2] - E^2[\bar{X}]$$

where $E[\bar{X}^2] = E\left[\frac{(X_1 + X_2)(X_1 + X_2)}{4}\right]$

$$= \frac{1}{4}(E[X_1^2] + E[X_2^2] + 2E[X_1]E[X_2])$$

Then

$$\sigma^2_{\bar{X}} = \frac{1}{4}E[X_1^2] + \frac{1}{4}E[X_2^2] + \frac{1}{2}E[X_1]E[X_2]$$

$$- \frac{1}{4}(E^2[X_1] + E^2[X_2] + 2E[X_1]E[X_2])$$

$$= \frac{1}{4}(E[X_1^2] - E^2[X_1]) + \frac{1}{4}(E[X_2^2] - E^2[X_2])$$

$$= \frac{1}{4}(\sigma^2_{X_1} + \sigma^2_{X_2})$$

In general, it can be shown that

$$E[\bar{X}] = \frac{E[X_1] + E[X_2] + \cdots + E[X_n]}{n}$$

and $\sigma^2_{\bar{X}} = \frac{1}{n^2}(\sigma^2_{X_1} + \sigma^2_{X_2} + \cdots + \sigma^2_{X_n})$

If, in addition, the random variables X_i have the same mean, μ, and variance, σ^2, then the expected value of the sample mean, $\bar{X}$, is equal to μ and the variance about $\bar{X}$, $\sigma^2_{\bar{X}}$, is equal to $\sigma^2_{\bar{X}}$ divided by the sample size, n. The sample variance is given by

$$s^2 = \frac{(X_1 - \bar{X})^2 + (X_2 - \bar{X})^2 + \cdots + (X_n - \bar{X})^2}{n-1} \tag{4.36}$$

The population mean is $E[X] = \mu$, and the population variance is given by σ^2. The sample mean and sample variance are called estimators of their corresponding population values. Furthermore, if X is normally distributed, so too is $\bar{X}$.

4.11 CENTRAL LIMIT THEOREM

The central limit theorem of probability states that if each of the n random variables X_i has a mean μ and variance σ^2, then the sample mean, $\bar{X}$, has a normal distribution with mean μ and variance σ^2/n as the sample size, n, becomes large. This allows us to use the properties of normal distributions when we take large sample sizes. The statistics gathered from the simulation of stochastic systems can be viewed as statistics gathered from a sampling process.

EXAMPLE 4.8 Suppose an urn contains five balls, two labelled 3, one labelled 2, and two labelled 1. The probability of obtaining a 3 is $\frac{2}{5}$, a 2 is $\frac{1}{5}$, and a 1 is $\frac{2}{5}$. If Y is the random variable denoting a number drawn, the theoretical (population) mean of the probability distribution is

$$\mu = E[Y] = 3\left(\frac{2}{5}\right) + 2\left(\frac{1}{5}\right) + 1\left(\frac{2}{5}\right) = 2$$

The theoretical variance is

$$\begin{aligned}
\sigma^2 &= E[(Y - \mu)^2] \\
&= E[Y^2 - 2Y\mu + \mu^2] \\
&= E[Y^2] - 2\mu^2 + \mu^2 \\
&= E[Y^2] - \mu^2 \\
&= 3^2\left(\frac{2}{5}\right) + 2^2\left(\frac{1}{5}\right) + 1^2\left(\frac{2}{5}\right) - 2^2 \\
&= \frac{4}{5}
\end{aligned}$$

Now suppose we conduct a series of experiments in which two balls are drawn at random from the urn. After each ball is drawn, it is replaced in the urn. This experiment is repeated four times, with the following results:

	First Ball (Y_1)	*Second Ball* (Y_2)	*Mean*
Repetition 1	3	2	2.5
Repetition 2	1	3	2.0
Repetition 3	2	1	1.5
Repetition 4	3	2	2.5
			8.5

From equation (4.35) we obtain the mean of the sample means:

$$\bar{X} = \frac{8.5}{4} = 2.125$$

and from equation (4.36) we obtain the sample variance of the sample means:

$$s^2 = \tfrac{1}{3}[(2.5 - 2.125)^2 + (2.0 - 2.125)^2 + (1.5 - 2.125)^2 + (2.5 - 2.125)^2]$$
$$= 0.229$$

The *mean of the sample means* is an estimate of the theoretical (population) mean of the random variable $\bar{Y}$, the sample mean. We also know that

$$E[\bar{Y}] = \mu$$

Similarly, the variance is

$$\sigma^2_{\bar{Y}} = \frac{\sigma^2}{2} = \frac{4/5}{2} = 0.4$$

If we had repeated this experiment, drawing three balls at random instead of two, then

$$\sigma^2_{\bar{Y}} = \frac{\sigma^2}{n} = \frac{4/5}{3} = 0.267$$

Upon more careful inspection of this example, we would have noticed that the mean of the repetitions is itself a random variable with its own distribution. This distribution is a *sampling distribution*. The sampling distribution also has a mean and a variance. The sample mean $E[\bar{Y}]$ of the sample means $(Y_1 + Y_2)/2$ is an estimate of the theoretical (population) mean of the random variable Y:

$$E[\bar{Y}] = E[Y] = \mu$$

Similarly, the variance of the sampling distribution is equal to the population variance divided by n^2:

$$\sigma^2_{\bar{Y}} = \frac{\sigma^2}{n^2}$$

Often we do not know σ^2 and use s^2 instead to compute

$$\sigma^2_{\bar{Y}} = \frac{s^2}{n^2}$$

EXAMPLE 4.9 This example is designed to clarify some sampling concepts. Consider the following 200 integers as our population:

104	108	87	111	130	95	108	104	108	99	105	89	102	104
99	97	100	77	89	88	86	95	77	108	120	100	106	109
93	92	103	107	101	105	105	89	90	94	100	102	104	85
93	103	107	97	107	94	107	99	95	114	88	125	86	90
102	99	98	111	105	96	97	103	102	92	106	85	92	97
88	80	110	97	91	110	111	108	113	97	115	95	87	97
110	111	80	98	103	114	99	103	83	77	101	113	93	102
110	114	93	105	95	92	89	111	109	100	107	97	91	110
108	86	104	95	97	98	105	100	96	98	106	90	99	108
102	98	93	91	100	96	92	95	99	95	110	104	84	127
91	102	100	92	106	99	89	94	99	93	110	107	88	108
110	104	90	100	93	91	108	96	101	98	93	98	93	104
114	97	89	108	112	115	101	99	109	103	102	109	97	89
88	111	103	85	95	101	119	92	96	92	98	87	116	98
90	81	103	113										

The histogram of this population is shown in Figure 4.13. The population appears to be normally distributed, and we compute its mean to be

$$\mu = 99.60$$

and its standard deviation to be

$$\sigma = 9.37$$

Now suppose we randomly drew the following 10 numbers from the population:

108 77 89 88 106 106 92 86 93 110

FIGURE 4.13 Histogram of Population for Example 4.9

Midpoint	Count	
75	3	◆◆◆
80	3	◆◆◆
85	11	◆◆◆◆◆◆◆◆◆◆◆
90	30	◆◆◆◆◆◆◆◆◆◆◆◆◆◆◆◆◆◆◆◆◆◆◆◆◆◆◆◆◆◆
95	37	◆◆◆◆◆◆◆◆◆◆◆◆◆◆◆◆◆◆◆◆◆◆◆◆◆◆◆◆◆◆◆◆◆◆◆◆◆
100	40	◆◆
105	32	◆◆◆◆◆◆◆◆◆◆◆◆◆◆◆◆◆◆◆◆◆◆◆◆◆◆◆◆◆◆◆◆
110	29	◆◆◆◆◆◆◆◆◆◆◆◆◆◆◆◆◆◆◆◆◆◆◆◆◆◆◆◆◆
115	10	◆◆◆◆◆◆◆◆◆◆
120	2	◆◆
125	2	◆◆
130	1	◆

The sample mean is

$$\bar{X} = 95.50$$

and the sample standard deviation is

$$s = 11.24$$

Thus, $\bar{X}$ and s are estimates of μ and σ, respectively. At this point, we might say that $\bar{X}$ approximates μ. However, before we conclude this, let us randomly draw another 10 samples from the population:

97 105 89 86 95 100 110 107 101 102

The sample mean here is

$$\bar{X} = 99.20$$

and the sample standard deviation is

$$s = 7.63$$

We can repeat this many more times and develop the sampling distribution of $\bar{X}$.

The mean of the two sample means is

$$m_{\bar{X}} = \frac{95.50 + 99.20}{2} = 97.35$$

and $s_{\bar{X}} = 2.62$

$$\sigma_{\bar{X}} = \frac{\sigma}{n} = \frac{9.37}{10} = 0.937$$

Now let us see the effects of sample size. Suppose we randomly sampled the following 20 integers from the population:

100	110	98	92	104	106	83	93	97	100
103	102	91	98	108	97	107	103	119	98

Then

$$\bar{X} = 100.45$$

and $s = 7.80$

Compare the value of $\bar{X}$ to μ. What can we conclude from these results? Let us repeat this process by randomly sampling another 20 numbers:

100	104	96	108	93	111	77	93	105	89
125	108	108	99	107	107	103	99	95	84

Then

$$\bar{X} = 100.55$$

and $s = 10.58$

If we continued to repeat this sampling and calculation again and again, we would get a sampling distribution. For our two repetitions of 20 samples,

$$m_{\bar{X}} = \frac{100.45 + 100.55}{2} = 100.5$$

$$s_{\bar{X}} = 0.07$$

$$\sigma_{\bar{X}} = \frac{9.37}{20} = 0.469$$

Let us increase our sample size by randomly drawing the following 50 numbers:

114	99	89	88	105	100	114	108	103	93
98	109	102	110	104	97	100	95	108	108
95	92	100	111	111	104	91	85	97	96
111	87	105	92	91	98	102	111	120	103
91	103	100	97	99	104	127	97	97	119

Then

$$\bar{X} = 101.60$$

$$s_{\bar{X}} = 8.97$$

$$\sigma_{\bar{X}} = \frac{9.37}{50} = 0.18$$

What can we conclude from our results now? Let us increase the sample one more time by randomly drawing the following 100 numbers:

77	111	89	108	89	98	110	113	99	111
105	97	87	98	103	108	90	92	104	110
93	100	99	110	104	102	108	97	130	106
88	108	104	115	86	110	105	110	102	88
110	109	109	99	98	77	114	89	96	93
102	110	99	106	91	119	92	97	80	88
110	111	108	97	101	98	96	94	87	107
92	108	85	102	114	97	102	125	99	92
103	84	108	104	101	100	113	89	95	92
105	93	100	95	98	103	99	95	95	114

Then

$$\bar{X} = 100.53$$

$$s_{\bar{X}} = 9.7$$

$$\sigma_{\bar{X}} = \frac{9.37}{100} = 0.0937$$

Figure 4.14 shows the relationship between sample size and the sampling distributions of sample means.

FIGURE 4.14 Relationship between Sample Size and Sampling Distribution of Sample Means

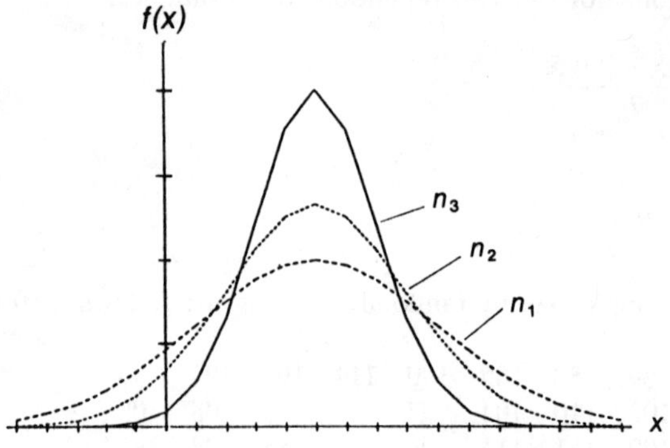

4.12 REGRESSION ANALYSIS

In most processes of interest to us as simulation analysts, variables are functionally related. That is, the values of some variables depend on the value of others. Regression analysis is used to obtain the relationship between a dependent variable and one or more independent variables. Algebraically, the relationship is approximated by a function that involves a set of parameters called regression coefficients. When a function is linear in the parameters, the regression is said to be linear. We should choose the form of the regression equation on the basis of the relationship, rather than on the basis of statistics. We can then use an equation derived through regression analysis to derive the regression coefficients, enabling us to predict the behavior of the dependent variable from a set of values of the independent variables.

The least squares method is the most widely used regression analysis technique. The least squares coefficients of the simple linear regression equation,

$$y = B_0 + B_1 x + \text{error} \tag{4.37}$$

are given by

$$B_1 = \frac{\left(n \sum_{i=1}^{n} x_i y_i\right) - \left(\sum_{i=1}^{n} x_i\right)\left(\sum_{i=1}^{n} y_i\right)}{\left(n \sum_{i=1}^{n} x_i^2\right) - \left(\sum_{i=1}^{n} x_i\right)^2} \tag{4.38}$$

and
$$B_0 = \left(\frac{1}{n} \sum_{i=1}^{n} y_i\right) - B_1\left(\frac{1}{n} \sum_{i=1}^{n} x_i\right) \tag{4.39}$$

where n = the number of samples

y_i = the ith sample of the dependent variable

x_i = the ith sample of the independent variable

FIGURE 4.15 Least Squares Linear Regression

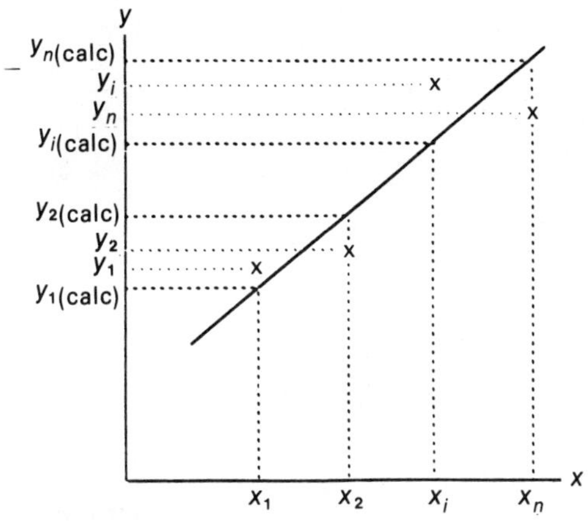

These choices of B_0 and B_1 minimize $\sum_{i=1}^{n} (y_i - y_{i(\text{calc})})^2$, where $y_{i(\text{calc})}$ is the value calculated from equation (4.37) with $x = x_i$ and error = 0. See Figure 4.15.

A multiple linear regression equation has the following form:

$$y = B_0 + B_1x_1 + B_2x_2 + \cdots + B_px_p + \text{error} \tag{4.40}$$

where y = the dependent variable

$x_1, x_2, \ldots, x_p$ = the independent variables

Another form of the regression equation is called the polynomial regression and is given by

$$y = B_0 + B_1x + B_2x^2 + \cdots + B_px^p \tag{4.41}$$

EXAMPLE 4.10 Obtain a least squares fit for the following data:

x_i	y_i	x_iy_i	x_i^2
0	1	0	0
1	2	2	1
3	4	12	9
5	6	30	25
9	13	44	35

Solution Using equations (4.38) and (4.39), we obtain the following:

$$B_1 = \frac{4(44) - 9(13)}{4(35) - (9)^2} = 1$$

$$B_0 = \frac{1}{4}(13) - 1\left(\frac{9}{4}\right) = 1$$

Therefore, the regression equation is

$$y = 1 + x$$

EXAMPLE 4.11 Obtain a linear least squares regression for the following data:

Sample i	*Country*	*GNP (\$Billion)* (y)	*Television Sets per 1000 Population* (x)
1	United States	1709.4	571
2	Japan	620.2	239
3	West Germany	494.8	311
4	France	350.3	309
5	United Kingdom	230.3	317
6	Italy	181.5	220
7	The Netherlands	94.8	274
8	Brazil	147.8	96
9	Soviet Union	1029.9	217

Source: *Statistical Abstract of the United States*, 1980, p. 922.

Solution

$$\sum_{i=1}^{9} y_i = 4859 \qquad \sum_{i=1}^{9} x_i = 2554 \qquad \sum_{i=1}^{9} x_i y_i = 1{,}763{,}008 \qquad \sum_{i=1}^{9} x_i^2 = 855{,}634$$

and thus

$$B_1 = \frac{9(1{,}763{,}008) - (2554)(4859)}{9(855{,}634) - (2554)^2} = 2.94$$

$$B_0 = \frac{1}{9}(4859) - 2.94\left(\frac{1}{9}\right)(2554) = -293.1$$

Therefore, the regression equation is

$$y = -293 + 2.94x$$

4.13 CORRELATION ANALYSIS

Correlation analysis indicates how well a regression fits the sampled data. It says nothing about cause and effect. In short, regression is the process of fitting a curve to data, and correlation is a measure of the quality of that fit.

The correlation coefficient for a simple linear regression is given by

$$r = \frac{n\left(\sum_{i=1}^{n} x_i y_i\right) - \left(\sum_{i=1}^{n} x_i\right)\left(\sum_{i=1}^{n} y_i\right)}{\sqrt{\left[n\left(\sum_{i=1}^{n} x_i^2\right) - \left(\sum_{i=1}^{n} x_i\right)^2\right]\left[n\left(\sum_{i=1}^{n} y_i^2\right) - \left(\sum_{i=1}^{n} y_i\right)^2\right]}} \tag{4.42}$$

The correlation coefficient can have values between -1 and 1 inclusive. If $r = -1$, the regression equation is a perfect fit and the dependent variable decreases as the independent variable increases. When $r = 0$, there is no correlation at all and the equation is not a good fit. If $r = 1$, the regression equation also fits the data perfectly and the dependent variable, y, increases as the independent variable, x, increases. The coefficient of determination, r^2, is defined as the ratio of explained variations from the mean, $\sum_{i=1}^{n} (y_{i(\text{calc})} - \bar{y})^2$, to the total variation from the mean, $\sum_{i=1}^{n} (y_i - \bar{y})^2$. The unexplained variation from the mean is $\sum_{i=1}^{n} (y_i - y_{i(\text{calc})})^2$. Here, $y_{i(\text{calc})}$ is the value of y calculated from the regression equation, y_i are the sample points at sample point i, and $\bar{y} = (1/n) \sum y_i$.

EXAMPLE 4.12 Obtain the correlation coefficient and the coefficient of determination for the data in Example 4.10.

Solution Using equation (4.42),

$$r = \frac{4(44) - (9)(13)}{\sqrt{[4(35) - (9)^2][4(57) - (13)^2]}} = \frac{59}{\sqrt{(59)^2}} = 1$$

This result indicates that we have a perfect fit, with $B_1 > 0$ and $r^2 = 1$.

EXAMPLE 4.13 Obtain the correlation coefficient and the coefficient of determination for the data in Example 4.11.

Solution Again, using equation (4.42),

$$r = \frac{9(1{,}763{,}008) - (2554)(4859)}{\sqrt{[9(855{,}634) - (2554)^2][9(4{,}851{,}739) - (4859)^2]}}$$

$$= 0.711$$

$$r^2 = 0.505$$

MINITAB, SPSS, BMDP, GAUSS, and SAS are some special-purpose packages that are available to the scientist to calculate the statistics of various random variables. Now that we have studied the basic tools for probability and statistics, we will use them when we develop simulation models in later chapters. We will also use these tools to determine the number of simulation runs needed to establish confidence in our results. The regression and correlation analyses will help us determine the relationships of interaction between entities in the system. Regression analysis can also be used as an external description for systems.

4.14 SUMMARY

In this chapter, we studied the fundamental concepts of probability theory and used the idea of relative frequency to derive basic axioms. Random variables, along with distribution and density functions, were introduced.

We learned the characteristics and statistics of four distributions (uniform, exponential, Poisson, and normal) which will be used later in the text when we simulate stochastic systems such as queueing systems and production systems. They are also the basis for the design of most of today's communication systems. In the next chapter, we will be interested in generating random variables having these distributions. Both regression and correlation analysis provide us with the ability to relate system variables to one another. Commercially available statistical packages can perform these analyses and are designed to relieve us of tedious computation. The topics covered in this chapter will also be used when we address the number of simulation experiments needed to develop confidence in the simulation results.

4.15 REVIEW PROBLEMS

1. Define a random variable and give some examples.
2. What is the distribution function? What does it tell us?
3. Draw the distribution function for the rolling of a die.
4. Describe the density function. What information does it convey?
5. The following are population and total employment data from 1960 to 1985 for the United States:

Year	*Population (Thousands)*	*Employment (Thousands)*
1976	218,035	90,420
1977	220,239	93,673
1978	222,585	97,679
1979	225,055	100,429
1980	227,757	100,907
1981	230,138	102,042
1982	232,138	101,194
1983	234,799	102,510
1984	237,019	106,702
1985	239,283	108,856

Source: *Statistical Abstract of the United States*, 1987, pp. 8, 374.

Using a statistical package:

a. Obtain a linear regression between time and population.
b. Repeat with a polynomial regression.
c. Find the correlation between population and employment.

6. The average monthly marriage rate in France for the years 1816–1865 is given below.

January	128
February	152
March	52
April	111
May	97
June	116
July	94
August	72
September	90
October	101
November	131
December	56
	1200

Source: M.G. Mulhall. *Dictionary of Statistics* (London: Associated Book Publishers (U.K.) Ltd., 1909). Reprinted with permission.

Obtain a least squares linear regression for the marriage rate as a function of time. What is the correlation coefficient? What is the coefficient of determination? Is it a good fit? Justify your answer.

7. The following data were compiled for the State of Connecticut for the year 1774:

Town	*Males over 70, Married*	*Males over 70, Single*	*Town*	*Males over 70, Married*	*Males over 70, Single*
1	11	5	16	13	3
2	20	10	17	28	13
3	29	7	18	13	3
4	20	5	19	22	19
5	37	8	20	13	5
6	21	5	21	12	6
7	35	17	22	6	4
8	3	17	23	35	9
9	42	20	24	15	10
10	15	8	25	48	44
11	23	10	26	33	10
12	39	8	27	20	6
13	14	3	28	19	8
14	15	5	29	34	5
15	12	6	30	14	8

continues

Continued

Town	*Males over 70, Married*	*Males over 70, Single*	*Town*	*Males over 70, Married*	*Males over 70, Single*
31	49	13	52	17	8
32	55	23	53	26	2
33	21	11	54	8	2
34	26	5	55	35	3
35	22	13	56	11	13
36	14	6	57	7	1
37	30	11	58	7	3
38	19	9	59	7	0
39	9	3	60	7	0
40	20	6	61	9	1
41	43	8	62	10	5
42	10	4	63	8	2
43	7	4	64	19	6
44	38	14	65	4	1
45	19	5	66	11	1
46	21	1	67	19	9
47	17	8	68	3	5
48	36	14	69	9	1
49	43	9	70	1	0
50	17	13	71	22	16
51	13	4			

Source: Connecticut Colony Public Records, vol. 14, pp. 485–91; *A Century of Population Growth, 1790–1900* (Baltimore: Genealogical Publishing Co., 1967). Reprinted with permission.

a. Assume that the number of married males over 70 is a random variable. Obtain the probability density function, $f(x)$, from the population of people over 70. What is the probability of meeting between five and six married males over 70 years old in a town?

b. Repeat part (a) using the single population as the random variable. What is the probability of meeting between three and four single males over 70 years old in a town?

8. Generate 12 random numbers in the range (0, 1) and form their sum. Repeat the process 500 times. Calculate the mean value of the sums. Calculate the standard deviation about the mean of the sums. Obtain the probability density function of the sums. What theorem, among those studied, is relevant here?

9. Suppose that a coin is tossed four times. Plot the probability distribution function for the number of heads. Repeat for the number of tails. Repeat the problem with five tosses.

10. Derive equations (4.38) and (4.39).

11. Apply the least squares method to find the coefficients of the multiple regression equation

$$y = B_0 + B_1x_1 + B_2x_2 + \text{error}$$

12. Electrical engineering students often obtain the frequency response of electrical circuits in the laboratory. Paul and Susan obtained the following data for the RC circuit shown in Figure 4.16.

ω	$T(\omega)$
100	0.71
200	0.45
300	0.32
400	0.24
500	0.20
600	0.16
700	0.14
800	0.12
900	0.11
1000	0.10
1100	0.09
1200	0.08
1300	0.08
1400	0.07
1500	0.07
1600	0.06
1700	0.06
1800	0.05
1900	0.05
2000	0.05

FIGURE 4.16 Problem 12

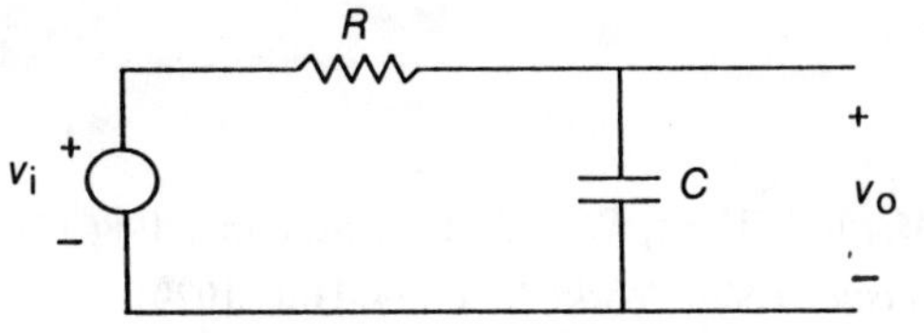

It can be easily shown that the theoretical values of RC can be computed using

$$\left|\frac{V_o(j\omega)}{V_i(j\omega)}\right| = T(\omega) = \frac{1}{\sqrt{(RC\omega)^2 + 1}}$$

a. Develop a linear least squares model for $T(\omega)$.

b. Develop a polynomial least squares model for $T(\omega)$. The model has the form

$$T(\omega) = a_0 + a_1\omega + a_2\omega^2 + \cdots + a_n\omega^n + \text{error}$$

c. Can you suggest a more appropriate model?

d. Using the experimental data, obtain the least squares fit for parts (a) through (c).

e. Find the value of RC.

f. Plot the measured data and the least squares data. Discuss your results. You may use statistical packages to solve this problem.

13. Assume the probability density function shown in Figure 4.17.

a. Find the mean value of X.

b. Find the mean value of X when $0 \le X \le 1$.

FIGURE 4.17 Problem 13

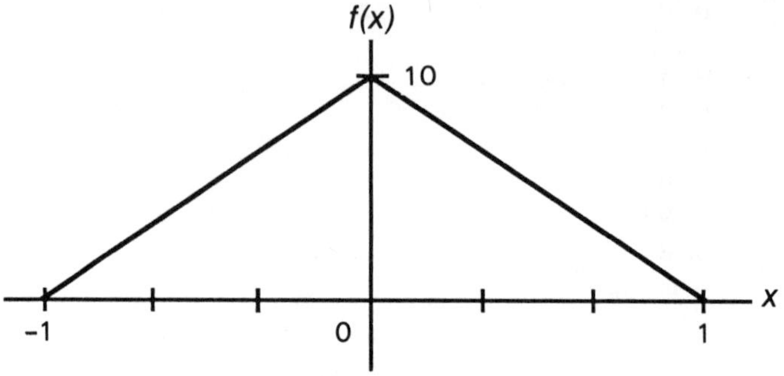

14. Let X be a Poisson random variable with parameter λ. Show that $E[X] = \lambda$.

15. Assume a Gaussian density function. Suppose that we choose a capacitor C from a box of capacitors with population mean $\mu = 1000\ \mu F$ and standard deviation $\sigma = 10\ \mu F$. Compute the probability that C will have a value of $1000\ \mu F \pm 10\%$.

4.16 REFERENCES

1. Aptech Systems, *Gauss Manual* (Kent, WA: Aptech Systems, 1986).
2. Blalock, H.M., *Social Statistics* (New York: McGraw-Hill, 1979).
3. Bortkiewicz, L., *Das Gesetz der Kleinen Zahlen* (Leipzig: Teubner, 1888).
4. Coleman, J., *Introduction to Mathematical Sociology* (Free Press of Glencoe; London: Collier-Macmillan, 1964).
5. Daellenbach, H.G.; George, J.A.; and McNickle, D.C., *Introduction to Operations Research Techniques* (Boston: Allyn & Bacon, 1983).
6. Dixon, W.J., ed., *Biomedical Computer Programs* (Berkeley, CA: University of California Press, 1975).
7. Dixon, W.J., ed., *BMDP Statistical Software* (Berkeley, CA: University of California Press, 1981).
8. Draper, N.R., and Smith, H., *Applied Regression Analysis* (New York: Wiley, 1981).
9. Ens, P.G., *Business Statistics: Methods and Applications* (Homewood, IL: Richard D. Irwin, 1985).
10. Feller, W., *An Introduction to Probability Theory and Its Applications* (New York: Wiley, 1957).
11. Frank, C.R., *Statistics and Economics* (New York: Holt, Rinehart, & Winston, 1971).

12. Freund, J.E., and Walpole, R.E., *Mathematical Statistics* (Englewood Cliffs, NJ: Prentice-Hall, 1987).

13. Halley, G., *Introduction to Business Statistics* (San Francisco: Holden-Day, 1968).

14. Hastings, N.A.J., and Peacock, J.B., *Statistical Distributions* (New York: Wiley, 1975).

15. Helwig, J.T., *SAS Introductory Guide* (Raleigh, NC: SAS Institute, 1979).

16. Kaplan, R.M., *Basic Statistics for the Behavior Sciences* (Boston: Allyn & Bacon, 1987).

17. Larsen, Richard J., and Marx, Morris L., *An Introduction to Mathematical Statistics and Its Applications* (Englewood Cliffs, NJ: Prentice-Hall, 1981).

18. Mood, A.M.; Graybill, F.A.; and Boes, D.C., *Introduction to the Theory of Statistics*, 3rd ed. (New York: McGraw-Hill, 1974).

19. Papoulis, A., *Probability, Random Variables, and Stochastic Processes*, 2nd ed. (New York: McGraw-Hill, 1984).

20. Parzen, E., *Stochastic Processes* (San Francisco: Holden-Day, 1967).

21. Render, B., and Stair, R.M., Jr., *Quantitative Analysis for Management* (Boston: Allyn & Bacon, 1985).

22. Ryan, Barbara; Joiner, F.; Brian, L.; and Ryan, Thomas A., Jr., *Minitab Handbook*, 2nd ed. (Boston: Duxbury, 1985).

23. SPSS Inc., *SPSS User's Guide* (Chicago, IL: SPSS Inc., 1986).

5 The Monte Carlo Method and Simulation

5.1 INTRODUCTION

Computer simulation of stochastic processes depends on the ability of the simulation program to produce a sequence of "random" numbers having required probability distributions. The Monte Carlo simulation method, for example, imitates a process step by step, but requires suitable random numbers to simulate the process, which itself proceeds randomly. There are various ways to obtain "random" numbers, computer methods and random number tables being the most popular[12,13,20]. The method we choose as best depends on the system being simulated and the computer facilities available.

Computer methods usually rely on a uniformly distributed random number generator, which can be used to generate random variables of any desired distribution. Because computers are finite-state machines and have finite word length, we cannot expect to find a random number generator algorithm that gives a *perfect* uniform distribution. The most desirable characteristics of (computer) random number generators are as follows:

1. The program must execute quickly.
2. The program should have the capability of producing a long sequence of random numbers before it begins to repeat the sequence (cycle). The length of the sequence is called a *period*. The period always exists because the digital computer is a finite-state machine.
3. The program should be short in order to minimize computer memory requirements.
4. The program should be capable of reproducing the sequence. Because of this capability, all algorithms that produce "random" numbers are really generating numbers in a predictable manner, and thus the numbers are called *pseudorandom*. We usually supply the starting number, called the *seed*, which determines the sequence of the pseudorandom numbers.

The ability to reproduce the random number sequence provides us with the ability to repeat the simulation for further study. The history and methods of generating random numbers are well documented in references 11, 12, 13, 16, 23, and 24.

In this chapter, we will present a uniformly distributed random number generator, methods for testing this and other generators, and methods of generating random numbers with nonuniform distributions. We will use the random number generators to simulate a stochastic queueing system. We will also present a method for determining the number of simulation runs needed to establish a certain confidence level.

5.2 UNIFORMLY DISTRIBUTED RANDOM NUMBER GENERATORS

We call the most widely used method for generating pseudorandom numbers the *mixed congruential* method. Theoretical and empirical investigations[12,13] have shown that the recursive formula

$$r_{i+1} = (\alpha r_i + \beta)\text{MOD } N \qquad \textbf{(5.1)}$$

generates uniformly distributed random numbers between zero and $N-1$, with $N = 2^{20} = 1{,}048{,}576$, $\alpha = 1909$, and $\beta = 221{,}571$. The MOD operation is defined as the integer remainder of $(\alpha r_i + \beta)/N$.

The period of the pseudorandom sequence is $N = 2^{20} = 1{,}048{,}576$, and the limits on the seed are $1 \le r_0 < N$.

EXAMPLE 5.1 Let us generate two uniformly distributed random numbers using the mixed congruential method. First, we randomly choose the seed to be 10. The first random number will be

$$\begin{aligned} r_1 &= [(1909)(10) + 221{,}571]\text{MOD } 1{,}048{,}576 \\ &= 240{,}661 \text{ MOD } 1{,}048{,}576 \\ &= 240{,}661 \end{aligned}$$

and

$$\begin{aligned} r_2 &= [(1909)(240{,}661) + 221{,}571] \text{ MOD } 1{,}048{,}576 \\ &= 459{,}643{,}420 \text{ MOD } 1{,}048{,}576 \\ &= 367{,}132 \end{aligned}$$

EXAMPLE 5.2 Let us generate two uniformly random numbers within the interval (0,1). Using the results of Example 5.1—$r_1 = 240{,}661$ and $r_2 = 367{,}132$—we divide by $N = 1{,}048{,}576$ and obtain

$$r_1 = 0.2295$$

and $$r_2 = 0.3501$$

We call uniformly distributed random numbers in the interval (0,1) *standard uniformly distributed numbers*.

EXAMPLE 5.3 Let us generate uniformly distributed random integers over the interval [1,10]. Here we will use the standard uniformly distributed number of Example 5.2 and let

$$X = \text{Trunc}\,[(0.2295) \times 10] = 2$$

The random integer will then be $X + 1$, or in this case, 3. Similarly, we generate the second random number by $X = \text{Trunc}\,[(0.1388) \times 10] = 1$, which we calculate to be 2.

Since the generation of pseudorandom numbers is machine dependent (computer word size and implementation of arithmetic operations), most computing facilities provide their users with an optimized built-in function or subprogram that generates uniformly distributed random numbers.

5.3 TESTING RANDOM NUMBER GENERATORS

Kendall and Babington-Smith have proposed the following tests for both randomness and uniformity of distribution of random numbers:[11]

1. *Frequency.* We subdivide the total range of random values into L subintervals of equal length and compute the proportion of the random numbers that fall into each subinterval. The proportions should be equal.

2. *Serial relationship.* In this case, we subdivide the interval into L subintervals and compute the proportion of the random numbers falling into two consecutive intervals for all subintervals. These proportions should be equal to $1/L$.

3. *Poker relationship.* We convert the random numbers, which are decimal based, to binary. By comparing successive and nonoverlapping sets of binary numbers, we tabulate the numbers of ones or zeros for each set. Each tabulation corresponds to a single card in a five-card hand. In a sense, we can think of each interval of five sets as a "poker hand" of zeros and ones. We expect the frequency of each poker hand to conform with the expected frequency of the corresponding hand, assuming that each card is equally likely to be drawn.

4. *The gap test.* Here, we divide the interval into subintervals and compute the length of the gaps in the sequence of random numbers. The gap length is considered to be zero if two consecutive random numbers fall in the same interval. The gap length is one if in any subsequence the first and third numbers fall in the same intervals and the second number does not. In general, the gap length is defined as the number of random numbers that occur between random numbers falling in the same interval. The length of the gaps can be $1, 2, \ldots, n, \ldots$. Usually, we take all gap lengths that are greater than n and group them as though they were of length $n + 1$. The test is complete when we compare the theoretical frequencies of the gap lengths to the observed frequencies of the gap lengths.

5.4 THE CHI-SQUARE TEST

The chi-square test is a frequency type of test for random number generators. We will first describe the chi-square distribution and then describe the test.

If $Y_1, Y_2, \ldots, Y_n$ are n random variables, each independently and normally distributed with a mean of zero and theoretical variance σ^2, the statistic

$$\chi^2 = \frac{\sum_{i=1}^{n} Y_i^2}{\sigma^2}$$

has a chi-square distribution with n degrees of freedom. Figure 5.1 shows the probability density functions for the chi-square distribution with $n = 2$ and $n = 6$.

As mentioned earlier, one specific test of the many available for testing random number generators is the chi-square test. To use the chi-square test, we divide the range of the random variable into L mutually exclusive and exhaustive subintervals. Then, from the probability distribution function, we compute the theoretical probabilities $p_i = 1, 2, \ldots, L$ associated with the subintervals such that

$$\sum_{i=1}^{L} p_i = 1 \tag{5.2}$$

We then compute the relative frequencies of a sequence of M random numbers g_i computed for the corresponding subintervals and calculate the quantity

$$\chi_c^2 = \frac{M \sum_{i=1}^{L} (g_i - p_i)^2}{p_i} \tag{5.3}$$

FIGURE 5.1 Density Functions for the Chi-Square Distribution

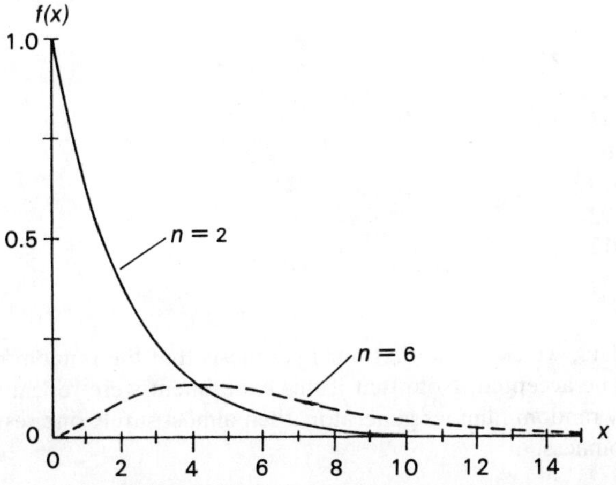

where the subscript c means "computed." We should keep in mind that $Mp_i \geq 5$ for all i. Our computation will result in

χ_c^2 being a random variable, since different sequences of M random numbers would lead to varying χ_c^2 values

The χ_c^2 values being larger whenever any one of the differences $(g_i - p_i)$ is large

A low value of χ_c^2 indicates an acceptable sequence of M pseudorandom numbers, whereas a high value is unacceptable.

The probability distribution function depends on L, the number of subintervals into which the random variables are placed. Thus, there is an entire family of these random variables and there exists an appropriate probability distribution function for each integral value of $L (L \geq 2)$. The probability distribution functions are known as the chi-square distributions of $(L-1)$ degrees of freedom whenever L subintervals are used. These probability distribution functions have been tabulated[15].

EXAMPLE 5.4 ***Chi-Square Test of a Random Number Generator*** Let the maximum number = 10,000, $L = 10$ (9 degrees of freedom), $M = 10{,}000$, and $p_i = 0.1$ (uniform distribution). Calculate χ_c^2 for five sequences of length M using the congruential method described earlier.

Solution From a chi-square distribution table (see Section 5.14), we chose the criterion $\chi_c^2 = 21.66$ as acceptable because we would state that the computed χ_c^2 is significantly large whenever it exceeds 21.66. This criterion indicates that values of $\chi_c^2 > 21.66$ will occur with probability 0.01, assuming that the generator is indeed random. For the five sequences generated, the value of χ_c^2 did not exceed 21.66. Our results for this experiment are tabulated in the table below. The seeds were obtained from the random numbers generated by the same congruential method.

Seed	χ_c^2
2,733	11.944
9,396	3.46
24,402	3.618
2,478	5.492
1,211	16.012

From these data, we conclude that our hypothesis that the random number generator is uniform should be accepted. Note that if this experiment were repeated a large number of times on a truly random number generator, then almost surely one result would seem to indicate nonrandomness.

5.5 SCATTER DIAGRAMS

The scatter diagram method is another effective technique for testing a uniform random number generator. For example, let r_n be the nth pseudorandom number generated in the sequence by the random number generator, and let r_{n+1} be the $(n+1)$th number. The scatter diagram is a plot of random number r_n versus random number r_{n+1}. For a good random number generator, no pattern should be discernible. While seemingly crude, this is an effective method because the eye can visualize patterns that are hard to detect mathematically. A scatter diagram is shown in Figure 5.2.

In the problems at the end of the chapter, we will see the patterns resulting from the scatter diagrams of exponential and normally distributed random numbers. We suggest looking at a scatter diagram of a random number generator that produces numbers serially in order to see the pattern, which can help us test the quality of the generators.

FIGURE 5.2 Scatter Diagram for a Uniform Random Number Generator

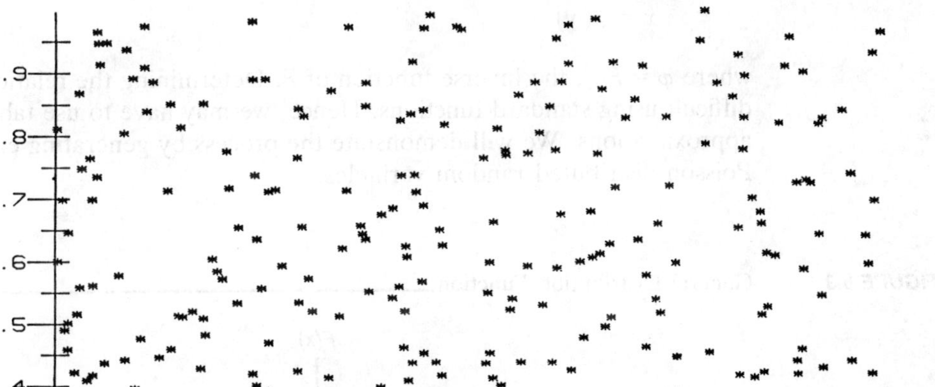

5.6 PRINCIPLES OF RANDOM PROCESS GENERATORS

The simulation of stochastic processes often requires the generation of nonuniform random variables. We can use uniform random number generators to design a variety of process generators with specific distribution functions.

The basic principle that we will use to generate a random variable with a desired probability distribution function can be illustrated with the aid of Figure 5.3. First, we generate a uniformly distributed random variable between zero and one and set $F(x)$ equal to this number. (Remember that $0 \leq F(x) \leq 1$.) Next, we find this value of $F(x)$ on Figure 5.3 and move horizontally across the graph until we intersect the probability distribution function. We then find the x-coordinate for this point and record the value of x as the value of the random variable generated. Through repetition of this process, we find other values of the random variable. This process can be described mathematically as follows.

Let Y be a uniformly distributed random variable over the interval (0,1). Then

$$p(\mathrm{Y} \leq y) = y \tag{5.4}$$

where $0 < y < 1$. Let $y = F(x)$. $F(x)$ is the desired distribution function. Then

$$F(x) = p(\mathrm{X} \leq x) = p(\mathrm{Y} \leq y) = y = p(\mathrm{Y} \leq F(x))$$

and

$$x = \phi(y)$$

where $\phi = F^{-1}$, the inverse function of F. Determining the relationship ϕ may be difficult using standard functions. Hence, we may have to use tabular methods or approximations. We will demonstate the process by generating exponentially and Poisson distributed random variables.

FIGURE 5.3 General Distribution Function

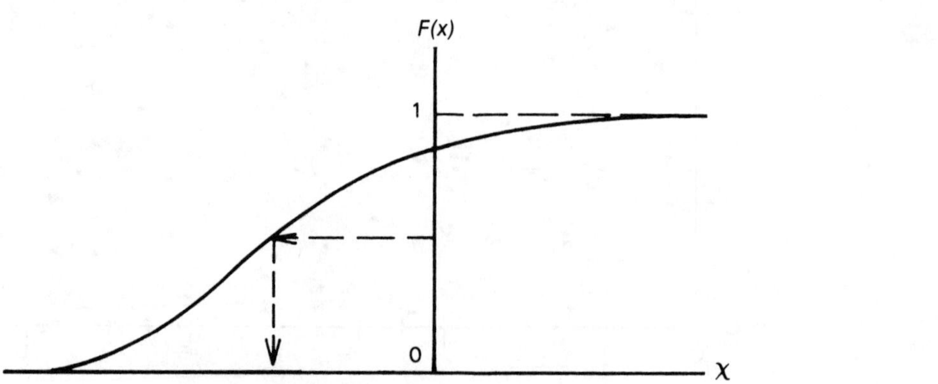

5.7 EXPONENTIAL PROCESS GENERATOR

An exponentially distributed random variable with parameter λ has a probability distribution function of

$$F(x) = 1 - e^{-\lambda x} \tag{5.5}$$

where $0 \le x < \infty$. Parameter λ is given by

$$\lambda = \frac{1}{E[X]} \tag{5.6}$$

where $E[X]$ is the expected value of the random variable X.

Generation of an exponentially distributed random variable X proceeds as follows. First, we generate a uniformly distributed random variable Y over the interval (0,1). Since $0 < F(x) < 1$, we let $F(x) = y$. From equation (5.5), $y = 1 - e^{-\lambda x}$, we solve for x and obtain

$$x = -\frac{1}{\lambda}\log_e(1 - y)$$

Since Y is a random variable with a uniform distribution, $(1 - Y)$ is also a random variable with the same distribution as Y, and

$$x = -\frac{1}{\lambda}\log_e(y) \tag{5.7}$$

Section 5.15 presents uniformly and exponentially distributed random numbers.

The random sequence of x's obtained when y is used instead of $(1 - y)$ in equation (5.7) is called the *antithetic series*. (An example of an exponential series of random numbers and its antithetic series are shown on p. 123.) Antithetic series are often used to reduce variance in the estimation of parameters in Monte Carlo simulations and will be expanded upon later in the chapter.

EXAMPLE 5.5 Let us generate two exponentially distributed random numbers with parameter $\lambda = 2$, using a standard uniform distribution. Assume that our standard uniform numbers are 0.1 and 0.7. Then

$$x_1 = -\frac{1}{2}\log_e(1 - 0.1) = -\frac{1}{2}(-0.105) = 0.053$$

$$x_2 = -\frac{1}{2}\log_e(0.3) = -\frac{1}{2}(-1.204) = 0.602$$

If we used the antithetic series, we would get

$$x_3 = -\frac{1}{2}\log_e(0.1) = -\frac{1}{2}(-2.303) = 1.151$$

$$x_4 = -\frac{1}{2}\log_e(0.7) = -\frac{1}{2}(-0.357) = 0.178$$

With the same technique, Table 5.1 can be used to design other popular random process generators, such as the Pareto, Rayleigh, and Weibull.

TABLE 5.1 Table for Generating Random Variables

Distribution	*Probability Distribution Function*	*Range of Random Variable*	*Random Variable*
Uniform	$\frac{1}{b-a}(x-a)$	$a<x<b$	$x = a + \text{NUMBER}\,(b-a)$
Gamma	$1 - e^{-\lambda x}\sum_{m=0}^{k-1}\frac{(\lambda x)^m}{m!}$	$0<x<\infty$	$x = -\frac{1}{\lambda}\log_e\left[\prod_{i=1}^{k}\text{NUMBER}(i)\right]$
Gaussian	$\int_{-\infty}^{x}\frac{1}{\sqrt{2\pi\sigma^2}}\exp\left[-\frac{(x-\mu)^2}{2\sigma^2}\right]dx$	$-\infty<x<+\infty$	$x = \lambda + \frac{r-0.5}{\lvert r-0.5\rvert}$ $\sigma\,\frac{V-2.511517+0.802835V+0.010328V^2}{1+1.432788V+0.189269V^2+0.001308V^3}$ $V = -2\ln 0.5(1-\lvert 1-2r\rvert)$ $r = \text{NUMBER}$ (see reference 23)
Pareto	$1 - x^{-\lambda}$	$x \geq 1$	$x = \text{NUMBER}^{-1/\lambda}$
Rayleigh	$1 - \exp(-\lambda x^2)$	$x>0$	$x = \left[-\frac{1}{\lambda}\log_e(\text{NUMBER})\right]^{1/2}$
Weibull	$1 - \exp(-\lambda x^b)$	$x>0$	$x = \left[-\frac{1}{\lambda}\log_e(\text{NUMBER})\right]^{1/b}$

Note: NUMBER is a uniformly distributed random number between 0 and 1.

5.8 POISSON PROCESS GENERATOR

Across all the sciences, the Poisson distributed random variable has probably received more attention than any other random variable. Its applications in queueing theory, engineering, inventory control, sociology, and many other areas are well documented.[1,2,4,8,22,23] To better understand its great utility, we will examine a method that uses a Poisson distribution and another parameter, called the average frequency of events, λ.

We introduced the Poisson distribution in Section 4.9. Using equation (4.29), we can compute the probability of no events occurring in time interval Δt to be

$$p(0) = \frac{(\lambda\Delta t)^0}{0!}\,e^{-\lambda\Delta t}$$
$$= e^{-\lambda\Delta t}$$

The probability of at least one event occurring in Δt is

$$p(\text{at least one event}) = 1 - p(0) = 1 - e^{-\lambda\Delta t}$$

Hence, we should generate a random number in the range (0,1) and if the number is less than or equal to $1 - e^{-\lambda\Delta t}$, we can say that at least one event has occurred. If Δt is taken to be small, then we can say that only one event has occurred.

It might be more appropriate if we develop a random number from 1 to $1/(1-e^{-\lambda\Delta t})+1$ truncated to an integer. If this number coincides with a previously chosen number, the event will occur in Δt seconds. Because of possible idiosyncracies in the random number generator, we should develop random numbers of four times this range and let four possible numbers occurring indicate that this event will occur. If $\lambda\Delta t > 0.1$, we should use another method. In the generation of a Poisson random variable, the interoccurrence times are exponentially distributed.

EXAMPLE 5.6 Let us generate Poisson distributed random numbers with parameter $\lambda = 1$. We recall that $\lambda = 1$ means that, on the average, one event occurs per unit time. Let us first choose Δt such that $\lambda\Delta t < 0.1$, which leads to $\Delta t < 0.1$. Let us choose $\Delta t = 0.05$. Then

$$\frac{1}{1-e^{-\lambda\Delta t}}+1=\frac{1}{0.049}+1=20.4+1\cong 21$$

and

$$4(21)=84$$

Let us arbitrarily choose the four numbers to compare with our random number to be 5, 17, 29, and 57. Now, let us generate a random number between 1 and 84. Let us suppose that the first number we generate is 12. Then we compare 12 with 5, 17, 29, and 57. Since 12 is not equal to any of these four numbers, we conclude that no events occurred at $\Delta t = 0.05$ unit of time.

Now we generate a second random number, 25. We can say that no events have occurred in $2\Delta t = 0.1$ unit of time. If the next number we generate is 17, we can conclude that one event occurred in 0.15 unit of time. We would repeat this process until we exceed one unit of time. The sum of events minus one will be our first Poisson random variable.

We can also generate Poisson distributed random numbers from a uniformly distributed number by successive multiplication. If we wish to generate a Poisson distributed random number with mean λ, we multiply n uniform random numbers until the product is less than $e^{-\lambda}$. The value of the Poisson random variable x is $n-1$. This can be represented mathematically by

$$\prod_{i=1}^{n} r_i < e^{-\lambda}$$

EXAMPLE 5.7 Let us generate a Poisson distributed random number with mean $\lambda = 3$, using the multiplication method. Here, $e^{-\lambda} = e^{-3} = 0.05$. If the first (uniformly distributed) random number is 0.5, then $0.5 > 0.05$ and we must generate another uniformly distributed number. Suppose that we generate 0.02 as our next number. Then $(0.5)(0.02) = 0.01 < 0.05$ and our Poisson distributed number would be 1.

5.9 APPLICATIONS: MONTE CARLO SIMULATION

Consider the following Monte Carlo simulation of a queue at a professor's office during registration week. Suppose that n students are given appointments at time $t = 0$ and the rest of the students (m in all) are scheduled at half-hour intervals. Assume that the consulting time is exponentially distributed with mean = 30 minutes. The question is how to schedule student appointments so that there is a balance between the average waiting times of the students and the waiting time of the professor.[3]

Let us simulate the process for two cases, $n = 1$ and $n = 2$, using the following abbreviations:

- X = a random number chosen from an exponential distribution with mean 100. X can be obtained from a table such as that shown on p. 123.
- S_{arr} = the arrival time of a student according to the appointment. Note that we have used the random variable notation here although in this case the arrival time of the students is fixed.
- T_{beg} = the beginning time for the consultation
- T_{end} = the completion time for the consultation
- CT = the consulting time for the consultation, an exponentially distributed random number with mean 30. Values for CT are obtained from CT = (30/100)X where X is chosen from the table on p. 123.

If we assume that students waiting in line are consulted on a first-come, first-served basis, then with the above information we have all that we need to perform the simulation. The fundamental equation for the simulation is

$$T_{end} = T_{beg} + CT$$

Table 5.2 shows the results of two simulation experiments for $m = 10$.

TABLE 5.2 Simulation Results for Two Experiments

	n = 1					*n* = 2	
Student No.	S_{arr}	T_{beg}	X	CT	T_{end}	S_{arr}	T_{end}
1	0*	0	199	60	60	0*	60
2	30	60	286	86	146	0	146
3	60	146	55	17	163	30	163
4	90	163	280	84	247	60	247
5	120	247	8	2	249	90	249
6	150	249	186	56	305	120	305
7	180	305	77	23	328	150	328
8	210	328	81	24	352	180	352
9	240	352	127	38	390	210	396
10	270	390	38	11	401	240	401
	1350			401	2641	1080	2647

* Student did not wait.

From Table 5.2,

	Professor Waited $\left(T_{end} - \sum CT\right)$	***Total Waiting Time for All Students*** $\left(\sum T_{end} - \sum S_{arr} - \sum CT\right)$
$n = 1$	$401 - 401 = 0$	$2641 - 1350 - 401 = 890$
$n = 2$	$401 - 401 = 0$	$2647 - 1080 - 401 = 1166$

The above simulation used the same consulting times for $n = 1$ and $n = 2$. This assumption usually saves computation time. It also results in a reduced variance in the estimate of the change in waiting times as n varies.

One way of explaining variance reduction is as follows. Let us define N_1 as the random variable for the waiting time with $n = 1$. The estimate of the mean of N_1 is $\overline{N}_1$. Similarly, we define N_2 as the random variable for waiting time with $n = 2$, with the estimate of the mean equal to $\overline{N}_2$. The change in the estimate of mean waiting time between the two cases is given by

$$\overline{D} = \overline{N}_1 - \overline{N}_2$$

Variance reduction is important when we compare the two cases because we want to determine the estimate of the relative difference, $\overline{D}$, rather than the absolute values of $\overline{N}_1$ and $\overline{N}_2$. If we use two random number sequences to obtain N_1 and N_2, then we can compute the variance of D as

$$\sigma^2[D] = \sigma^2[N_1] + \sigma^2[N_2]$$

However, if we use the same random number sequence for both cases, then both cases will be compared under the same conditions. Using the same random number sequence establishes a correlation between the two random variables N_1 and N_2. Now we will be estimating a different random variable D*, which has the same mean as D, that is,

$$\overline{D^*} = \overline{D}$$

and whose variance is

$$\sigma^2[D^*] = \sigma^2[N_1] + \sigma^2[N_2] - 2\sigma^2[N_1, N_2]$$

$\sigma^2[D^*]$ is less than $\sigma^2[D]$ when N_1 and N_2 are correlated.

Table 5.3 shows the same procedure using Y the antithetic series of the exponentially distributed random numbers shown on p. 124.

TABLE 5.3 Simulation Results Using Antithetic Series

	$n = 1$					$n = 2$	
Student No.	S_{arr}	T_{beg}	Y	CT	T_{end}	S_{arr}	T_{end}
1	0*	0	15	5	5	0*	5
2	30*	30	6	2	32	0*	7
3	60*	60	5	2	62	30*	32
4	90*	90	86	26	116	60*	86
5	120*	120	6	2	122	90*	92
6	150*	150	253	76	226	120*	196
7	180	226	17	5	231	150	201
8	210*	231	62	19	250	180	220
9	240*	250	59	18	268	210	238
10	270	270	33	10	280	240*	250
	1350			165	1592	1080	1372

* Student did not wait.

From Table 5.3,

	Professor Waited	*Total Waiting Time for All Students*
$n = 1$	$280 - 165 = 116$	$1592 - 135 - 165 = 77$
$n = 2$	$250 - 165 = 85$	$1327 - 1080 - 16 = 82$

A summary of the pair of simulations is given by the set of means:

	Professor Waited	*Students Waited*
$n = 1$	$(0 + 116)/2 = 58$	$(890 + 77)/2 = 484$
$n = 2$	$(0 + 85)/2 = 42.5$	$(1159 + 82)/2 = 620.5$

Note, however, that you should not draw a firm conclusion from one or two experiments but should carry out more experiments if you want to recommend a certain value of n. Because this system can be readily simulated with a general-purpose simulation language (e.g., GPSS, SIMSCRIPT, SLAM, SIMAN, or GASP), by running several simulations you could develop a value for n in which you could be confident.

5.10 REDUCTION OF VARIANCE

The previous queueing example used an antithetic pair of series and then calculated the mean for each pair. We can use the antithetic series because the effect of chance on the means is less than the effect of chance in the original simulation. Two techniques for variance reduction are

Use of antithetic series

Use of the same numbers in corresponding situations

EXAMPLE 5.8 In 10 simulations of the queueing system depicted in Figure 5.4, the total time in the system—the time from arrival to departure—was found to be

199 286 294 55 280 8 186 77 81 127

The mean of these values is 159.3 and the standard deviation is 104.76. If we used the antithetic series of the random number generator, the total time in the system for the 10 simulations would be

15 6 5 85 253 17 62 59 33 115

The mean of these values is 65 and the standard deviation is 75.42. The series of means is thus

107.0 146.0 149.5 70.5 266.5 12.5 124.0 68.0 57.0 121.0

The mean of this series is 112.2 and the standard deviation is 69.3. If we supplemented the first series with 10 additional simulations with random numbers from the random number generator, the 10 additional results for total time in the system would be

38 38 11 232 188 114 75 34 1 95

The mean of the series of 20 values is 120.6 and the standard deviation is 97.6. Using the antithetic series produces the same decimal accuracy with $(69.3/97.6)^2 \cong 1/2$ of the work required if completely random numbers are used. This quantitative work is uncertain, since it is obtained from a small number of simulations.

Suppose there are two ways of carrying out an experiment. Experiment 1 has a variance σ_1^2 and experiment 2 has variance σ_2^2. If experiment 1 is repeated n_1 times and experiment 2 is repeated n_2 times, the same accuracy is obtained if

$$\frac{n_1}{n_2} = \frac{\sigma_1^2}{\sigma_2^2}$$

This relationship implies that if a variance reduction by a factor l can be achieved by a particular way of carrying out the experiment, then the number of experiments needed can also be reduced by l.

FIGURE 5.4 Simple Queueing System

5.11 CONFIDENCE INTERVALS AND NUMBER OF SIMULATIONS

By now it should be clear that a simulation of a stochastic process is itself a stochastic process. Therefore, a natural question is how much confidence we can have in the results of our simulations. Depending on the seed, the desired outputs form an ensemble. Thus, determining the required number of simulations is similar to deciding how many samples are needed in any statistical experiment. Therefore, we must initially define

The quantity to be estimated
The allowable error in the estimate
The confidence level associated with the allowable error

And standard methods described in statistical books can be applied to determine the required number of simulation runs.

G.J. Hahn suggests the following criterion.[7] Suppose that E is the maximum allowable error in estimating a proportion of a parameter p; that V is the desired confidence level that the estimated proportion p does not differ from p' by more than $\pm E$; and that P' is an initial estimate of p. Then, the number of simulation experiments is given by

$$n = \frac{P'(1-P')}{E^2} Z^2 \left(\frac{1+V}{2}\right) \tag{5.8}$$

where $Z[(1+V)/2]$ denotes the $100[(1+V)/2]$ percent point of a standard normal distribution.

The qth percentile of the distribution of any random variable X is the value $x_q/100$ defined by

$$p\left(\mathrm{X} \leq \frac{x_q}{100}\right) = \frac{q}{100}$$

Thus, for $V = 0.9$, we use

$$\frac{1+V}{2} = 0.95$$

and

$$Z\left(\frac{1+V}{2}\right) = 1.644$$ (See the table of percentiles for the standard normal distribution on p. 127)

For example, if $E = 0.05$, $V = 0.9$, $P' = 0.8$, and $Z[(1+V)/2] = 1.644$, then

$$n = \frac{0.8(1-0.8)(1.644)^2}{(0.05)^2} = 173$$

The estimate p' of the true proportion p obtained from the 173 simulation runs will not differ from the true proportion p by more than 0.05, with confidence level 0.90. This method should not be used when np or $n(1-p)$ is less than 5.

If, instead of estimating a proportion p, we desire an estimate $(\overline{X})$ of a true average (μ) of a variable, then the number of required simulations is

$$n = \left[\frac{Z\left(\frac{1+V}{2}\right)\sigma'}{E} \right]^2$$

where σ' is an initial estimate of the variable's standard deviation. We can make a conservative estimate using a large σ'.

5.12 SUMMARY

In this chapter, we introduced the basic tools for simulation and stochastic systems. We presented methods for generating random variables from several distributions; in particular, we covered uniformly, Poisson, and exponentially distributed random variable generators. The problems that follow will take you beyond these into normally distributed random numbers. We also introduced a simple (queueing system) simulation. Confidence intervals and the number of simulations needed to establish confidence were also addressed. Generation of these random variables will provide us with the means of simulating stochastic discrete event systems, as we will see in Chapter 6.

5.13 REVIEW PROBLEMS

1. What is the difference between a random number and a pseudorandom number? What are the advantages of using pseudorandom numbers? The disadvantages?
2. List some methods of generating random numbers.
3. What are the desirable characteristics of a random number generator?
4. Write and execute a computer program using the congruential method to produce a uniformly distributed random number such that $1 \leq$ random number $\leq$ maximum number $\leq 1{,}048{,}576$, where the maximum number is determined by the particular application.
5. Produce 1000 uniformly distributed random numbers using your computing facility's random number generators?
6. What are the four schemes suggested by Kendall and Babington-Smith for testing uniform random number generators
7. Use the chi-square criterion to test your computing facility's random number generator for 9 degrees of freedom. Choose the criterion $\chi_c^2 = 19.0$. What is the probability that $\chi_c^2 > 19.0$?
8. Plot a scatter diagram for a uniform random number generator. Do you detect any pattern(s)?

9. It has been observed that in the time period 1500–1931, 299 wars were started. Assuming that the interval between wars is exponentially distributed, simulate the process and compare your results with the observed value. How many simulation experiments did you carry out? Calculate the mean time between wars and compare it with the observed value.

10. In the period 1820–1929, the following frequency table was compiled for the outbreak of wars:

Number of outbreaks in a year (i):	0	1	2	3	4	Total
Year with i outbreaks:	65	35	6	4	0	110

Assume that the process is Poisson distributed. Simulate the process and compare your results with the above data.

11. What is an antithetic series? When and why should it be used? Did you apply an antithetic series in problems 9 and 10?

12. L. Bortkiewicz found the average number of deaths due to horse kicks in the Prussian army per corps-year for 10 corps was 0.61 over a period of 20 years.[1] The compiled data are as follows.

Number of deaths:	0	1	2	3	4
Number of corps with i deaths per year:	109	65	22	3	1

Assume that the process is Poisson distributed. Simulate the process and compare your results with the above data. How many computer simulation experiments did you perform?

13. What is the relationship between an exponential and a Poisson distribution?

14. What type of events have a Poisson distribution?

15. The distribution of annual incomes of individuals in a large country is said to be Pareto distributed. Generate 500 Pareto distributed numbers. What are the characteristics of this distribution? The distribution is said to have a long tail. Can this distribution be fitted to predict the number of papers authored by scientists or the number of paintings produced by artists?

16. Normally distributed variables are often associated with characteristics of people and weather. Generate 500 normally distributed numbers to represent the weight or height of students in your class.

17. Plot a scatter diagram for an exponential distribution with $\lambda = 1$. Do you see any patterns? Repeat with $\lambda = 10$.

18. Assuming that the initial estimate of a parameter p is 0.5, how many simulation runs are needed to obtain the value of p to within 0.01, with a 0.95 confidence level?

19. How can the variance of the estimate of a parameter in a simulation experiment be reduced?

20. Let there be k possible exclusive responses to a question or to a grouping in social categories. Let $p_1, p_2, \ldots, p_k$ be the probabilities of responding to the first, second, . . . , kth possible response. For a population of 1000, what responses will occur most often in each of the following cases?

    a. $k = 2$, $p_1 = p_2 = 0.5$
    b. $k = 2$, $p_1 = 0.6$, $p_2 = 0.4$
    c. $k = 3$, $p_1 = 0.5$, $p_2 = 0.3$, $p_3 = 0.2$

21. Suppose that a sociologist is interested in the number of suicides in a year in several towns with similar characteristics. What type of distribution would you use to model the phenomenon?

22. Assume that we have a table of pairs (x_i, y_i). Write a computer program that will calculate the value of y, given any value of x, such that $x_1 \le x \le x_n$. Use a linear interpolation. For example, if $x_1 < x < x_2$, then

$$y = y_1 + \frac{x - x_1}{x_2 - x_1}(y_2 - y_1)$$

23. It has been shown by R.W. Hamming (*Numerical Methods for Scientists and Engineers* [New York: McGraw-Hill, 1962], p. 34) that

$$Y = \frac{\sum_{i=1}^{k} X_i - \frac{k}{2}}{\left(\frac{k}{12}\right)^{1/2}}$$

has a normal distribution with mean zero and standard deviation one as $k \to \infty$, where the X_i are uniform random numbers. Small values of k give good accuracy. Write a program to generate normally distributed random numbers using $k = 12$. What are the mean and standard deviation?

24. Here we suggest another method for generating Poisson distributed random variables with mean λ. The method generates exponentially distributed random numbers with mean one and adds them until their sum is greater than λ. The number of times the exponentially distributed number is generated to meet this condition represents a Poisson distributed number. Use this method to generate 50 Poisson numbers. From this data, estimate $p(0)$, $p(1)$, $p(2)$, $p(3)$, and $p(4)$. You may select a value for λ.

25. In a simple point-to-point computer communication network, a packet is transmitted from the source host to the destination host as shown in Figure 5.5. The probability that a packet could become garbled along links 1 and 2 is p, and the probability that it becomes garbled along links 3 and 4 is q. If garbled, the packet will have to be retransmitted from the source node. The probability that it travels along link 1 is r.

- **a.** Compute the probability of a packet reaching the destination node without error.
- **b.** Compute the average number of times a packet must be transmitted before it reaches its destination.
- **c.** Simulate the system above and compare your simulation results with the computations in parts (a) and (b).

FIGURE 5.5 Problem 25

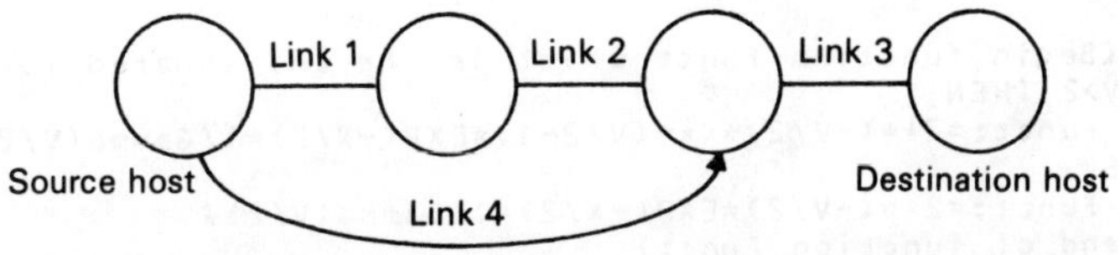

5.14 APPENDIX: CHI-SQUARE DISTRIBUTIONS

Pascal Program for Calculating the 99% Points for Chi-Square Distributions with 2 to 20 Degrees of Freedom

```
PROGRAM ChiSquareTable(input,output);


   {program to calculate the 99% point for Chi}
 {square distributions with degrees of freedom from 2 to 20}
          {using Simpson's integration method}

CONST

   VMax=20;
   Percentage=99;

VAR {of the main program Chi_Table}

   Ulim,Llim,Last,Next,ErrorAllowed:    REAL;
   I,J,K,L,V,M:INTEGER;
   Values:ARRAY[0..10] of REAL;

FUNCTION Int(Top,Bottom,Error:REAL;V:INTEGER):REAL;

VAR {of the function Int}
   S1,S2:REAL;

FUNCTION Funct(X,V:REAL):REAL; {this is the function being integrated}

   FUNCTION Gamma(N:REAL):REAL;

   VAR {of the function Gamma}
       K:INTEGER;

   BEGIN {Begin function Gamma}
       K:=TRUNC(2*N+0.1);
       IF (K) MOD 2<>0 THEN
           BEGIN
               IF N>1.5 THEN
                   Gamma:=Gamma(N-1)*(N-1)
               ELSE
                   Gamma:=0.886226925;
           END
       ELSE
           BEGIN
               IF N>1 THEN
                   Gamma:=Gamma(N-1)*(N-1)
               ELSE
                   Gamma:=1
           END;
   END;

BEGIN {Begin function Funct which is the chi squared function}
   IF V>2 THEN
       Funct:=2**(-V/2)*X**(V/2-1)*EXP(-X/2)*1/Gamma(V/2)
   ELSE
       Funct:=2**(-V/2)*EXP(-X/2)*1/Gamma(V/2);
END; {end of function Funct}
```

Pascal Program for Calculating the 99% Points for Chi-Square Distributions with 2 to 20 Degrees of Freedom (Continued)

```
FUNCTION Simpson(T,B:REAL;V:INTEGER):REAL;

VAR {of the function Simpson}

   Midpoint:REAL;


BEGIN {begin Simpson}
   Midpoint:=(T+B)/2;
   Simpson:=((T-B)/6)*(Funct(B,V)+4*Funct(Midpoint,V)+Funct(T,V))
END; {end Simpson}

BEGIN {begin Int}
   S1:=Simpson(Top,Bottom,V);
   S2:=Simpson(Top,(Top+Bottom)/2,V)+Simpson((Top+Bottom)/2,Bottom,V);
   IF (S1-S2) < 10*Error THEN
       INT:=(1/15)*(16*S2-S1)
   ELSE
       Int:=Int(Top,(Top+Bottom)/2,Error/2,V)+Int((Top+Bottom)/2,Bottom,Error/2,V);
END;

BEGIN {begin main program ChiSquareTable}

{in each iteration the program integrates from the previous point}
{to the present one and then adds the previously calculated value}
{of the integral from zero to the previous point.  When the value}
{is greater than .99 the iterations stop and the point is written}
{to the terminal along with the number of degrees of freedom used}
                {for that set of iterations}


    WRITE('              Degrees of Freedom        ');
    WRITELN('       99% Point        ');
        WRITELN;
    FOR V:=2 TO VMax DO
        BEGIN
            I:=1;
            Last:=0;
            REPEAT
                Ulim:=I/100;
                Llim:=(I-1)/100;
                ErrorAllowed:=0.0000000000000001;
                Next:=Last+Int(Ulim,Llim,ErrorAllowed,V);
                IF Next<Percentage/100 THEN
                    BEGIN
                        I:=I+1;
                        Last:=Next;
                    END

            UNTIL (I>5000) OR (Next>Percentage/100);
            WRITE('                          ',V);
            IF V<10 THEN
                WRITE(' ');
            WRITELN('                             ',(I-1)/100:4:2);
            WRITELN;
        END;
END.
```

Chi-Square Table for 99% Points

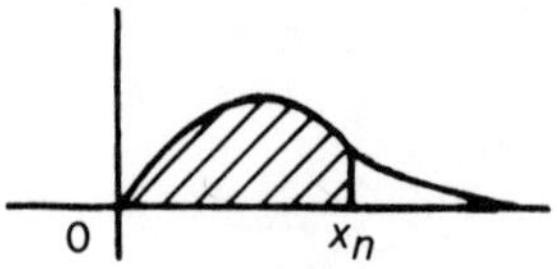

Shaded area = 0.99
$P(\chi^2 < x_n) = 0.99$

Degrees of Freedom (n)	*99% Point (x_n)*
2	9.21
3	11.34
4	13.27
5	15.08
6	16.81
7	18.47
8	20.09
9	21.66
10	23.20
11	24.72
12	26.21
13	27.68
14	29.14
15	30.57
16	31.99
17	33.40
18	34.80
19	36.19
20	37.56

5.15 APPENDIX: RANDOM NUMBER TABLES

100 Uniformly Distributed Random Numbers

R

.8633	.4201	.5359	.9481	.8471	.8659	.7185	.1550	.9393	.7307	.0793	.5121
.8441	.9426	.9020	.1378	.8027	.6532	.6815	.0859	.3180	.6585	.3193	.3266
.1056	.1141	.6124	.6063	.0391	.5395	.9733	.0343	.5259	.5006	.2893	.2115
.0091	.7287	.9302	.0048	.7358	.4104	.2738	.5587	.3836	.1791	.0255	.9196
.7437	.4919	.6578	.8535	.2686	.9094	.1118	.9549	.6192	.7011	.6651	.4861
.9489	.2354	.6689	.3098	.9414	.5568	.2177	.1160	.1138	.3597	.5268	.8994
.9897	.3672	.1687	.8565	.5583	.4232	.1159	.4364	.0784	.6386	.6768	.8994
.0095	.4334	.8624	.9469	.2451	.4250	.8723	.7862	.9108	.0208	.7181	.7637
.2036	.9429	.2528	.5537								

100 Uniformly Distributed Random Numbers (Continued)

Middle of Interval	*Number of Observations*
0.0	7 *******
0.1	10 **********
0.2	8 ********
0.3	8 ********
0.4	9 *********
0.5	8 ********
0.6	9 *********
0.7	14 **************
0.8	5 *****
0.9	19 *******************
1.0	3 ***

1 − R

.1367	.9141	.6164	.1465	.0586	.5750	.6734	.9909	.4605	.8882	.1006	.7964
.4640	.8209	.3349	.5636	.1277	.6820	.8859	.0698	.0450	.9216	.5666	.7472
.1529	.5139	.0103	.2138	.3415	.7107	.3937	.2642	.3614	.2819	.1559	.1341
.0267	.7646	.8313	.0531	.7885	.2563	.5896	.7823	.6328	.5799	.0980	.9657
.3808	.9792	.4417	.2815	.2713	.3422	.8840	.0892	.9206	.0519	.1973	.2989
.4732	.5768	.8450	.4741	.9952	.7314	.6902	.4879	.8944	.3648	.7261	.2970
.9905	.0571	.4994	.9745	.0906	.8841	.0607	.0574	.3876	.4413	.8852	.2363
.1376	.4463	.0804	.0511	.4432	.2693	.6807	.8622	.9609	.6403	.3232	.1435
.7549	.3185	.5081	.3311								

Exponentially Distributed Random Numbers and Their Antithetic Series

MEAN = 100

$$\text{(a) } X = \frac{|\log(1-R)|}{0.01}$$

199	286	294	55	280	8	186	77	81	127	38
38	11	232	188	114	75	34	1	95	162	362
48	3	136	266	4	32	97	109	297	107	133
12	12	75	458	111	31	25	8	113	1	18
284	12	242	127	23	198	82	206	131	72	54
29	28	17	107	40	286	296	201	9	69	24
12	15	101	3	20	252	130	93	78	82	121
67	68	0	53	310	45	121	27	192	240	12
102	230	144	2	57	81	37	46	57	194	55
154										

continues

Exponentially Distributed Random Numbers and Their Antithetic Series (Continued)

(b) Antithetic series: $Y = \dfrac{|\log R|}{0.01}$

15	6	4	86	6	253	17	62	59	33	115
114	225	10	17	38	64	124	470	49	22	3
96	367	30	7	324	130	48	41	5	42	31
219	217	64	1	40	131	152	255	39	466	178
6	215	9	33	159	15	58	14	31	67	87
138	141	186	42	112	6	5	14	245	69	155
217	198	45	337	172	8	32	50	62	58	36
72	71	534	89	5	102	35	145	16	9	215
45	11	27	387	83	59	117	100	84	15	86
24										

Note: To obtain random numbers with the density function $\lambda e^{-\lambda x}$, divide the values by 100λ.

5.16 APPENDIX: NORMAL DISTRIBUTIONS

Pascal Program for Computing the Area under the Normal Curve from $-\infty$ to 3.4

```
PROGRAM Inter(input,output);

{a program to calculate the area under the normal curve from}
   {negative infinity to I/100 where I goes from 0 to 349}


VAR {of the main program Inter}
   Ulim,Llim, Err:     Real;
   I,J,K,L,M:INTEGER;
   Values:ARRAY[0..350] of REAL;

FUNCTION Int(Top,Bottom,Error:REAL):REAL;

VAR {of the function Int}
   S1,S2:REAL;

FUNCTION Funct(X:REAL):REAL; {this is the function being integrated}

BEGIN
   Funct:=EXP(-(X*X)/2)/SQRT(2*3.141592654)
END; {end of function Funct}

FUNCTION Simpson(T,B:REAL):REAL;

VAR {of the function Simpson}
   Midpoint:REAL;
```

Pascal Program for Computing the Area under the Normal Curve from $-\infty$ to 3.4 (Continued)

```
BEGIN {begin Simpson}
   Midpoint:=(T+B)/2;
   Simpson:=((T-B)/6)*(Funct(B)+4*Funct(Midpoint)+Funct(T))
END; {end Simpson}

BEGIN {begin Int}
   S1:=Simpson(Top,Bottom);
   S2:=Simpson(Top,(Top+Bottom)/2)+Simpson((Top+Bottom)/2,Bottom);
   IF (S1-S2) < 10*Error THEN
       INT:=(1/15)*(16*S2-S1)
   ElSE
       Int:=Int(Top,(Top+Bottom)/2,Error/2)+Int((Top+Bottom)/2,Bottom,Error/2);
END; {end Int}

BEGIN {begin main program Inter}

{in each iteration the program integrates from the previous point}
{to the present one and then adds the previously calculated value}
{of the integral from negative infinity to the previous point}

   Values[0]:=0.5;
   FOR I:=1 TO 349 DO
       BEGIN
           Ulim:=I/100;
           Llim:=(I-1)/100;
           Err:=0.00000000001;
           Values[I]:=INT(Ulim,Llim,Err)+Values[I-1];
       END;

 {Now output the results in a table}

   WRITELN;
   WRITELN;
   WRITELN;
   WRITE('                                    ');
   WRITELN;
   FOR J:=0 TO 6 DO
       BEGIN
           WRITELN;
           WRITE('                  ');
           FOR L:=0 TO 3 DO
               WRITE((J/2+L/10):2:2,'            ');
           WRITELN((J/2+0.4):2:2);
           WRITELN;
           FOR I:=0 TO 9 DO
               BEGIN
                   WRITE('       ',(I/100):2:2,'       ');
                   FOR K:=0 TO 3 DO
                       WRITE(Values[J*50+10*K+I]:4:4,'          ');
                   WRITELN(Values[J*50+10*(K+1)+I]:4:4);
               END;
           IF (J=3) THEN {a loop to create a margin at the bottom}
               BEGIN     {of the first page of the table}
                   FOR M:=1 TO 14 DO
                       WRITELN
               END;
       END;
END. {of the main program Inter}
```

Area under the Normal Distribution Curve from $-\infty$ to α

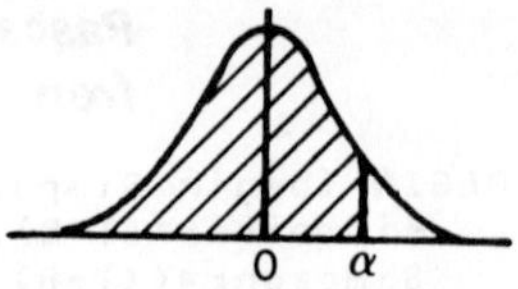

α	0.00	0.01	0.02	0.03	0.04	0.05	0.06	0.07	0.08	0.09
0.0	0.5000	0.5040	0.5080	0.5120	0.5160	0.5199	0.5239	0.5279	0.5319	0.5359
0.1	0.5398	0.5438	0.5478	0.5517	0.5557	0.5596	0.5636	0.5675	0.5714	0.5753
0.2	0.5793	0.5832	0.5871	0.5910	0.5948	0.5987	0.6026	0.6064	0.6103	0.6141
0.3	0.6179	0.6217	0.6255	0.6293	0.6331	0.6368	0.6406	0.6443	0.6480	0.6517
0.4	0.6554	0.6591	0.6628	0.6664	0.6700	0.6736	0.6772	0.6808	0.6844	0.6879
0.5	0.6915	0.6950	0.6985	0.7019	0.7054	0.7088	0.7123	0.7157	0.7190	0.7224
0.6	0.7257	0.7291	0.7324	0.7357	0.7389	0.7422	0.7454	0.7486	0.7517	0.7549
0.7	0.7580	0.7611	0.7642	0.7673	0.7704	0.7734	0.7764	0.7794	0.7823	0.7852
0.8	0.7881	0.7910	0.7939	0.7967	0.7995	0.8023	0.8051	0.8078	0.8106	0.8133
0.9	0.8159	0.8186	0.8212	0.8238	0.8264	0.8289	0.8315	0.8340	0.8365	0.8389
1.0	0.8413	0.8438	0.8461	0.8485	0.8508	0.8531	0.8554	0.8577	0.8599	0.8621
1.1	0.8643	0.8665	0.8686	0.8708	0.8729	0.8749	0.8770	0.8790	0.8810	0.8830
1.2	0.8849	0.8869	0.8888	0.8907	0.8925	0.8944	0.8962	0.8980	0.8997	0.9015
1.3	0.9032	0.9049	0.9066	0.9082	0.9099	0.9115	0.9131	0.9147	0.9162	0.9177
1.4	0.9192	0.9207	0.9222	0.9236	0.9251	0.9265	0.9279	0.9292	0.9306	0.9319
1.5	0.9332	0.9345	0.9357	0.9370	0.9382	0.9394	0.9406	0.9418	0.9429	0.9441
1.6	0.9452	0.9463	0.9474	0.9484	0.9495	0.9505	0.9515	0.9525	0.9535	0.9545
1.7	0.9554	0.9564	0.9573	0.9582	0.9591	0.9599	0.9608	0.9616	0.9625	0.9633
1.8	0.9641	0.9649	0.9656	0.9664	0.9671	0.9678	0.9686	0.9693	0.9699	0.9706
1.9	0.9713	0.9719	0.9726	0.9732	0.9738	0.9744	0.9750	0.9756	0.9761	0.9767
2.0	0.9772	0.9778	0.9783	0.9788	0.9793	0.9798	0.9803	0.9808	0.9812	0.9817
2.1	0.9821	0.9826	0.9830	0.9834	0.9838	0.9842	0.9846	0.9850	0.9854	0.9857
2.2	0.9861	0.9864	0.9868	0.9871	0.9875	0.9878	0.9881	0.9884	0.9887	0.9890
2.3	0.9893	0.9896	0.9898	0.9901	0.9904	0.9906	0.9909	0.9911	0.9913	0.9916
2.4	0.9918	0.9920	0.9922	0.9925	0.9927	0.9929	0.9931	0.9932	0.9934	0.9936
2.5	0.9938	0.9940	0.9941	0.9943	0.9945	0.9946	0.9948	0.9949	0.9951	0.9952
2.6	0.9953	0.9955	0.9956	0.9957	0.9959	0.9960	0.9961	0.9962	0.9963	0.9964
2.7	0.9965	0.9966	0.9967	0.9968	0.9969	0.9970	0.9971	0.9972	0.9973	0.9974
2.8	0.9974	0.9975	0.9976	0.9977	0.9977	0.9978	0.9979	0.9979	0.9980	0.9981
2.9	0.9981	0.9982	0.9982	0.9983	0.9984	0.9984	0.9985	0.9985	0.9986	0.9986
3.0	0.9987	0.9987	0.9987	0.9988	0.9988	0.9989	0.9989	0.9989	0.9990	0.9990
3.1	0.9990	0.9991	0.9991	0.9991	0.9992	0.9992	0.9992	0.9992	0.9993	0.9993
3.2	0.9993	0.9993	0.9994	0.9994	0.9994	0.9994	0.9994	0.9995	0.9995	0.9995
3.3	0.9995	0.9995	0.9995	0.9996	0.9996	0.9996	0.9996	0.9996	0.9996	0.9997
3.4	0.9997	0.9997	0.9997	0.9997	0.9997	0.9997	0.9997	0.9997	0.9997	0.9998

Percentiles for the Standardized Normal Distribution

p	0.00	0.01	0.02	0.03	0.04	0.05	0.06	0.07	0.08	0.09
0.00		−2.3263	−2.0537	−1.8808	−1.7507	−1.6449	−1.5548	−1.4758	−1.4051	−1.3408
0.10	−1.2815	−1.2265	−1.1750	−1.1264	−1.0803	−1.0364	−0.9945	−0.9541	−0.9154	−0.8779
0.20	−0.8416	−0.8064	−0.7722	−0.7388	−0.7063	−0.6745	−0.6433	−0.6128	−0.5828	−0.5534
0.30	−0.5244	−0.4958	−0.4677	−0.4399	−0.4125	−0.3853	−0.3585	−0.3319	−0.3055	−0.2793
0.40	−0.2533	−0.2275	−0.2019	−0.1764	−0.1510	−0.1257	−0.1004	−0.0753	−0.0502	−0.0250
0.50	0.0000	0.0250	0.0502	0.0753	0.1004	0.1257	0.1510	0.1764	0.2019	0.2275
0.60	0.2533	0.2793	0.3055	0.3319	0.3585	0.3853	0.4125	0.4399	0.4677	0.4958
0.70	0.5244	0.5534	0.5828	0.6128	0.6434	0.6745	0.7063	0.7388	0.7722	0.8064
0.80	0.8416	0.8779	0.9154	0.9541	0.9945	1.0364	1.0803	1.1264	1.1750	1.2265
0.90	1.2815	1.3408	1.4051	1.4758	1.5548	1.6449	1.7507	1.8808	2.0537	2.3263

Note: Entries provide values of α for which

$$\text{area} = \int_x^\infty \frac{1}{\sqrt{2\pi}} \exp(-x^2)\, dx$$

Source: G. Authur Mirham, *Simulation: Statistical Foundations and Methodology* (New York: Academic Press, 1972), p. 499. Reprinted by permission.

5.17 REFERENCES

1. Bortkiewicz, L., *Das Gesetz der Kleinen Zahlen* (Leipzig: Teubner, 1888).
2. Coleman, J., *Introduction to Mathematical Sociology* (Free Press of Glencoe; London: Collier-Macmillan, 1964).
3. Dahlquist, G., and Bjorck, A., *Numerical Methods* (Englewood Cliffs, NJ: Prentice-Hall, 1974).
4. Gerlough, P.L., and Barnes, F.C., *The Poisson and Other Probability Distributions in Highway Traffic* (Saugatuck, CT: ENO Foundation for Transporation, 1971).
5. Gordon, Geoffrey, "GPSS: Sound Recording," recorded at the History of Programming Language Conference, ACM Sig Plan, Los Angeles, June 1978.
6. Gordon, Geoffrey, *System Simulation* (Englewood Cliffs, NJ: Prentice-Hall, 1978).
7. Hahn, G.J., "Sample Sizes for Monte Carlo Simulations," *IEEE Transactions on Systems, Man, and Cybernetics*, vol. SMC-2, no. 5 (November 1972): 678–80.
8. Haight, F.A., *Handbook of the Poisson Distribution* (New York: Wiley, 1967).
9. IBM, *GPSS—General Purpose Simulation System* V, GH20-0825-0, IBM Corporation, White Plains, NY, 1970.
10. IBM, *General Purpose Simulation System* V *User's Manual*, SH20-0851-2, IBM Corporation, White Plains, NY, 1977.
11. Kendall, D.G., and Babington-Smith, B., "Randomness and Random Sampling Numbers," *Journal of the Royal Statistical Society* 101 (1938): 147–66.
12. Knuth, D.E., *The Art of Computer Programming*, vol. 2 (Reading, MA: Addison-Wesley, 1967).
13. Kochenburger, R.J., and Turcio, C.J., *Computers in Modern Society* (Carlsbad, CA: Hamilton, 1974).

14. Kreutzer, W., *System Simulation: Programming Styles and Languages* (Reading, MA: Addison-Wesley, 1986).
15. Lindgren, B.W., *Statistical Theory* (New York: Macmillan, 1976).
16. Martin, F.F., *Computer Modeling and Simulation* (New York: Wiley, 1968).
17. Mihram, G.A., *Simulation: Statistical Foundations and Methodology* (New York: Academic Press, 1972).
18. Pritsker, A.A., and Kiviat, P.J., *Simulation with GASP II: A FORTRAN Based Simulation Language* (Englewood Cliffs, NJ: Prentice-Hall, 1969).
19. Pritsker, A. Alan B., and Young, R.E., *Simulation with GASP: APL/Based Continuous/ Discrete Simulation Language* (New York: Wiley, 1975).
20. RAND Corporation, *A Million Random Digits with 100,000 Normal Deviates* (Santa Monica, CA: Rand Corporation, 1955).
21. Ripley, Brian D., *Stochastic Simulation* (New York: Wiley, 1987).
22. Saaty, T.L., *Elements of Queueing Theory* (New York: McGraw-Hill, 1983).
23. Schmidt, J.W., and Taylor, R.E., *Simulation and Analysis of Industrial Systems* (Homewood, IL: Richard D. Irwin, 1970).
24. Sowey, E.R., "A Chronological and Classified Bibliography on Random Number Generation and Testing," *International Statistical Review*, 40, no. 3 (December 1972): 335–71.

6 Discrete-Event Simulation

6.1 INTRODUCTION

We are now seeing the application of discrete-event simulation in the design of computer systems, computer communication networks, and automated factories, among other areas. Discrete-event simulation languages, developed in the 1960s, enjoy widespread use and acceptance because of the growing popularity of personal computers (PCs). Using the graphics capabilities of modern systems, some of these simulation packages now include animation. We will begin the discussion by providing a perspective on simulation languages, including discrete-event, continuous-time, and combined discrete-continuous. (Chapter 9 focuses on continuous-time languages.) Next we will learn how discrete-event simulations work and what some of the basic characteristics of discrete-event simulation languages are. We refer the reader to the references for details about specific languages.

The choice of a simulation language is often based on language availability, the available documentation, and the fluency of the user. These are not always the best criteria, for as we will see, depending on how we view a system, one language might be more appropriate than another.

6.2 A PERSPECTIVE ON SIMULATION LANGUAGES

General-purpose simulation languages have been designed with the nonprogrammer in mind. One goal of the designers of these general-purpose simulation languages is to make the languages applicable to a broad class of systems while maintaining a relatively fixed set of procedures for carrying out simulation automatically. With that goal in mind, language designers attempt

- To produce a generalized structure for designing simulation models
- To provide the special program formulations that simulation requires, particularly time-keeping and special accounting, to relieve researchers of detailed program construction and allow them to address the computer with the definition of the problem
- To make possible easy modification of the simulation program to reflect changes in the simulation model
- To make the language easy to use

In general, these objectives are achieved by introducing restrictions that do not allow optimum use of the computer. Compromises also must be made between the ease with which the simulation language can be used and the complexity of the system that can be simulated.

Today there exist over a hundred simulation languages, and the number is growing. The existence of so many simulation languages suggests that no one general-purpose language is adequate for all general simulation. The optimal choice for a particular simulation depends on the nature of the system under study.

The most popular simulation languages (packages) are GASP, GPSS, SIMSCRIPT, SIMAN, SIMULA, SLAM, ACSL, NDTRAN, DYNAMO, and CSMP. The first six are best suited for simulation of discrete-event systems, such as scheduling and queueing problems. GASP, a FORTRAN-based language, offers modular programs, allowing easy expansion and alteration of simulation programs to suit the needs of the analyst. GPSS, the most widely used package, is easier to use than SIMSCRIPT, according to one comparative study, but SIMSCRIPT can simulate more complex systems, according to another study. ACSL, NDTRAN, DYNAMO, and CSMP lend themselves best to simulation of complex feedback systems.

In spite of the availability of these simulation languages, most simulation problems are programmed using FORTRAN, ALGOL, Pascal, C, and PL/1. According to Katzke and Reitman, and Kay[16,17], [our comments in brackets], this is due in part to the following:

1. Computer facilities often do not have a simulation compiler or an interpreter available. [This situation should be disappearing soon because of the reduced prices and widespread use of PCs.]
2. Deadline pressure makes analysts feel that they do not have time to acquire skills in a simulation language. [Owing to the complexity of systems being simulated, however, simulation programming is only a small part of the total project.]
3. The problem might be too complex for the simulation language. [Most recent simulation packages are fully capable of simulating even the most complex systems.]
4. The system analyst might have to model his system according to a restrictive structure imposed by the simulation language. [Given the wide variety of simulation packages and their ability to call user-written subprograms if needed, this restriction does not apply in most situations.]
5. The problem would use too much memory when handled by the simulation language. [Memory is no longer a major cost of computing systems, and virtual operating systems are common for minicomputers.]
6. Execution time is usually longer when a simulation language is used. [Execution time is a function of the programmer's experience and expertise, and now computer time is cheaper relative to a person's time.]

It is our view that, over time, more analysts will recognize and make full use of the flexibility of newer and yet-to-be-developed simulation packages.

6.3 A DISCRETE-EVENT SIMULATION: PRINCIPLES OF OPERATION

In Chapter 5, we introduced a hand simulation of a single-server (the professor), single-queue (students waiting outside the professor's office) system. This is a discrete-event simulation. The arrival time of a student, the beginning time of a consultation, and the departure time of a student are system events. In general, the time at which the simulation is completed will be called a program event. Using our earlier notation, the states or conditions of the system can be defined as the status of the queue (number of students waiting) and the status of the professor (consulting with students; busy, e.g., preparing his lectures or his next research article; or idle). The system events can change the states of the system. However, the clock time at which we complete the simulation, a program event, does not change the states of the system. By definition of a discrete system, these events occur at specific moments in time. The actual time it took us to complete the simulation experiment is called the *run time*. Simulation languages have been designed to reduce human run time. In this chapter, we will discuss the methods used to simulate systems of a similar nature.

Let us review the steps we used in the hand simulation of Chapter 5. First, we started to keep track of time. Here we will call it *clock time* and measure it in units that are best suited to describe the behavior of the system under study. In our simulation, clock time was measured in minutes and initialized to zero. Let us define clock time, T_i, as the arrival time of the ith student and B_i as the time between arrival of the $(i-1)$th student and arrival of the ith student. In our case, B_i was specified to be 30 minutes, indicating that the interarrival time has a deterministic distribution. By knowing B_i, we could immediately generate the following list:

Student No.		$S_{arr} = T_i$
1	B_2	0
2		30
3		60
4		90
5	B_6	120
6		150
7		180
8		210
9		240
10		270

In our simulation, we then generated the consulting time required for each student, CT_i, using an exponential distribution with mean 30. Given this additional information, we could have generated Table 6.1. From these data, we can

TABLE 6.1 Arrival Times and Consulting Times

Student No.	$S_{arr} = T_i$	CT_i
1	0	60
2	30	86
3	60	17
4	90	84
5	120	2
6	150	56
7	180	23
8	210	24
9	240	38
10	270	11

easily perform our simulation. We will assume that the professor is initially idle. In general, if the professor is busy when a student arrives, the student waits at the end of the line (queue). When the professor completes the consulting period with a student, the student at the head of the queue enters the professor's office and begins his consulting. This procedure is referred to as a first-in, first-out (FIFO) queueing discipline. To simplify our example, we assume no limit on the size of the queue.

Let us begin the clock time at $CL = 0$. We notice that student 1 arrives at this time and an arrival event has occurred. We choose not to use the option of "tagging" the student with some identification, such as the time of arrival or student number or some other attribute. At this point, the student enters the professor's office, causing the start of a consulting event (the change of state of the professor, server, resource). We *scan* Table 6.1 to determine if another event has occurred simultaneously. We could keep track of the duration of the consulting time with the professor. The state and statistics of the system at this simulation clock time are

$CL = 0$
Number of students waiting in the queue = 0
Status of the professor = busy
Student 1 arrived at $CL = 0$

What do we do next? We have a choice. We can increment the clock time by one and see if an event has occurred. If an event has occurred, we would update our states and statistics, increment our clock time, and repeat the steps. If nothing has occurred, we would increment the clock time and repeat the steps. This type of simulation is called a *fixed-increment* or *synchronous time slice discrete-event simulation* and could require much computation.

A more popular method in discrete-event simulation languages involves scanning the tables, determining the next *imminent* event, and updating the simulation clock time to the time of the event. This type of simulation is called *next-event* or *asynchronous discrete-event simulation*. In our case, we see that the consulting time for the first student is 60 minutes; therefore, the departure time of the student would be the clock time at entry, designated T_{beg} in Chapter 5, plus the consultation time. This is the event at which the status of the professor changes to idle. However, we also determined that the second student arrives at $CL = 30$. The next imminent event is therefore an arrival event. At this moment, the states, statistics, and clock time of the system are updated to

$CL = 30$

Number of students waiting in the queue = 1

Status of the professor = busy

Student 1 arrived at $CL = 0$

Student 2 arrived at $CL = 30$

$$\text{Average number in the queue} = \frac{0}{30} = 0$$

$$\text{Utilization of the professor} = \frac{30}{30} \times 100\% = 100\%$$

We scan Table 6.1 again and determine that student 1 completes his consultation and simultaneously student 3 arrives. Therefore, two types of events occur simultaneously. We must choose which event we will consider first in order to update our states. Let us decide to always consider the arrival events first and the completion of a consultation event second in the case of simultaneous events. The choice to handle arrivals first has implications. Consider the case in which the professor is completing a consultation with a student. When a second student arrives for consultation and learns that the professor is busy, he may choose to wait or he may leave, deciding to come back at a time when the professor is free. We use the term *balk* to describe the tendency to choose not to join the queue or to leave the queue before being serviced. The states, statistics, and clock time at this point are updated as follows.

$CL = 60$

Number of students waiting in the queue = 2

Status of the professor = busy

Student 3 arrived at $CL = 60$

$$\text{Average number in the queue} = \frac{1 \times 30}{60} = \frac{1}{2}$$

Utilization of the professor = 100%

Again, we scan our tables and determine that a student has finished his consultation period. In these steps, we are trying to schedule the next event.

We look at the queue and find that a student is waiting; therefore, the student at the head of the line moves to the professor's office immediately, and our update is

$CL = 60$

Number of students in the queue = 1

Status of the professor = busy

Average number in the queue = $\frac{1}{2}$

Utilization of the professor = 100%

Maximum number of students waiting = 2

Minimum number of students waiting = 0

We scan the tables again and repeat the steps.

We terminate the simulation after all 10 students have finished their consultations. This is a program event and allows us to print the statistics we have gathered.

Let us reflect on the method just described. First, it required the generation of tables, which would be very large if we attempted to simulate the system with many more students and professors. From a computer system point of view, we have used a lot of storage (computer memory). We should ask whether we could do it another way. Of course the answer is always yes in a textbook! All we need is a slight modification of our solution; but before we do that, let us discuss the organization of commercially available next-event, time-advance simulation languages.

In the first step, we assign a value to the simulation clock to specify its starting time. Normally the clock is set to zero ($CL = 0$). Then we specify the states of the system, such as the queue is empty and the professor is idle. The statistical data are then initialized so that the waiting time in the queue and the utilization of the professor are zero. Next we indicate what type of event will occur first in the simulation, in our case a student's arrival rather than a completion of service. All of this is commonly done by an *initialization* subprogram.

The second step is to determine the next event type, event i, in order to advance the simulation clock time. This determination is often accomplished by a *timing subprogram*. In our case, a student arrives, and the event type is an arrival event. The simulation clock is advanced to $CL = 0$ and we proceed to a subprogram to process the results of event type i (arrival). This is usually called an *event subprogram*.

In the third step, the event subprogram updates the system states and statistical variables, and then generates future events by adding them to a list of events to be scanned later by the timing subprogram.

The fourth step is to determine whether the simulation process is complete. This determination can be based on a specification to stop the simulation when the clock time exceeds a certain value or when a certain number of specific types of events occur, such as the departure of 10 students. If the termination criterion is not satisfied, program execution is transferred back to the second step. When the termination criterion is satisfied, the fifth major step computes the final statistical variables of interest, such as the average waiting time in queue, the utilization of the professor, the final value of the clock time, the number of arrivals, and the number of departures. The data are printed and histograms are plotted.

Some simulation languages modify the third step to reduce the table size, by generating the time of enough future events to continue the simulation and adding the times at which the events occur to the *events list* or *calendar*. If we return to our example of a single student arriving at the professor's office and use the additional information, then

W_i = the waiting time in the queue of the *i*th student

D_i = the time at which the *i*th student departs from the professor's office $= T_i + W_i + CT_i$

E_i = the time at which any event *i* occurs

In general, T_i, B_i, W_i, D_i, and E_i will be random variables.

Figure 6.1 demonstrates the sequence of steps needed to create an events list. First, we initialize the system states, events, and simulation clock time as shown in Figure 6.1(a). Next, we call the timing routine and determine that a student arrives at simulation clock time = 0, as shown in Figure 6.1(b). This determines the simulation clock time of the first event, T_i, and the arrival event, E_i. The timing routine generates the consulting time for the first student as shown in Figure 6.1(c). At this point, it is not clear when the next event will occur. The timing routine generates the interarrival time, B_2. Now we can compare $T_1 + B_2$ to $T_1 + CT_1$. Since $T_1 + B_2 < T_1 + CT_1$, we know that an arrival event has occurred. This result is shown in Figure 6.1(d). To determine the time of the next event, the timing routine will generate the interarrival time, B_3. To determine the next imminent event, we compare $T_2 + B_3$ to $T_1 + CT_1$. In our case, these two quantities are equal ($T_3 = T_4$), which means that the two events occur simultaneously, as shown in Figure 6.1(e). This process is repeated until the simulation is complete.

In our simulation, we have identified the observable data to be the students' arrival times, the queue length, and the professor's utilization. We initially assumed the professor to be idle, the queue to be empty, and the statistics to be zero. We defined the arrival rate as deterministic with a mean time between arrivals of 30 minutes and the consulting time to be exponentially distributed with a mean of 30 minutes. We reached our stopping or terminal criterion when 10 students left the professor's office. We collected the data on the professor's utilization and the queue length. Therefore, we expect to provide and obtain similar data from simulation languages.

FIGURE 6.1 Progression of Simulation Clock Time Using Next-Event Simulation Language

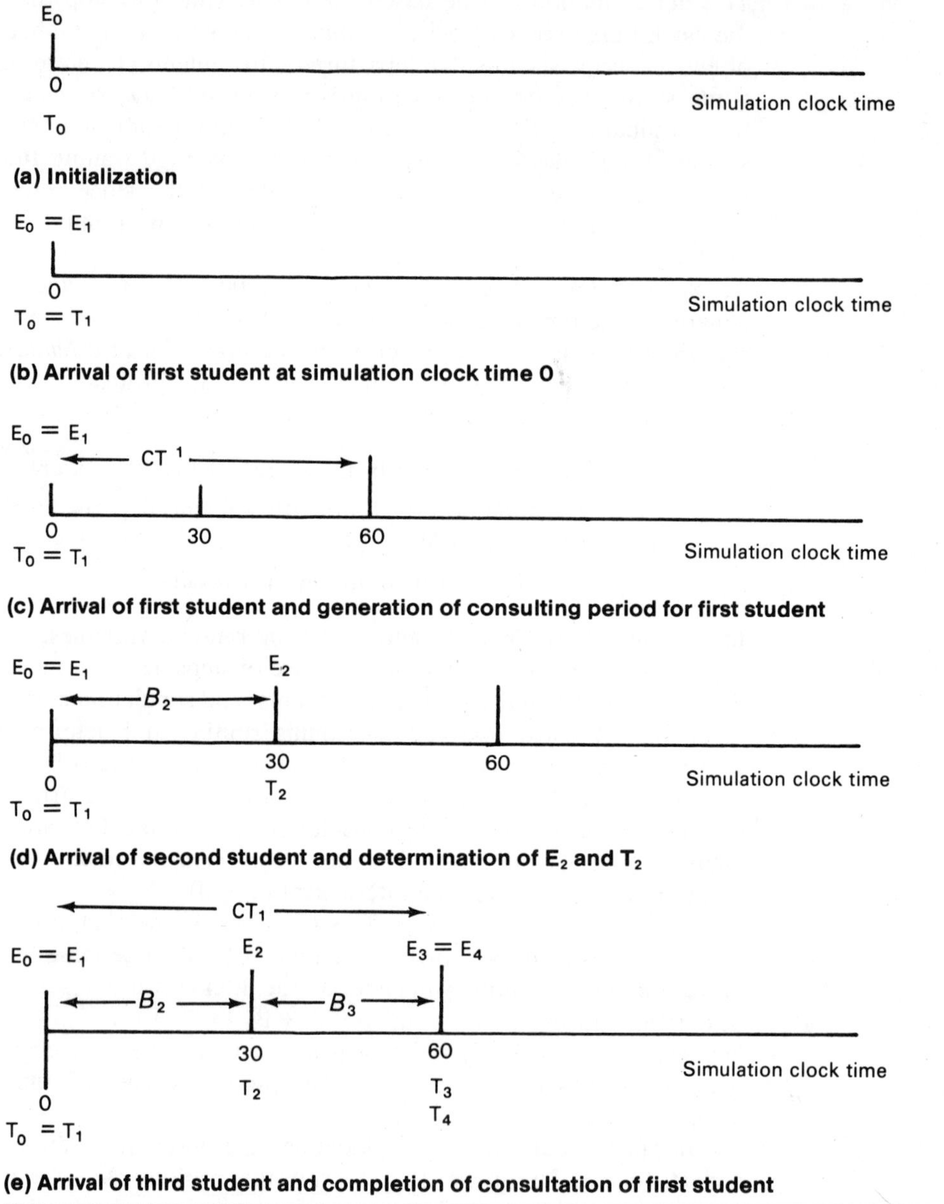

6.4 WORLD VIEWS

Let us look at our example in greater detail in order to define some important terms. In simulation language terminology, our *system* is a collection of students, called *entities* or *transactions*, passing through a sequence of *actions*. The transactions have characteristics called *attributes*. For example, "time of arrival" can be an attribute of a transaction. The transactions arrive from a *source* and depart to a *sink*. This type of system is an *open queueing* system because the transactions depart from the system forever. If the transactions remain in the system, then we have a *closed queueing* system. The professor is a *resource* or *server* unit, performing some service for the transactions. The line in which the transactions wait their turn to use the resource is the *queue*. In summary, our simulation included the following classes of *objects*: sources, sinks, transactions, servers, and queues.

Simulation languages are classified according to the manner in which these objects are viewed. The two common points of view for implementing simulation languages are the *material-oriented* and the *machine-oriented* world views.[12] In the material-oriented view, the behavior of a system is viewed from the perspective of the transactions, where the transactions direct themselves through the system. The server and the queues are viewed as passive objects to be used by the transactions as they see fit. In this case, the times at which events occur are attributes of the transactions, and the resources are acquired, held, and released by the transactions. In the machine-oriented approach, the world is viewed from the resource's perspective, and the simulation follows the operation of the resource on the transaction. The terms *machine* and *material* originated when discrete-event simulations were used to simulate manufacturing systems.

Simulation languages can be further classified as *event*, *process*, and *activity*. In our queueing example, we used the next-imminent-event sequencing to perform our simulation and to update the clock. This is an event-type simulation language. Examples are GASP, SIMSCRIPT, and SIMPAS. A SIMSCRIPT simulation program of our example is shown in Program 6.1.

PROGRAM 6.1 SIMSCRIPT Program

```
PREAMBLE

   PROCESSES INCLUDE ARRIVAL.GENERATOR, GO.HOME, AND STUDENT

   RESOURCES INCLUDE PROFESSOR

   DEFINE APPOINTMENT.LENGTH, DELAY.IN.QUEUE, MEAN.CONSULTING.TIME,
      AND MEAN.INTERARRIVAL.TIME AS REAL VARIABLES

   DEFINE NUM.PROFESSORS AS AN INTEGER VARIABLE

   DEFINE .MINUTES TO MEAN UNITS
```

continues

PROGRAM 6.1 SIMSCRIPT Program (Continued)

```
    TALLY MAX.DELAY.IN.QUEUE AS THE MAXIMUM,
       MEAN.DELAY.IN.QUEUE AS THE MEAN, AND NUM.CONSULTED
       AS THE NUMBER OF DELAY.IN.QUEUE

    ACCUMULATE AVG.NUMBER.IN.QUEUE AS THE AVERAGE AND
       MAX.NUMBER.IN.QUEUE AS THE MAXIMUM OF N.Q.PROFESSOR

    ACCUMULATE UTIL.PROFESSOR AS THE AVERAGE OF N.X.PROFESSOR

END ''PREAMBLE
MAIN

    CALL READ.DATA
    CALL INITIALIZE

    START SIMULATION

    CALL REPORT

END ''MAIN
ROUTINE READ.DATA

    READ APPOINTMENT.LENGTH, MEAN.INTERARRIVAL.TIME,
       MEAN.CONSULTING.TIME AND NUM.PROFESSORS

    PRINT 7 LINES THUS


CONSULTING MODEL -- INPUT PARAMETERS
____________________________________

    SKIP 3 LINES


    PRINT 9 LINES WITH NUM.PROFESSORS, MEAN.INTERARRIVAL.TIME,
       MEAN.CONSULTING.TIME, AND APPOINTMENT.LENGTH THUS

NUMBER OF PROFESSORS   **

INTERARRIVAL TIMES ARE DETERMINISTIC WITH MEAN ***.* MINUTES

CONSULTING TIMES ARE EXPONENTIAL WITH MEAN  ***.* MINUTES

LAST APPOINTMENT IS  BEFORE ****.* MINUTES

    START NEW PAGE
```

PROGRAM 6.1 SIMSCRIPT Program (Continued)

```
END ''READ.DATA
ROUTINE INITIALIZE

   LET TIME.V = 0

   CREATE EVERY PROFESSOR(1)
   LET U.PROFESSOR(1) = NUM.PROFESSORS
   ACTIVATE AN ARRIVAL.GENERATOR NOW
   ACTIVATE A GO.HOME IN APPOINTMENT.LENGTH .MINUTES

END''INITIALIZE
PROCESS STUDENT

   DEFINE TIME.OF.ARRIVAL AS A REAL VARIABLE

   LET TIME.OF.ARRIVAL = TIME.V
   REQUEST 1 PROFESSOR(1)
   LET DELAY.IN.QUEUE = TIME.V - TIME.OF.ARRIVAL
   WORK EXPONENTIAL.F(MEAN.CONSULTING.TIME,1) .MINUTES
   RELINQUISH 1 PROFESSOR(1)

END''STUDENT
PROCESS GO.HOME

   INTERRUPT ARRIVAL.GENERATOR

END''GO.HOME
PROCESS ARRIVAL.GENERATOR

    WHILE TIME.V < APPOINTMENT.LENGTH
    DO
       WAIT MEAN.INTERARRIVAL.TIME .MINUTES
       ACTIVATE A STUDENT NOW
    LOOP

END '' ARRIVAL.GENERATOR
ROUTINE REPORT

   PRINT 6 LINES THUS

SIMULATION RESULTS (ALL TIMES ARE IN MINUTES)
_____________________________________________

   SKIP 3 LINES

   PRINT 3 LINES WITH TIME.V THUS

SIMULATION ENDED AT ****.*

   SKIP 3 LINES

   PRINT 8 LINES WITH NUM.CONSULTED, MEAN.DELAY.IN.QUEUE, AND
     MAX.DELAY.IN.QUEUE THUS

TOTAL NUMBER OF          MEAN TIME      MAX TIME
STUDENTS CONSULTED       IN QUEUE       IN QUEUE
```

continues

PROGRAM 6.1 SIMSCRIPT Program (Continued)

```
                  ___________________      ___________     ________
                                                .
                    ***              ****.*              ****.*

    SKIP 3 LINES

    PRINT 7 LINES WITH UTIL.PROFESSOR(1)/NUM.PROFESSORS,
       AVG.NUMBER.IN.QUEUE(1), AND MAX.NUMBER.IN.QUEUE(1) THUS

PROFESSOR            AVERAGE NUMBER OF    MAXIMUM NUMBER OF
UTILIZATION          STUDENTS IN QUEUE    STUDENTS IN QUEUE
___________          _________________    _________________

    ***.*                ****.*                 *****.

END '' REPORT
```

In a process technique simulation language, a sequence of events is considered a *process*. In our example, we have one process that each student steps through; in other simulations, there might be many processes. The sequence of steps for all students is the same. Upon arrival, each student schedules the arrival of the next student and places himself in the queue. If the professor is idle, the student seizes the professor. After consultation, the student releases the professor. This information is then transmitted to other students. When we have multiple processes, each of the processes interacts with all the others. Process technique simulation languages include GPSS, SIMULA, SIMSCRIPT, and SLAM. A SLAM simulation program of our system is shown in Program 6.2.

In the activity method, the simulation tracks the beginning and ending times of the consulting period, or the "activity," of the students. This technique is implemented in GPSS and CSL. An example using GPSS is shown in Program 6.3.

PROGRAM 6.2 SLAM Program

```
;EXAMPLE SIMULATION OF STUDENTS AT PROFESSOR'S OFFICE
GEN,SIMULATION THEORY,PROFESSOR EXAMPLE,6/18/87,2;
LIMITS,1,1,50;
NETWORK;
;HERE WE INTRODUCE ONE ATTRIBUTE
       CREATE,30,0,1;     STUDENT ARRIVALS,1ST. STUDENT ARRIVES AT T=0
        QUEUE(1);
        ACTIVITY(1)/1,EXPON(30),0;
;HERE WE COLLECT STATISTICS
       COLCT,INT(1),TIME IN SYSTEM,5/0/10;
        TERMINATE,10;
       ENDNETWORK;
INIT,0,10000;
SIMULATE;
FIN;
```

PROGRAM 6.3 GPSS Program

```
  1     FUNCTION    RN1,C12              FUNCTION FOR EXPONENTIAL DISTRIBUTION
0.0,0.00/0.1,3.16/0.2,6.69/0.3,10.70/0.4,15.33/0.5,20.79/0.6,27.50
0.7,36.12/0.8,48.28/0.9,69.08/0.95,89.87/0.99,138.16
*                   ABOVE IS AN EXPONENTIAL DISTRIBUTION WITH MEAN 30 MINUTES

        GENERATE    30                   CREATE STUDENT ARRIVALS EVERY 30 MINUTES
*
        QUEUE       1                    QUEUE FOR PROFESSOR
*
        SEIZE       1                    CONSULTATION BEGINS IF PROFESSOR IS FREE
*
        MARK                             MARK THE BEGINNING OF THE CONSULTATION
*
        ADVANCE     1,FN1                CONSULTING TIME IS EXPONENTIALLY
*                                        DISTRIBUTED, CONSULTING TIME IS COMPUTED
*                                        FROM FUNCTION 1
*
        RELEASE     1                    CONSULTING PERIOD ENDS
*
        TABULATE    2                    TABULATE CONSULTING TIME
*
 EXIT   TERMINATE   1                    STUDENT GOES BACK TO DORM
*
  2     TABLE       M1,0,10,5            DEFINE TABULATION INTERVALS
*
        START       10                   SIMULATE FOR TEN STUDENTS
*
        END
```

6.5 DISCRETE-EVENT SIMULATION OF A DISTRIBUTED COMPUTING SYSTEM

Let us consider the following distributed computing system. Assume that jobs are submitted to the front-end computer, which distributes the jobs to one of four computers as shown in Figure 6.2. By simulation, we will compare the performance of two strategies for distributing jobs. One strategy is to distribute jobs in a round-robin fashion and the other is to distribute jobs to the computer with the shortest queue. In case of ties, the job is assigned to the lowest-numbered computer.

Assume that the jobs arrive at the front end according to a Poisson distribution with parameter $\lambda = 3$ jobs per unit time. Let us also assume that

The queue length for the four computers is infinite

All computers have the same processing speed of 20,000 lines per second

The number of lines per job is exponentially distributed with a mean of 30,000 lines

Let us consider the average time in the system as our performance measure. Which strategy yields a lower average time in the system? Here we will present two SLAM simulation programs and evaluate the output results. Program 6.4 is for the round-robin strategy. Figure 6.3 shows a SLAM summary report. We see

FIGURE 6.2 Distributed Computing System

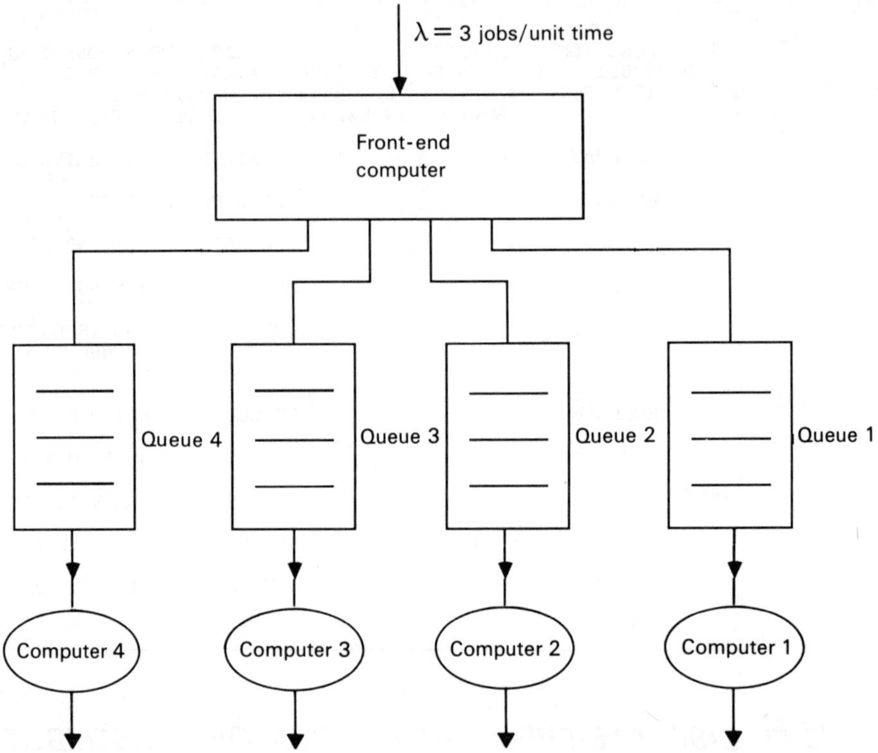

PROGRAM 6.4 SLAM Simulation Program Using the Round-Robin Strategy

```
;
GEN, KANWAL, COMPUTER SYSTEM EXAMPLE, 2/23/87; 1;
LIMITS, 4, 2, 500;
NETWORK;
      CREATE, EXPON(0.33),, 1, 500;            CREATE JOB ARRIVALS
      ASSIGN, ATRIB(2)=EXPON(30000)/20000;     ASSIGN JOB SERVICE TIME
      SELECT, CYC,,, QUE1, QUE2, QUE3, QUE4;   SELECT COMPUTERS IN
;                                              ROUND ROBIN FASHION
QUE1  QUEUE(1);                                WAIT FOR COMPUTER 1
      ACTIVITY(1)/1, ATRIB(2),, FINAL;         COMPUTER 1
QUE2  QUEUE(2);                                WAIT FOR COMPUTER 2
      ACTIVITY(1)/2, ATRIB(2),, FINAL;         COMPUTER 2
QUE3  QUEUE(3);                                WAIT FOR COMPUTER 3
      ACTIVITY(1)/3, ATRIB(2),, FINAL;         COMPUTER 3
QUE4  QUEUE(4);                                WAIT FOR COMPUTER 4
      ACTIVITY(1)/4, ATRIB(2);                 COMPUTER 4
FINAL GOON;
      COLCT, INT(1), TIME IN SYSTEM, 15/3/5;   COLLECT STATISTICS ON
;                                              TIME SPENT IN SYSTEM
      TERMINATE;
      ENDNETWORK;
INITIALIZE, 0, 1000;
SIMULATE;
FIN;
```

FIGURE 6.3 SLAM Summary Report

STATISTICS FOR VARIABLES BASED ON OBSERVATION

	MEAN VALUE	STANDARD DEVIATION	COEFF. OF VARIATION	MINIMUM VALUE	MAXIMUM VALUE	NUMBER OF OBSERVATIONS
TIME IN SYSTEM	.2566E+02	.1715E+02	.6684E+00	.2591E-01	.5969E+02	500

FILE STATISTICS

FILE NUMBER	ASSOC NODE LABEL/TYPE	AVERAGE LENGTH	STANDARD DEVIATION	MAXIMUM LENGTH	CURRENT LENGTH	AVERAGE WAITING TIME
1	QUE1 QUEUE	16.7688	9.7598	36	0	28.6562
2	QUE2 QUEUE	5.1712	4.9714	17	0	8.8370
3	QUE3 QUEUE	14.0617	8.7086	30	0	24.0301
4	QUE4 QUEUE	20.3621	10.9159	35	0	34.7968
5	CALENDAR	4.4300	.9290	6	0	.6416

SERVICE ACTIVITY STATISTICS

ACTIVITY INDEX	START NODE OR ACTIVITY LABEL	SERVER CAPACITY	AVERAGE UTILIZATION	STANDARD DEVIATION	CURRENT UTILIZATION	AVERAGE BLOCKAGE	MAXIMUM IDLE TIME/SERVERS	MAXIMUM BUSY TIME/SERVERS	ENTITY COUNT
1	COMPUTER 1	1	.9763	.1521	0	.0000	1.5124	198.4478	125
2	COMPUTER 2	1	.7750	.4176	0	.0000	39.7871	110.1001	125
3	COMPUTER 3	1	.9749	.1564	0	.0000	2.5896	189.6689	125
4	COMPUTER 4	1	.9620	.1912	0	.0000	5.7852	205.2807	125

in Figure 6.3 that the mean time in the system for a job is 25.7 seconds, and the standard deviation is 17.2 seconds. The minimum time in the system is 0.03 second and the maximum value is 59.7 seconds. We also see that the average utilizations of computers 1, 2, 3, and 4 are 0.98, 0.78, 0.98, and 0.96, respectively. Figure 6.4 shows a histogram of the mean time in the system for the 500 observations.

FIGURE 6.4 SLAM Simulation Histogram for Mean Time in the System

```
OBSV     RELA     CUML       UPPER
FREQ     FREQ     FREQ   CELL LIMIT    0         20        40        60        80      100
                                       +    +    +    +    +    +    +    +    +    +    +
  53     .106     .106   .3000E+01     +*****                                            +
  67     .134     .240   .8000E+01     +********   C                                     +
  39     .078     .318   .1300E+02     +****           C                                 +
  29     .058     .376   .1800E+02     +***               C                              +
  37     .074     .450   .2300E+02     +****                 C                           +
  40     .080     .530   .2800E+02     +****                     C                       +
  62     .124     .654   .3300E+02     +******                          C                +
  22     .044     .698   .3800E+02     +**                                C              +
  50     .100     .798   .4300E+02     +*****                                  C         +
  41     .082     .880   .4800E+02     +****                                       C     +
  28     .056     .936   .5300E+02     +***                                           C  +
  31     .062     .998   .5800E+02     +***                                              C
   1     .002    1.000   .6300E+02     +                                                 C
   0     .000    1.000   .6800E+02     +                                                 C
   0     .000    1.000   .7300E+02     +                                                 C
   0     .000    1.000   .7800E+02     +                                                 C
   0     .000    1.000      INF        +                                                 C
 ---                                   +    +    +    +    +    +    +    +    +    +    +
 500                                   0         20        40        60        80      100
```

STATISTICS FOR VARIABLES BASED ON OBSERVATION

	MEAN VALUE	STANDARD DEVIATION	COEFF. OF VARIATION	MINIMUM VALUE	MAXIMUM VALUE	NUMBER OF OBSERVATIONS
TIME IN SYSTEM	.2566E+02	.1715E+02	.6684E+00	.2591E-01	.5969E+02	500

PROGRAM 6.5 SLAM Simulation Program Using the Shortest-Queue Strategy

```
;
GEN, KANWAL, COMPUTER SYSTEM EXAMPLE, 2/23/87,1;
LIMITS, 4, 2, 500;
NETWORK;
      CREATE, EXPON(0.33),, 1, 500;             CREATE JOB ARRIVALS
      ASSIGN, ATRIB(2)=EXPON(30000)/20000;      ASSIGN JOB SERVICE TIME
      SELECT, SNQ,,, QUE1, QUE2, QUE3, QUE4;    SELECT COMPUTER WITH
;                                               SHORTEST LINE
QUE1  QUEUE(1);                                 WAIT FOR COMPUTER 1
      ACTIVITY(1)/1, ATRIB(2),, FINAL;          COMPUTER 1
QUE2  QUEUE(2);                                 WAIT FOR COMPUTER 2
      ACTIVITY(1)/2, ATRIB(2),, FINAL;          COMPUTER 2
QUE3  QUEUE(3);                                 WAIT FOR COMPUTER 3
      ACTIVITY(1)/3, ATRIB(2),, FINAL;          COMPUTER 3
QUE4  QUEUE(4);                                 WAIT FOR COMPUTER 4
      ACTIVITY(1)/4, ATRIB(2);                  COMPUTER 4
FINAL GOON;
      COLCT, INT(1), TIME IN SYSTEM, 11/3/5;    COLLECT STATISTICS ON
;                                               TIME SPENT IN SYSTEM
      TERMINATE;
      ENDNETWORK;
INITIALIZE, 0, 1000;
SIMULATE;
FIN;
```

Program 6.5 simulates the distribution of jobs to the computer with the shortest queue. The results are shown in Figure 6.5. The mean time in the system for a job is 24.4 seconds, and the standard deviation is 13.7 seconds. The minimum time in the system is 0.03 second, and the maximum value is 57.2 seconds. We also see that the average utilizations of computers 1, 2, 3, and 4 are 0.91, 0.94, 0.98, and 0.91, respectively.

FIGURE 6.5 SLAM Summary Report and Histogram of Mean Time in the System

STATISTICS FOR VARIABLES BASED ON OBSERVATION

	MEAN VALUE	STANDARD DEVIATION	COEFF. OF VARIATION	MINIMUM VALUE	MAXIMUM VALUE	NUMBER OF OBSERVATIONS
TIME IN SYSTEM	.2438E+02	.1371E+02	.5625E+00	.2591E-01	.5722E+02	500

FILE STATISTICS

FILE NUMBER	ASSOC NODE LABEL/TYPE	AVERAGE LENGTH	STANDARD DEVIATION	MAXIMUM LENGTH	CURRENT LENGTH	AVERAGE WAITING TIME
1	QUE1 QUEUE	13.1682	8.4741	29	0	18.7942
2	QUE2 QUEUE	13.6714	8.2301	29	0	23.1027
3	QUE3 QUEUE	14.2321	7.8176	28	0	25.6947
4	QUE4 QUEUE	12.9071	8.2785	28	0	24.7853
5	CALENDAR	4.4799	1.0251	6	0	.6403

SERVICE ACTIVITY STATISTICS

ACTIVITY INDEX	START NODE OR ACTIVITY LABEL	SERVER CAPACITY	AVERAGE UTILIZATION	STANDARD DEVIATION	CURRENT UTILIZATION	AVERAGE BLOCKAGE	MAXIMUM IDLE TIME/SERVERS	MAXIMUM BUSY TIME/SERVERS	ENTITY COUNT
1	COMPUTER 1	1	.9056	.2924	0	.0000	19.1488	189.5358	148
2	COMPUTER 2	1	.9396	.2382	0	.0000	10.1434	192.2010	125
3	COMPUTER 3	1	.9794	.1419	0	.0000	1.6035	200.2125	117
4	COMPUTER 4	1	.9051	.2931	0	.0000	12.8646	187.3396	110

FIGURE 6.5 SLAM Summary Report and Histogram of Mean Time in the System (Continued)

```
 OBSV      RELA      CUML        UPPER
 FREQ      FREQ      FREQ     CELL LIMIT     0         20        40        60        80        100
                                             +    +    +    +    +    +    +    +    +    +    +
   44      .088      .088     .3000E+01      +****                                             +
   39      .078      .166     .8000E+01      +****   C                                         +
   35      .070      .236     .1300E+02      +***        C                                     +
   37      .074      .310     .1800E+02      +****          C                                  +
   50      .100      .410     .2300E+02      +*****              C                             +
   76      .152      .562     .2800E+02      +********                   C                     +
   96      .192      .754     .3300E+02      +**********                           C           +
   44      .088      .842     .3800E+02      +****                                     C       +
   37      .074      .916     .4300E+02      +****                                         C   +
   20      .040      .956     .4800E+02      +**                                             C +
   14      .028      .984     .5300E+02      +*                                               C+
    8      .016     1.000     .5800E+02      +*                                                C
    0      .000     1.000        INF         +                                                 C
  ---                                        +    +    +    +    +    +    +    +    +    +    +
  500                                        0         20        40        60        80        100
```

STATISTICS FOR VARIABLES BASED ON OBSERVATION

	MEAN VALUE	STANDARD DEVIATION	COEFF. OF VARIATION	MINIMUM VALUE	MAXIMUM VALUE	NUMBER OF OBSERVATIONS
TIME IN SYSTEM	.2438E+02	.1371E+02	.5625E+00	.2591E-01	.5722E+02	500

This is a heavily loaded system, as shown by the very high average utilization figures. Based on a comparison of mean time values, it appears that the shortest-queue strategy performs slightly better than the round-robin strategy. As mentioned in Chapter 5, this simulation is only one observation for estimating the mean time in the system. To obtain more confidence in these results, we should repeat the simulation using different random number generator seeds.

6.6 DISCRETE-EVENT SIMULATION OF A COMPUTER COMMUNICATION NETWORK[30,31]

Telephone companies in the United States and Europe are converting their communication systems from analog to digital. The goal is to give telephone users a digital communication link for both voice and data through a simple set of standardized interfaces with integrated switch digital networks (ISDNs). The stored program–controlled (SPC) switches supply the switching needed for ISDNs. Some SPCs use time-division switching, with signals assigned to different time slots. In satellite networks, earth stations transmit to the satellite at assigned time slots. This arrangement provides total connectivity between earth stations and is called fixed time-division multiple access (F-TDMA). Each of the N earth stations is assigned one slot in each frame of N slots. When the assignment pattern is the same for each frame, it is called synchronous time-division multiple access (S-TDMA). Simulation is often used to study the performance of these systems. This example demonstrates the use of SLAM for simulating TDMA systems when the transmitter's ON-OFF periods are exponentially distributed.

Consider a simple case of an S-TDMA system where each frame is divided into 10 one-second slots. Only one packet can be transmitted in one slot, as shown in Figure 6.6. The eighth slot is allotted for transmission of messages from

FIGURE 6.6 Synchronous Time-Division System

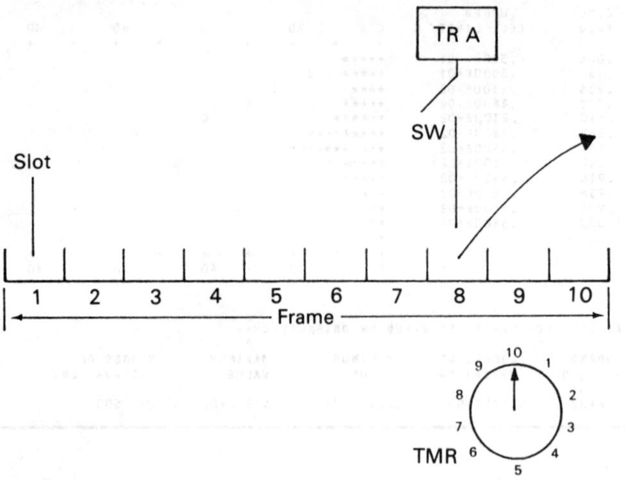

a transmitter, TR A. In this model, a timer—TMR—is used to identify the slot in each frame.

The transmitter, TR A, could be switched on and off at user-selected intervals by means of a switch SW. The messages are forwarded to the SPC buffer. They are then forwarded to the receiver on a first-in, first-out (FIFO) basis from the buffer at the time slot assigned for TR A. A message does not wait at the buffer if it is generated during the slot assigned for TR A. There are many approaches to model this transmission process, three of which are considered here.

Model 1

Program 6.6 represents the first SLAM model. Steps 3 through 39 are the SLAM network statements. The slot assigned for transmitter TR A (SLT8) is considered a "resource," and gate GTE1 defines slot position.

Timer TMR (step 7, labelled TMR) determines the end point of slot 8 (at the ninth second within a frame). Gate timer (step 10, labelled CGT) CGT opens the transmission channel for TR A at the beginning of slot 8 and closes it at the end point of slot 8. In this case, the closing occurs after one second.

Statements 17 through 28 of Program 6.6 demonstrate the unique characteristics of this model in obtaining the ON-OFF periods for TR A. First, the CREATE node in step 17 creates entities (messages) from time zero and produces a total of 10 entities at one-second intervals. This represents a 10-second ON duration of TR A. Then TR A is OFF until the 20th second, when the second CREATE node (step 19) starts its message creations. Thus, the six CREATE nodes in series give the ON cycles for the transmitter. This model has a disadvantage when we try to modify the ON-OFF process to random intervals, since the termination of each CREATE activity at the CREATE nodes is based

PROGRAM 6.6 SLAM Simulation Program for Model 1

```
03.00   NETWORK;
04.00        RESOURCE/SLT8(1),1;
05.00        GATE/GTE1,CLOSE,2;
06.00   ;
07.00   TMR  CREATE,10,9,,,1;    DEFINE SLOT END TIME
08.00        TERM;
09.00   ;
10.00   CGT  CREATE,10,8,,,1;     GATE TIMER
11.00        OPEN,GTE1,1;
12.00        ACT/1,1;
13.00        CLOSE,GTE1,1;
14.00        TERM;
15.00   ;
16.00   ;***TRANSMITTER SECTION***
17.00        CREATE,1,0,1,10,1;        I ON PERIOD
18.00        ACT,,,LB;
19.00        CREATE,1,20,1,10,1;       II ON PERIOD
20.00        ACT,,,LB;
21.00        CREATE,1,40,1,10,1;       III ON PERIOD
22.00        ACT,,,LB;
23.00        CREATE,1,60,1,10,1;       IV ON PERIOD
24.00        ACT,,,LB;
25.00        CREATE,1,80,1,10,1;       V ON PERIOD
26.00        ACT,,,LB;
27.00        CREATE,1,100,1,10,1;      VI ON PERIOD
28.00        ACT,,,LB;
28.50   ;***MULTIPLEXER SECTION***
29.00   LB   GOON,1;
30.00        COLCT,ATRIB(1), TR A,50/0/1,1;
31.00   RSC  AWAIT(1),SLT8/1,1;  MULTIPLEXER BUFFER
32.00   GTE  AWAIT(2),GTE1,1;
33.00        ACT/2,REL(TMR);
34.00        CLOSE,GTE1,1;
35.00   TG1  FREE,SLT8/1,1;     RECEIVER
36.00        COLCT,INT(1),RECEIVER,50/0/1,1;
37.00        TERM;
38.00   ;
39.00     ENDNET;
42.00
```

on an entity count, not on the duration of CREATE node operation. An entity captures a resource and waits at AWAIT node GTE (step 32), if necessary, until a slot is available. Step 33 represents the slot time after which the slot is closed (step 34). This closure of slot SLT8, and the fact that only one resource was defined (step 4), ensures that only one message is transmitted in a slot, regardless of its arrival time. After the slot is utilized, the resource unit is freed (step 35) and collected into the receiver (steps 36 and 37). The COLCT statement is for purposes of statistics collection.

Model 2

In our second model, represented by Program 6.7, the series of CREATE nodes used for the ON-OFF operation of transmitter TR A is eliminated. The main

PROGRAM 6.7 SLAM Simulation Program for Model 2

```
03.00   NETWORK;
04.00             RESOURCE/SLT8(1),1;
05.00             GATE/GTE1,CLOSE,2;
06.00             GATE/GTE2,OPEN,3;
07.00   ;
08.00   TMR       CREATE,10,9,,,1;  DEFINE SLOT END TIME
09.00             TERM;
10.00   ;
11.00   CGT       CREATE,10,8,,,1;  GATE TIMER
12.00             OPEN,GTE1,1;
13.00             ACT/1,1;
14.00             CLOSE,GTE1,1;
15.00             TERM;
16.00   ;
17.00   ;***TRANSMITTER ON-OFF SWITCH***
18.00   DGT       CREATE,20,10,,,1;
19.00             CLOSE,GTE2,1;
20.00             ACT/3,10;
21.00             OPEN,GTE2,1;
21.00             TERM;
22.00   ;
23.00   LB2       TERM;
24.00   ;***TRANSMITTER SECTION***
25.00             CREATE,1,0,1,,1;
26.00             ASSIGN,ATRIB(1)=TNOW,1;
27.00             COLCT,ATRIB(1),TR A,110/0/1,1;
28.00   DXT       AWAIT(3/1),GTE2,BALK(LB2),1;
29.00             ASSIGN,ATRIB(1)=TNOW,1;
30.00             COLCT,ATRIB(1),TR A OP,110/0/1,1;
31.00   RSC       AWAIT(1),SLT8/1,1;     BUFFER
32.00   GTE       AWAIT(2),GTE1,1;
33.00             ACT/2,REL(TMR);
34.00             CLOSE,GTE1,1;
35.00   TG1       FREE,SLT8/1,1;  RECEIVER
36.00             ASSIGN,ATRIB(2)=TNOW,1;
37.00             COLCT,INT(1),REC INTR,110/0/1,1;
38.00             COLCT,ATRIB(2),REC SLT,110/0/1,1;
39.00             TERM;
40.00   ;
42.00   FIN;
```

difference between this model and the previous one is the use here of an AWAIT node, labelled DXT, with balk facility. Recall that we defined balk as the choice, for whatever reasons, not to join the queue or to leave the queue before being serviced.

Steps 25 through 28 in Program 6.7 represent TR A controlled (the ON-OFF switching) by the timer labelled DGT. When DGT closes gate GTE2, the AWAIT node DXT will dump the incoming messages out of the system at the node labelled LB2 (step 23). This model is more versatile than model 1 since to modify the system for random ON-OFF periods, timer DGT would enable this switching on the basis of time rather than the entity count at the CREATE note in step 26. The rest of the steps of this model are similar to those of model 1.

A slight inaccuracy in the performance of the second model occurs at the beginning of every transmission cycle after the first. The AWAIT node DXT balks incoming messages only if at least one entity is already waiting in its file. When the OFF period ends, the waiting entity at DXT is immediately released. This extra release will cause a slight inaccuracy in the arrivals expected at the SPC buffer. The inaccuracy is usually negligible since the discrepancy remains a constant one unit in each ON cycle. As the number of messages sent during the ON cycle increases, the effect of this extra one packet becomes less noticeable.

Model 3

The third model, shown in Program 6.8, avoids both problems of the previous models. This is achieved in steps 20 through 23 which represent transmitter TR1. At the CREATE node labelled TR1, the entities (messages) are formed at the rate specified; in this case, one every second. The arrival time is then marked at the ASSIGN node, from which the message can take either one of two branches (ACT/3 or ACT/4 in steps 22 and 23) depending on the instant of message

PROGRAM 6.8 SLAM Simulation Program for Model 3

```
03.00   NETWORK;
04.00            RESOURCE/SLT8(1),1;
05.00            GATE/GTE1,CLOSE,2;
06.00   ;
07.00   TMR      CREATE,10,9,,,1;  SLOT END TIMER
08.00            TERM;
09.00   ;
10.00   CGT      CREATE,10,8,1,,1;  GATE TIMER
11.00            OPEN,GTE1,1;
12.00            ACT/1,1;
13.00            CLOSE,GTE1,1;
14.00   LB1      TERM;
15.00   ;
16.00   TLX      CREATE,20,10,,,1;  ON-OFF SWITCH
17.00            ACT/5,10;
18.00            TERM;
19.00   ;***TRANSMITTER SECTION***
20.00   TR1      CREATE,1,0,1,,1;
21.00            ASSIGN,ATRIB(1)=TNOW,1;
22.00            ACT/3,,NNACT(5).NE,0,LB1;
23.00            ACT/4;
24.00            COLCT,ATRIB(1),TR A ARV,200/0/1,1;
25.00   RSC      AWAIT(1),SLT8/1,1;  BUFFER
26.00   GTE      AWAIT(2),GTE1,1;
27.00            ACT/2,REL(TMR);
28.00            CLOSE,GTE1,1;
29.00   TG1      FREE,SLT8/1,1;  RECEIVER
30.00            ASSIGN,ATRIB(1)=TNOW,1;
31.00            COLCT,ATRIB(1),RECEIVER,200/6/1,1;
32.00            TERM;
33.00   ;
34.00            ENDNET;
```

arrival. The condition for ACT/3 is that there be at least one entity in ACT/5 (step 22), and ACT/3 operates only in 20-second intervals starting from the 10th second (step 16). At the beginning of the 10th second, an entity enters ACT/5 and stays there until the 20th second since the activity duration is 10 seconds (step 12). During this period—from the 10th second to the 20th second—messages from the node labelled TR1 will branch into ACT/3 (step 22), effectively dumping out of the system the messages arriving during this period. At the end of the 20th second, there are no more entities in ACT/4, which leads into the SPC buffer (steps 25 and 26). The process from here on is the same as for the previous two models.

Summary of Simulation Results

The following simulation results are from the final model described in Program 6.8. The simulation period was 500 time units. The "20" in the step labelled TLX has been modified to "EXPON(20)" for the exponentially distributed random ON-OFF periods in message transmission. The same technique is used in the step labelled TR1 (step 20) to get the exponentially distributed interarrival times for simulation. Ten random number streams with different seed values are available in SLAM, and these are picked in a random manner for each simulation run.

The simulation results are summarized in the graphs of Figures 6.7 through 6.9. These graphs are plotted from the SLAM summary report data for simulation runs in unit steps, from 5 to 13, for mean interarrival times.

Figure 6.7 describes the buffer characteristics for various interarrival rates. These include the maximum queue length experienced during the simulation (MXL), the length of the queue at the end of the simulation (CURL), and the

FIGURE 6.7 Buffer Characteristics

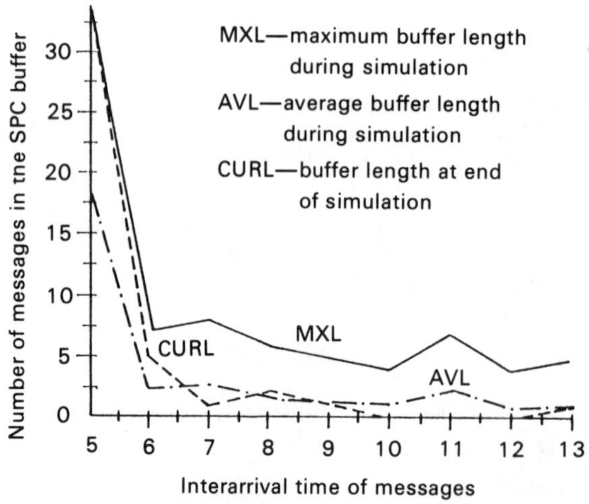

FIGURE 6.8 Slot Utilization Characteristic

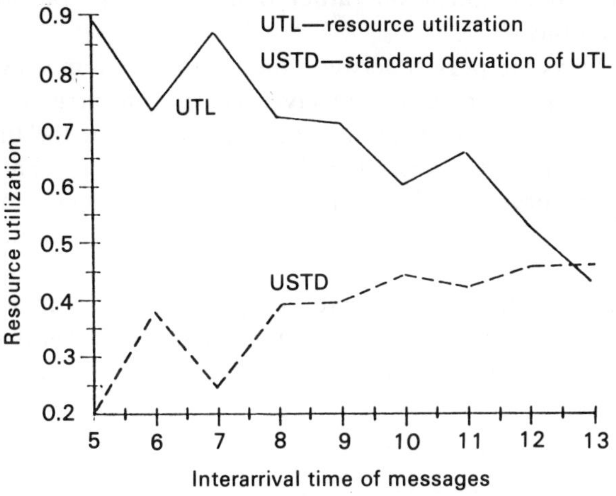

average queue length (AVL). As the arrival rate decreases, MXL, CURL, and AVL decrease. As seen in CURL, during the 500 time unit simulations, the number of messages that did not reach the receiver is very low for interarrival times greater than EXPON(7).

Figure 6.8 is a plot of the slot utilization factor (UTL) and the corresponding standard deviation (USTD) during each simulation for the different transmission rates. As expected, slot utilization decreases as the arrival rate decreases. This is an effective measure of the slot wastage in the synchronous time-division multi-

FIGURE 6.9 Maximum Message Transit Time Characteristic

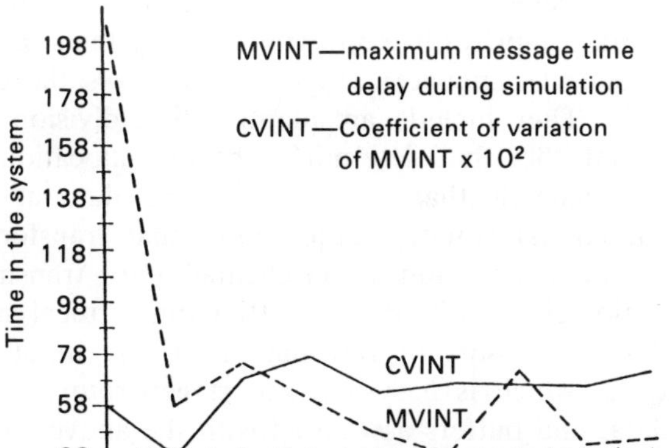

plexing process. The shape of the standard deviation plot implies an overall smooth degradation of utilization rather than the sharp transitions seen in the individual simulations.

Figure 6.9 shows the period between the creation of a message (MVINT) and the time the message reaches the receiver versus interarrival rates. MVINT is dependent to a large extent on the length of the queue at the time it comes into the buffer. The coefficient of variation (CVINT) plot implies that we could expect an overall smooth curve for the mean values of the message transit time.

Graphs of this nature could resolve problems in the design of TDMA communication channels. For example, Figure 6.7 shows that if the mean interarrival time is 6 or greater, there is no advantage in increasing it further, since the average queue length remains more or less minimal from this point on. Based on the application at hand, this would enable choosing the optimal buffer size for a transmitter.

6.7 DISCRETE-EVENT SIMULATION OF A FIVE-NODE COMPUTER NETWORK[30,31]

In this example, we present a SLAM network model for the simulation of a five-node computer network, shown in Figure 6.10. The input messages to the nodes are in packet form, and their interarrival times are exponentially distributed. The system uses a bus with time-division multiplexing to forward messages. The bus-scheduling algorithm for each node can be modified to obtain optimum network design. Design characteristics include a variable number of devices communicating within a node, variable buffer lengths, consideration of message processing times at buffers and devices, and collection of statistics at any point in the message flow. We will present sample simulation results.

The system described here is part of a data acquisition and processing network that communicates via adaptive time-division multiplexed (ATDM) channels. The network comprises five interconnected systems. Each system forms a node of a multiconnected network. Figure 6.10 shows the components of a system at each node. They include an adaptive time-division multiplexed bus (ATDMB); a bus controller that determines the slot allocation of time slots in each frame; source channels that provide incoming data at deterministic or random rates; bidirectional communication links that transfer data from one system (node) to another; communication channel 1 for transmitting processed reports; communication channel 2 for transmitting time-critical processed reports; processor 1 for data processing of raw data from the input channels of the network; processor 2, which is a system report generator; bulk memory that stores processed data; and buffers associated with the above components.

System operation can be described as follows. Messages arriving at each node are packetized, placed in the appropriate buffer, and forwarded to destination nodes at assigned time slots. The incoming messages to a node include messages from neighboring nodes along the corresponding links, messages generated at the

FIGURE 6.10 Network Node

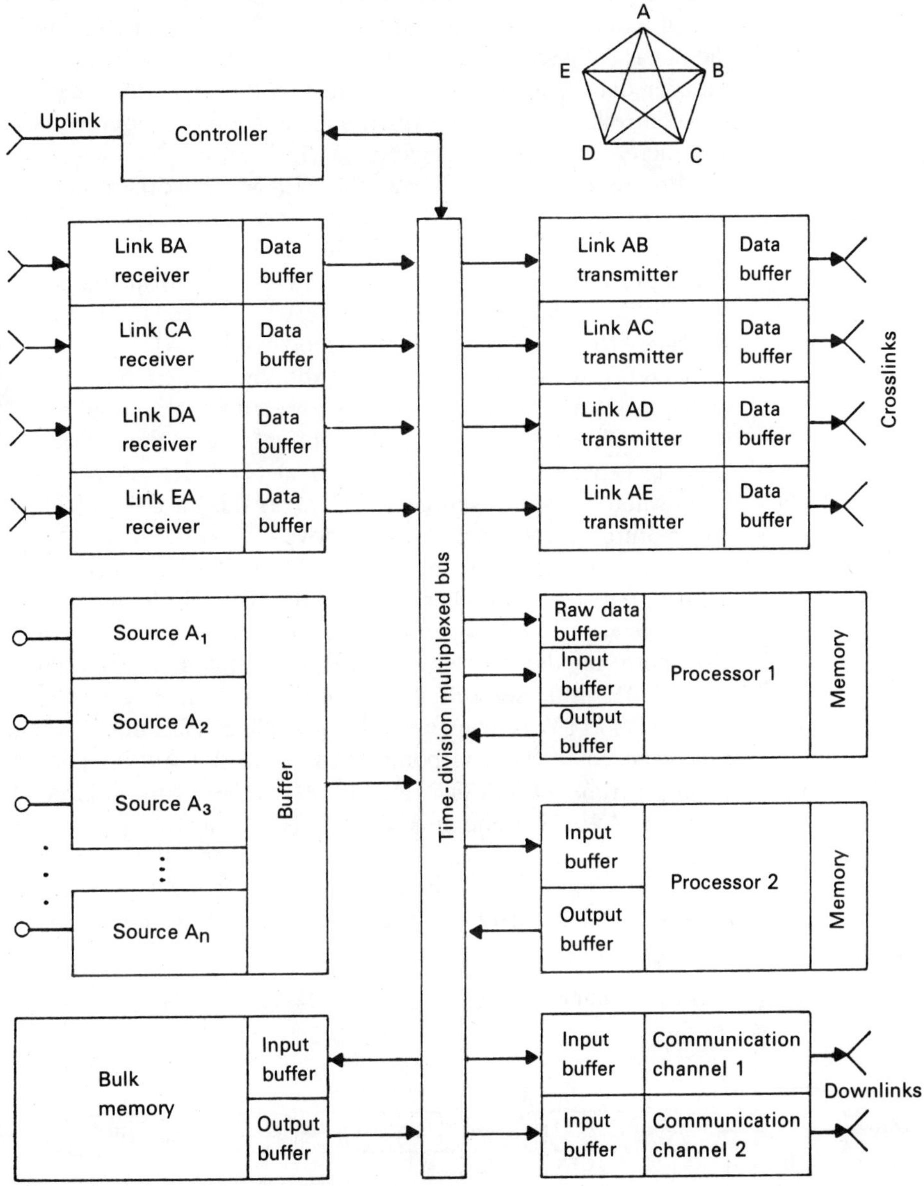

node itself, and messages to be transferred between the bulk memory and processors within the node.

When a packet arrives at a node, it joins the end of its queue (buffer) and awaits transmission to its destination on a first-in, first-out basis as soon as the assigned bus slot is available for it. The destination of a packet could be any other node or noninput device within the system. Only one packet can be transferred from one buffer to another during a single bus time slot.

In a time-division multiplexed bus (TDMB) system, each device connected to the bus is assigned a fixed portion of a time frame. During this period, the bus can transfer data from the output buffer to the input buffer of another device. The bus controller allows for dynamic assignments of slots to increase bus utilization and increase system throughput.

Programs 6.7 and 6.8 are the SLAM network models of one node of the five-node, fully connected computer network. This node will be explained in detail since the other nodes could be described using similar techniques. The nodes are linked together by ACTIVITIES to form the entire network.

The CREATE node labelled DV1 (device 1) in Figure 6.11 creates the messages to be transmitted. The message from DV1 goes to the GOON node labelled TRA which serves as a window to receive messages. The messages arriving at the GOON node are forwarded sequentially to the AWAIT node labelled GTA, representing the buffer where the messages wait in a queue (infinite capacity). Messages waiting at GTA are released on a first-in, first-out basis when a unit "resource"—SLT/1 in this case—is available. A message that has captured a resource at GTA moves on to the next AWAIT node. This node will release the message into the ACTIVITY REL(TMR) only when gate GTEA is open, which is at the beginning of the time slot assigned for the transmission of DV1 messages. The opening and closing of gate GTEA is controlled from elsewhere in the model, as will be explained in detail later. ACTIVITY REL(TMR) is a "server" or slot for the DV1 messages. The termination point of this ACTIVITY is controlled by the CREATE node labelled TMA which is synchronized to the end point of the time slot for DV1 messages. The message released from ACTIVITY REL(TMR) closes gate GTEA. This action ensures that the AWAIT node linked to GTEA will not release any more messages, thereby avoiding the risk of transmitting more than one message in a time slot. The following COLCT node collects statistics. The resource—SLT/1 in this case—is then freed at the FREE node, after which the message is terminated.

FIGURE 6.11 Message Multiplexer Model

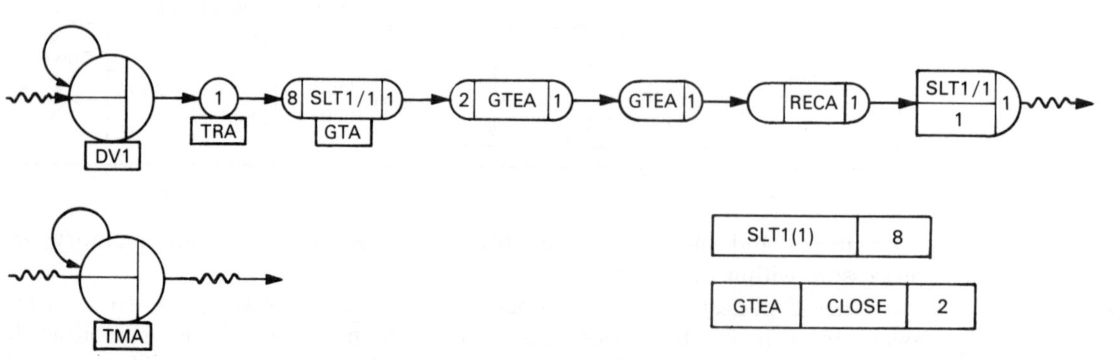

FIGURE 6.12 Slot Allocation Model

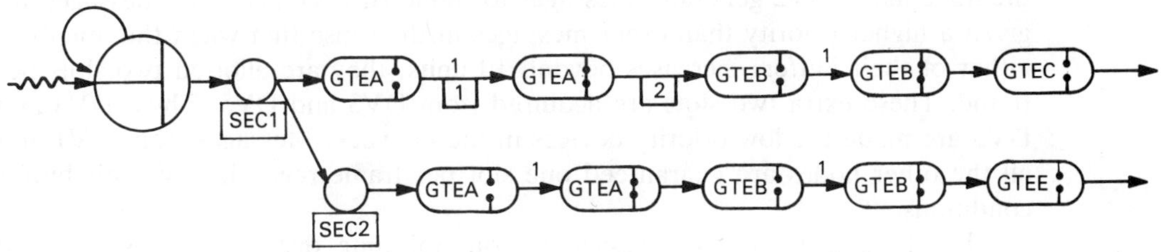

Termination is allowed only because the messages generated by DV1 are not to be sent outside node 1. For devices that transmit messages out of the node, the configuration can be the same as that of DV1, except that, instead of the TERMINATE node in the end, an ACTIVITY node is used to forward the message to the required node, where it is multiplexed again.

Figure 6.12 shows the slot allocation mechanism used in the model. There are two strings of OPEN and CLOSE nodes for gate GTEA, GTEB, and so on. An entity entering the upper string opens GTEA and begins ACTIVITY 1. This mechanism keeps GTEA open for one second, after which it is closed by the subsequent CLOSE node, representing the opening and closing of the first slot in the frame. Then the second slot, represented by GTEB, is opened. In this way, the time slots in a frame are opened and closed in a series, ensuring that no two slots are opened simultaneously.

The lower string of OPEN and CLOSE nodes is similar but the node order can be different. The slot allocation at the TDMB is thus made adaptive by choosing a particular string at the beginning of each frame period. The choice is made at the GOON node labelled SEC1 by branching of activities, conditioned upon the queue lengths at various buffers.

Figure 6.13 shows the network configuration of the system simulated for zero to 50 time units. The devices shown at node A are also available at all other

FIGURE 6.13 Network Configuration

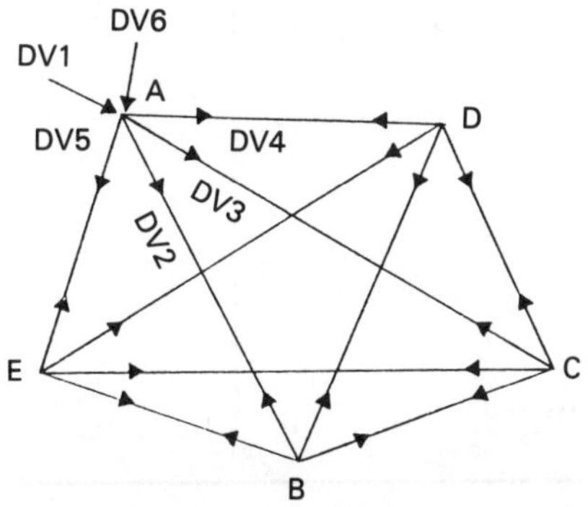

nodes. At node A, devices DV1 and DV6 generate messages whose destination is the node itself. DV2 generates messages for node B. DV6 and DV2 messages are given a higher priority than other messages in the sense that when the length of either of their buffers increases beyond 10 units, they are allotted two slots per frame. These extra two slots are acquired from DV3 and DV5. Thus, DV3 and DV5 are made the low-priority devices in the network. Messages from DV1 and all the other nodes are guaranteed one slot per frame regardless of their buffer conditions.

The following are the message interarrival time distributions used in the simulation for each device:

DV1: exponential distribution, mean = 1
DV2: exponential distribution, mean = 1
DV3: exponential distribution, mean = 1
DV4: exponential distribution, mean = 8
DV5: exponential distribution, mean = 5
DV6: exponential distribution, mean = 1

Statistics can be collected, using COLCT nodes, from virtually any point in the network. Some sample results from the SLAM summary report follow.

Figure 6.14 shows a histogram of the message generation at DV1 according to the interarrival distribution above. Figure 6.15 is a histogram of all messages arriving at node A, including DV1 messages as well as messages arriving from the other nodes. Figure 6.16 shows the message arrivals into the receiver buffer of node A. Since node A messages are guaranteed a slot in every frame, they arrive every six time units.

FIGURE 6.14 Histogram of Message Generation at DV1

```
 OBS RELA   UPPER
FREQ FREQ  CELL LIM 0         20        40        60        80        100
                    +    +    +    +    +    +    +    +    +    +    +
   1 .024  .000E+00 +*                                                +
   1 .024  .100E+01 +*C                                               +
   0 .000  .200E+01 + C                                               +
   1 .024  .300E+01 +*  C                                             +
   0 .000  .400E+01 +   C                                             +
   1 .024  .500E+01 +*   C                                            +
   0 .000  .600E+01 +    C                                            +
   3 .071  .700E+01 +****   C                                         +
   2 .048  .800E+01 +**        C                                      +
   0 .000  .900E+01 +          C                                      +
   2 .048  .100E+02 +**          C                                    +
   1 .024  .110E+02 +*            C                                   +
   2 .048  .120E+02 +**              C                                +
   0 .000  .130E+02 +                C                                +
   0 .000  .140E+02 +                C                                +
   0 .000  .150E+02 +                C                                +
   1 .024  .160E+02 +*               C                                +
   1 .024  .170E+02 +*                C                               +
   1 .024  .180E+02 +*                 C                              +
   1 .024  .190E+02 +*                  C                             +
   1 .024  .200E+02 +*                    C                           +
  23 .548    INF    +***************************                      C
 ---                +    +    +    +    +    +    +    +    +    +    +
  42                0         20        40        60        80        100
```

FIGURE 6.15 Histogram of All Messages Arriving at Node A

```
OBS  RELA   UPPER
FREQ FREQ  CELL LIM 0         20        40        60        80      100
                    +    +    +    +    +    +    +    +    +    +    +

 5   .064  .000E+00 +***                                              +
 6   .077  .100E+01 +****  C                                          +
 6   .077  .200E+01 +****      C                                      +
 5   .064  .300E+01 +***          C                                   +
 5   .064  .400E+01 +***             C                                +
 4   .051  .500E+01 +***                C                             +
 1   .013  .600E+01 +*                   C                            +
 6   .077  .700E+01 +****                  C                          +
 6   .077  .800E+01 +****                      C                      +
 0   .000  .900E+01 +                          C                      +
 3   .038  .100E+02 +**                          C                    +
 1   .013  .110E+02 +*                            C                   +
 2   .026  .120E+02 +*                             C                  +
 0   .000  .130E+02 +                              C                  +
 0   .000  .140E+02 +                              C                  +
 0   .000  .150E+02 +                              C                  +
 1   .013  .160E+02 +*                              C                 +
 1   .013  .170E+02 +*                              C                 +
 1   .013  .180E+02 +*                               C                +
 1   .013  .190E+02 +*                                C               +
 1   .013  .200E+02 +*                                C               +
23   .295    INF    +***************                                  C
---                 +    +    +    +    +    +    +    +    +    +    +
78                  0         20        40        60        80      100
```

Figure 6.17 shows the effect of high-priority arrivals (DV2) into their receiver buffers. After the 14th time unit, the arrival frequency has increased to two messages per frame because the queue length of a high-priority message has exceeded the 10-unit limit.

The collection of statistics at any point in the network model is possible with the use of COLCT nodes. ASSIGN nodes can tag the messages at the required points for later identification at COLCT. A COLCT node can also provide histograms on the statistics collected. These data could be used to determine the network specifications according to customer requirements.

FIGURE 6.16 Message Arrivals at the Receiver Buffer of Node A

```
OBS  RELA   UPPER
FREQ FREQ  CELL LIM 0         20        40        60        80      100
                    +    +    +    +    +    +    +    +    +    +    +
 0   .000  .000E+00 +                                                 +
 1   .111  .100E+01 +******                                           +
 0   .000  .200E+01 +     C                                           +

 0   .000  .300E+01 +     C                                           +
 0   .000  .400E+01 +     C                                           +
 0   .000  .500E+01 +     C                                           +
 0   .000  .600E+01 +     C                                           +
 1   .111  .700E+01 +******    C                                      +
 0   .000  .800E+01 +          C                                      +
 0   .000  .900E+01 +          C                                      +
 0   .000  .100E+02 +          C                                      +
 0   .000  .110E+02 +          C                                      +
 0   .000  .120E+02 +          C                                      +
 1   .111  .130E+02 +******          C                                +
 0   .000  .140E+02 +                C                                +
 0   .000  .150E+02 +                C                                +
 0   .000  .160E+02 +                C                                +
 0   .000  .170E+02 +                C                                +
 0   .000  .180E+02 +                C                                +
 1   .111  .190E+02 +******               C                           +
 0   .000  .200E+02 +                     C                           +
 5   .556    INF    +****************************                     C
---                 +    +    +    +    +    +    +    +    +    +    +
 9                  0         20        40        60        80      100
```

FIGURE 6.17 Effect of High Priority

```
OBS   RELA   UPPER
FREQ  FREQ  CELL LIM 0         20        40        60        80        100
                     +    +    +    +    +    +    +    +    +    +    +
 0    .000  .000E+00 +                                                 +
 0    .000  .100E+01 +                                                 +
 1    .067  .200E+01 +***                                              +
 0    .000  .300E+01 +  C                                              +
 0    .000  .400E+01 +  C                                              +
 0    .000  .500E+01 +  C                                              +

 0    .000  .600E+01 +  C                                              +
 0    .000  .700E+01 +  C                                              +
 1    .067  .800E+01 +***   C                                          +
 0    .000  .900E+01 +      C                                          +
 0    .000  .100E+02 +      C                                          +
 0    .000  .110E+02 +      C                                          +
 0    .000  .120E+02 +      C                                          +
 0    .000  .130E+02 +      C                                          +
 1    .067  .140E+02 +***      C                                       +
 0    .000  .150E+02 +         C                                       +
 0    .000  .160E+02 +         C                                       +
 1    .067  .170E+02 +***         C                                    +
 0    .000  .180E+02 +            C                                    +
 0    .000  .190E+02 +            C                                    +
 1    .067  .200E+02 +***             C                                +
10    .667     INF   +*********************************                C
---                  +    +    +    +    +    +    +    +    +    +    +
15                   0         20        40        60        80        100
```

6.8 DISCRETE-EVENT SIMULATION OF A MANUFACTURING PROCESS[1]

Let us consider a company that manufactures uPVC pipes to order only; no product is kept in stock. The process starts with a forecast of the raw material needs for the month ahead. An order is placed for the raw material, consisting of uPVC polymer and some additives. The company's policy is to keep the equivalent of one month's forecasted production in raw material as a buffer. The extrusion plant manager receives sales orders and schedules production.

A flow chart of the system is presented in Figure 6.18. The uPVC polymer is delivered in 25-kilogram (kg) bags and the additive in 50-kilogram barrels. Samples of both are taken and tested. If the samples are rejected, the whole delivery is sent back to the supplier. If the samples are accepted, the additives are stored as is, and the uPVC bags are transferred to an automatic debagging unit. From here, the polymer is conveyed to one of 47 storage silos, whose total capacity is 3000 tons of uPVC. From the silos, the polymer is transferred to the mixing plant by an automatic conveying system.

The additive and the polymer are combined according to designated proportions and sent to one of three mixers to produce a dry uPVC blend. This blend then undergoes a quality control check. Because of the nature of the tests, the results are not known until the end of the shift. Each of the 21 extrusion lines receives the dry blend automatically from one of the mixers. It is worthwhile to note that the polymer is conveyed in a closed system from bag to extruder.

Extruding uPVC pipes involves an extruder, die head, water bath, haul-off, printer, and saw. The polymer is extruded as a pipe through the die head, cooled in a water bath, pulled at a constant speed, labelled, and cut to the required length. Each line has a socketing machine which fits the pipe with either a rubber joint or solvent weld socket. After the socketing process, the pipes are inspected. Rejected pipes go to a crush silo and the polymer is recycled and fed to the

FIGURE 6.18 Flow Chart of a Manufacturing System

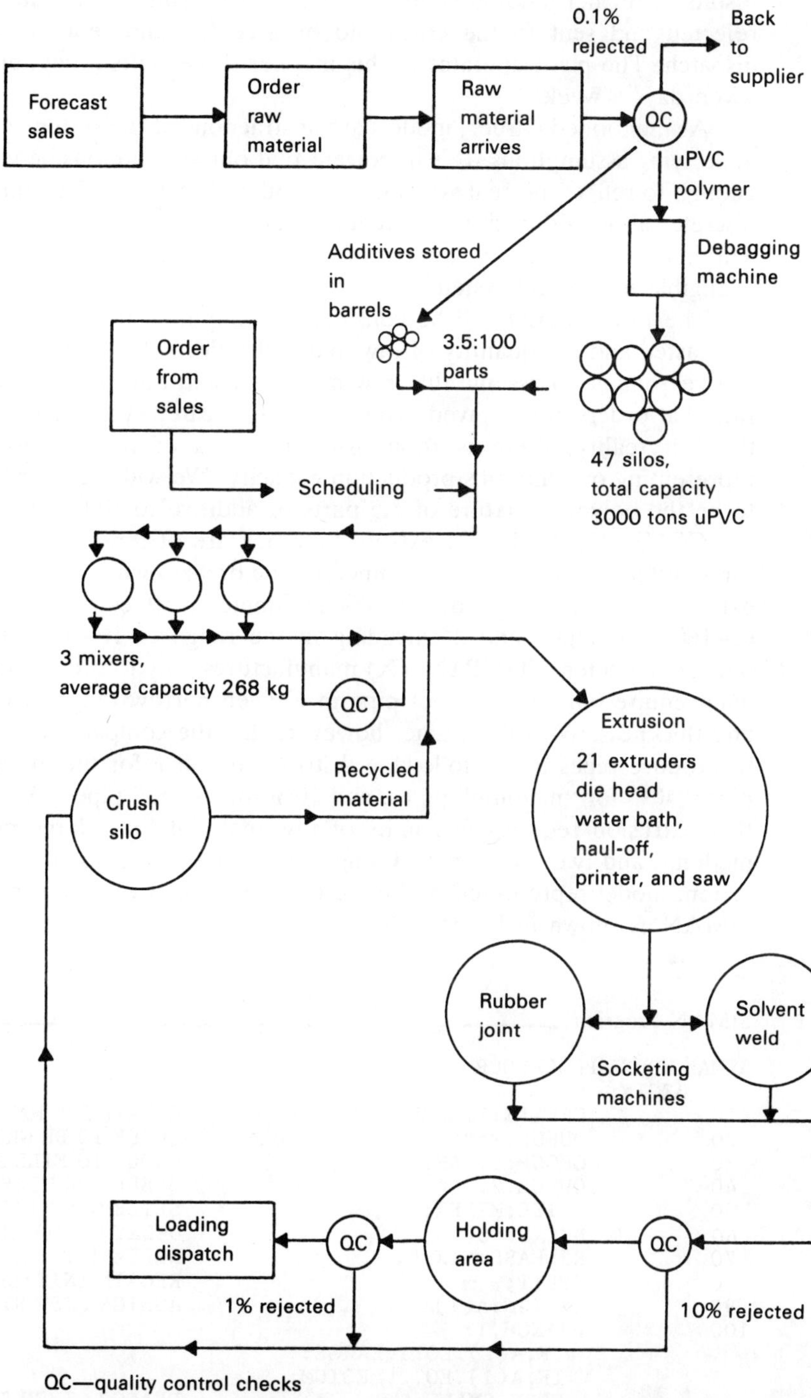

extruders. Tentatively accepted pipes are stacked in a holding area to await results of further laboratory testing. When the results are in, the pipes are either rejected and sent to the crush silo or accepted and sent on for loading and dispatch. The plant operates in this manner three shifts a day, eight hours each, seven days a week.

As mentioned earlier, models are abstractions of the system and require some underlying assumptions. It is important that our assumptions be valid and realistic enough to reflect the real system. Our model of this manufacturing system will be discrete, although much of the real system is of continuous flow. As a result, we will ignore the transfer times between stations in the model since they are negligible in the real system.

The raw material will be considered as input to the system. Since sales are forecasted and the quantity of raw material reflects the amount of pipes ordered, it is reasonable to assume that raw material could flow in the system at the same rate that orders are received. This assumption can be further justified by the fact that, in reality, there is room for 3000 tons of raw material in the silos, representing one month's production capacity. We will consider the raw material to be the polymer mixture of 3.5 parts of additive to 100 parts of uPVC.

COMPANY(X) has 21 extruders, which are of three types to accommodate three different ranges of pipe diameter. The breakdown is as follows: 10 machines extruding in the range of 200–400 millimeters, 10 extruding in the range of 63–160 millimeters, and 2 extruding in the range of 16–50 millimeters nominal outside diameter. COMPANY(X) manufactures only to order, and certain orders may require only one type of pipe in a given narrow range of outside diameter and thickness. We will assume, however, that the company uses its raw material to produce pipes in the following distribution: 50% for the extrusion of "large" pipes, 40% for "medium" pipes, and 10% for "small" pipes. We will also assume that extrusion requires five units of raw material for a large pipe, three for a medium, and two for a small. Using these assumptions about the system and the system model represented by Figure 6.18, we can develop a simulation model in SIMAN as shown in Program 6.9.

PROGRAM 6.9 SIMAN Program[1]

```
SIMAN MODEL PROCESSOR RELEASE
     BEGIN;
 10          CREATE:1:MARK(2);              CREATE RAW MATERIAL
 20          QUEUE,1;                       QUEUE TO BE GROUPED
 30          GROUP:5,LAST;                  GROUP TO FILL MIXER
 40          QUEUE,2;                       QUEUE FOR MIXER
 50          SEIZE:MIXER;                   SEIZE ONE MIXER
 60          DELAY:CO(1);                   DELAY BY MIXING TIME
 70          RELEASE:MIXER;                 RELEASE THE MIXER
 80          SPLIT;                         REGAIN INITIAL ENTITIES
 90          ASSIGN:A(1)=DP(3,1);           ASSIGN PIPE SIZE
100          BRANCH,1:
               IF,A(1).EQ.1,LARGE:
               IF,A(1).EQ.2,MEDIUM:
               ELSE,SMALL;                  BRANCH GIVEN SIZE
```

PROGRAM 6.9 SIMAN Program (Continued)

```
110 CRUSH    QUEUE,3;                          QUEUE FOR CRUSH SILO
120          SEIZE:CRUSHER;                    SEIZE THE CRUSH SILO
130          DELAY:CO(2);                      DELAY BY CRUSHING TIME
140          RELEASE:CRUSHER;                  RELEASE THE CRUSH SILO
150          BRANCH,1:
               IF,A(1).EQ.1,L1:
               IF,A(2).EQ.2,M1:
               ELSE,S1;                        BRANCH GIVEN SIZE
160 LARGE    QUEUE,4;                          QUEUE TO BE COMBINED
170          COMBINE:5,LAST;                   COMBINED TO MAKE PIPE
180 L1       QUEUE,5;                          QUEUE FOR LG EXTRUDER
190          SEIZE:EXTRUD1;                    SEIZE ONE LG EXTRUDER
200          DELAY:UN(4,2);                    DELAY BY EXTRUSION TIME
210          RELEASE:EXTRUD1;                  RELEASE LARGE EXTRUDER
220          QUEUE,6;                          QUEUE FOR SOCKETING
230          SEIZE:WELD1;                      SEIZE SOCKETING MACHINE
240          BRANCH,1:
               WITH,.15,SOLV1:
               ELSE,RUBBER1;                   BRANCH GIVEN JOINT TYPE
250 SOLV1    DELAY:UN(5,3):NEXT(END1);         DELAY FOR SOLVENT WELD
260 RUBBER1  DELAY:UN(6,4);                    DELAY FOR RUBBER JOINT
270 END1     RELEASE:WELD1;                    RELEASE MACHINE
280          COUNT:1,1:NEXT(PASS);             COUNT LG PIPES EXTRUDED
290 MEDIUM   QUEUE,7;                          QUEUE TO BE COMBINED
300          COMBINE:3,LAST;                   COMBINED TO MAKE PIPE
310 M1       QUEUE,8;                          QUEUE FOR MD EXTRUDER
320          SEIZE:EXTRUD2;                    SEIZE ONE MD EXTRUDER
330          DELAY:UN(7,5);                    DELAY BY EXTRUSION TIME
340          RELEASE:EXTRUD2;                  RELEASE MEDIUM EXTRUDER
350          QUEUE,9;                          QUEUE FOR SOCKETING
360          SEIZE:WELD2;                      SEIZE SOCKETING MACHINE
370          BRANCH,1:
               WITH,.15,SOLV2:
               ELSE,RUBBER2;                   BRANCH GIVEN JOINT TYPE
380 SOLV2    DELAY:UN(8,6):NEXT(END2);         DELAY FOR SOLVENT WELD
390 RUBBER2  DELAY:UN(9,7);                    DELAY FOR RUBBER JOINT
400 END2     RELEASE:WELD2;                    RELEASE MACHINE
410          COUNT:2,1:NEXT(PASS);             COUNT MD PIPES PRODUCED
420 SMALL    QUEUE,10;                         QUEUE TO BE COMBINED
430          COMBINE:2,LAST;                   COMBINED TO MAKE PIPE
440 S1       QUEUE,11;                         QUEUE FOR SM EXTRUDER
450          SEIZE:EXTRUD3;                    SEIZE ONE SM EXTRUDER
460          DELAY:UN(10,8);                   DELAY BY EXTRUSION TIME
470          RELEASE:EXTRUD3;                  RELEASE SMALL EXTRUDER
480          QUEUE,12;                         QUEUE FOR SOCKETING
490          SEIZE:WELD3;                      SEIZE SOCKETING MACHINE
500          DELAY:UN(11,9);                   DELAY FOR SOLVENT WELD
510          RELEASE:WELD3;                    RELEASE MACHINE
520          COUNT:3,1;                        COUNT SM PIPES PRODUCED
530 PASS     BRANCH,1:
               WITH,.1,CRUSH:
               ELSE,HOLD;                      BRANCH AFTER INSPECTION
540 HOLD     DELAY:CO(12);                     DELAY BY LAB INSPECTION
550          BRANCH,1:
               WITH,.01,CRUSH:
             ELSE,EXIT;                        BRANCH AFTER LAB RESULT
560 EXIT     TALLY:A(1),INT(2):DISPOSE;        TALLY TIME IN SYSTEM
      END;
```

One unit of raw material, representing one kilogram of a mixture of uPVC polymer and additive, enters the system at the CREATE block every time unit. By use of the MARK option, arrival time to the system of each new unit of raw material is stored in the entity's second attribute. The entities queue in file 1 to be grouped. Upon arriving at the GROUP block, a unit of raw material is held until five units are in the block to form a temporary set. Five entities fill one mixer. The GROUP block creates a new entity representing a set from each five arriving entities to the block but does not destroy the members of the set. The original entities are removed from the preceding QUEUE block and are linked to and carried along by the representative of the set. The attributes of the set representative are set to the attributes of the last entity grouped.

The new sets queue in file 2, where a single set waits to seize one unit of the resource MIXER. Each set is delayed at the DELAY block by the constant processing time before releasing the resource at the RELEASE block. The SPLIT block is used to recover the individual units of raw material in the temporary set; hence, upon arrival to this block, the five units are recovered and continue their flow through the system.

The entity, in the form of a dry-blend polymer, enters the ASSIGN block, where its first attribute is set to a value from a uniform discrete probability distribution. The value assigned to attribute 1 of each entity, A(1), is between 1 and 3 and denotes the pipe category. Next, the entity enters the BRANCH block. Here category 1 entities are sent to the block labelled LARGE, category 2 entities are sent to MEDIUM, and category 3 entities are sent to SMALL.

Entities sent to the LARGE sequence of blocks queue in file 4 to be combined. The COMBINE block holds arriving entities in the preceding QUEUE block until five entities (enough to extrude one pipe) reside in the QUEUE block. The COMBINE block then generates and releases a new entity which represents the set, and destroys the members of the set residing in the QUEUE block. The attributes of the newly created entity are those of the last entity to be combined. The new entities then queue in file 5, wait to seize one unit of EXTRUD1, and are delayed by extrusion time according to a uniform distribution. After the process is completed, the resource is released at the RELEASE block. At this stage, the new entity is in the form of a pipe. The pipes queue in file 6 for the socketing machine; a pipe will seize one unit of the resource WELD1. The pipe will arrive at a probabilistic BRANCH block where a pipe can be directed to one of two DELAY blocks depending on the socket type. Both delays are according to a uniform distribution, but it takes more time for the rubber joint. After processing time, the resource is released, and a count of the number of pipes manufactured is done at the COUNT block. At this stage, the pipes are sent to the block labelled PASS.

When entities are sent to the block labelled MEDIUM, the same process as described above takes place, except that the number of entities required to be combined to form a pipe is different. The same is true for the SMALL sequence of blocks, but here there is only one type of socket available. Thus, at the end of all three categories, all the pipes of different sizes are directed toward the block labelled PASS.

The PASS sequence of blocks allows for quality control measures; here there is a site inspection of pipes, and samples are taken. When a pipe arrives at this probabilistic BRANCH block, it goes to the block labelled HOLD if accepted or to the block labelled CRUSH if rejected. The acceptable pipes going to HOLD are delayed by a constant period of time, allowing laboratory results to become known. When results are in, another probabilistic BRANCH block directs pipes toward EXIT if passed or CRUSH if rejected.

When the pipes are directed toward EXIT, they arrive at a TALLY block, where the tie-in system for each pipe category and the number of pipes observed in each category are recorded. Then all the entities (pipes) exit the system at the DISPOSE modifier.

Rejected pipes are directed toward the block labelled CRUSH. The pipes queue in file 3 for the crush silo, where they wait to seize the CRUSH. Each pipe is delayed by the processing time; then the resource is released. At this BRANCH block, the entity (recycled polymer with marked attributes) goes to its designated block according to its attributes. When the entities arrive at their appropriate block, they flow through the system as discussed earlier.

The experimental frame defines the experimental conditions under which the model is run. Program 6.10 shows a SIMAN experiment processor release. The PROJECT element is used to label the SIMAN summary report. The DISCRETE element defines variables associated with a discrete system, such as the maximum number of entities allowed in the system at any one time, and the maximum number of attributes associated with an entity. The TALLIES element defines the report identifier and output unit for recorded statistics and is employed when a TALLY block is used in the model. The RESOURCES element

PROGRAM 6.10 SIMAN Experiment File

```
   BEGIN;
10 PROJECT, PLANT REPRESENTATION,TALAL AGGAD ,3/6/85;
20 DISCRETE,100,4,100,100;
30 TALLIES:1,LARGE:2,MEDIUM:3,SMALL;
40 RESOURCES:1,MIXER,3:2,CRUSHER:3,EXTRUD1,10:4,WELD1,10:
          5,EXTRUD2,10:6,WELD2,10:7,EXTRUD3,2:8,WELD3,2;
50 PARAMETERS:1,1.0:2,1.0:2,1.0:3,.50,1,.90,2,1.0,3:4,1.0,1.0:
              5,1.0,1.0:6,1.0,1.0:7,1.0,1.0,:8,1.0,1.0:
              9,1.0,1.0:10,1.0,1.0:11,1.0,1.0:12,1.0;
60 REPLICATE,1,0,1000;
70 DSTAT:1,NQ(1),GROUPING QUEUE:2,NQ(2), MIXER QUEUE:
        3,NQ(3),CRUSHER QUEUE:4,NQ(4),COMB EXTRUD1:
        5,NQ(5),EXTRUD1 QUEUE:6,NQ(6),WELD1 QUEUE:
        7,NQ(7),COMB EXTRUD2:8,NQ(8),EXTRUD2 QUEUE:
        9,NQ(9),WELD2 QUEUE:10,NQ(10),COMB EXTRUD3:
        11,NQ(11),EXTRUD3 QUEUE:12,NQ(12),WELD3 QUEUE:
        13,NR(1),MIXER UTIL:14,NR(2),CRUSHER UTIL:
        15,NR(3),EXTRUD1 UTIL:16,NR(4),WELD1 UTIL:
        17,NR(5),EXTRUD2 UTIL:18,NR(6),WELD2 UTIL:
        19,NR(7),EXTRUD3 UTIL:20,NR(8),WELD3 UTIL:
80 COUNTERS:1,LARGE:2,MEDIUM:3,SMALL;
90 TRACE,0,60;
   END;
```

refers to the resource names mentioned in the model and their initial capacity. The PARAMETERS element specifies parameter sets associated with the random variable (in this case all values have been set to unity). The REPLICATE element specifies the number of simulation runs, the beginning time of the first run, the maximum length of the runs, and any initialization between runs. The DSTAT element is used to obtain statistics on variables in a discrete system. Here the number of entities residing in queue and the utilization of resources will be given. The COUNTERS element specifies the report identifier and count limit. The TRACE element is used to generate a detailed report of the processing of entities within the discrete model.

Now that a simulation model has been developed, its performance must be evaluated. Verification of the model is an essential step before any experimentation can take place and must (should) establish that the SIMAN program performs as intended.

To verify the model, creation of entities is made constant at one entity for every unit of time and all parameters are set to unity. The trace for the first 60 units of time, shown in Section 6.10, is part of the verification process. This is a detailed report of the processing of entities within the model. The trace indicates the entity flow through the system at each unit of time. Thus, it has been determined that the program processes the entities as intended.

The SIMAN simulation program generates an output file known as the SIMAN summary report. For this example, the full report is shown in Figure 6.19. The first section in the report contains statistics that summarize the observations for the tally variables. The second section contains statistics that summarize the observations recorded by SIMAN on discrete-change variables, including statistics on the number in the various queues and on the utilization of various resources. Finally, the output includes statistics given by the counters. These are observations of the number of pipes produced in each category.

The relevant part of the report for further verification is the observations for the tally variables, in particular the number of observations made for each category of pipe. This is the number of pipes in each category that are successfully dispatched (i.e., exit the system). The run ended after 1000 units of time elapsed; hence 1000 entities were created. Since 500 entities went to extrude large pipes, the expected observation is 100 pipes dispatched, actual observation 97. Since 400 entities went to extrude medium pipes, the expected observation is 133 pipes dispatched, actual observation 140. Finally, 100 entities went to extrude small pipes; therefore, the expected observation is 50 pipes dispatched, actual observation 43. The discrepancy between the numbers can be attributed to several factors. The first is the fact that the category to which an entity is assigned is set according to a discrete probability distribution; thus, the number of entities assigned to each category from 1000 is probabilistic but can be approximated. The second factor is that some entities are still in the system when the simulation is stopped. Also, some pipes are being rejected and recycled through the system. Finally, when the 280 observed pipes are split back to original entities, we find that 991 entities have been processed. Thus, given the evidence of both the trace and the SIMAN output, it seems reasonable to assume verification of the model.

FIGURE 6.19 SIMAN Summary Report

TALLY VARIABLES

NUMBER IDENTIFIER	AVERAGE	STANDARD DEVIATION	MINIMUM VALUE	MAXIMUM VALUE	NUMBER OF OBS.
1 LARGE	.6155E+01	.1734E+01	.4000E+01	.1200E+02	97
2 MEDIUM	.6343E+01	.1884E+01	.4000E+01	.1300E+02	140
3 SMALL	.6884E+01	.2301E+01	.4000E+01	.1200E+02	43

DISCRETE CHANGE VARIABLES

NUMBER IDENTIFIER	AVERAGE	STANDARD DEVIATION	MINIMUM VALUE	MAXIMUM VALUE	TIME PERIOD
1 GROUPING QUEUE	.2000E+01	.1414E+01	.0000E+00	.5000E+01	.1000E+04
2 MIXER QUEUE	.0000E+00	.0000E+00	.0000E+00	.0000E+00	.1000E+04
3 CRUSHER QUEUE	.0000E+00	.0000E+00	.0000E+00	.1000E+01	.1000E+04
4 COMB EXTRUD1	.2020E+01	.1371E+01	.0000E+00	.5000E+01	.1000E+04
5 EXTRUD1 QUEUE	.0000E+00	.0000E+00	.0000E+00	.0000E+00	.1000E+04
6 WELD1 QUEUE	.0000E+00	.0000E+00	.0000E+00	.0000E+00	.1000E+04
7 COMB EXTRUD2	.9800E+00	.8122E+00	.0000E+00	.3000E+01	.1000E+04
8 EXTRUD2 QUEUE	.0000E+00	.0000E+00	.0000E+00	.0000E+00	.1000E+04
9 WELD2 QUEUE	.0000E+00	.0000E+00	.0000E+00	.0000E+00	.1000E+04
10 COMB EXTRUD3	.5000E+00	.5000E+00	.0000E+00	.2000E+01	.1000E+04
11 EXTRUD3 QUEUE	.0000E+00	.0000E+00	.0000E+00	.0000E+00	.1000E+04
12 WELD3 QUEUE	.0000E+00	.0000E+00	.0000E+00	.0000E+00	.1000E+04
13 MIXER UTIL	.2000E+00	.4000E+00	.0000E+00	.1000E+01	.1000E+04
14 CRUSHER UTIL	.3700E-01	.1888E+00	.0000E+00	.1000E+01	.1000E+04
15 EXTRUD1 UTIL	.1050E+00	.3066E+00	.0000E+00	.1000E+01	.1000E+04
16 WELD1 UTIL	.1050E+00	.3066E+00	.0000E+00	.1000E+01	.1000E+04
17 EXTRUD2 UTIL	.1400E+00	.3611E+00	.0000E+00	.2000E+01	.1000E+04
18 WELD2 UTIL	.1400E+00	.3611E+00	.0000E+00	.2000E+01	.1000E+04
19 EXTRUD3 UTIL	.7200E-01	.2623E+00	.0000E+00	.2000E+01	.1000E+04
20 WELD3 UTIL	.7200E-01	.2623E+00	.0000E+00	.2000E+01	.1000E+04

COUNTERS

NUMBER IDENTIFIER	COUNT	LIMIT
1 LARGE	105	INFINITE
2 MEDIUM	140	INFINITE
3 SMALL	72	INFINITE

Further experimentation with the SIMAN program can now take place. As mentioned earlier, the aim will be to optimize measures of system behavior such as time, resource utilization, and production output. This will be done through the modification of parameters, resources, and rejection ratios in the system.

6.9 SUMMARY

In this chapter, we provided a perspective on simulation languages in general and focused on the principles of operation of discrete-event simulation languages. We presented the material-oriented and machine-oriented systems. We classified simulation languages according to event, process, and activity types, and gave examples of each. Four significant discrete-event simulations were presented.

6.10 REVIEW PROBLEMS

1. Define discrete-event simulation.
2. What is the difference between material-oriented and machine-oriented simulations?
3. Describe each of the three types of discrete-event simulation languages.
4. Simulate the student-professor queueing system of Chapter 5. Let the student arrival rate be Poisson distributed with a mean of two students per hour. Let the professor's service rate be exponentially distributed with a mean of two students per hour. Compute the average time the student spends in the system from arrival to departure. What is the average student waiting time?
5. Many airline ticket counters use a single queue with multiple ticket agents as shown in Figure 6.20. The customers are served on a first-come, first-served basis and proceed to the first available agent. Let the customers' arrival rate be Poisson distributed with a mean of 40 customers per hour.

FIGURE 6.20 Problem 5

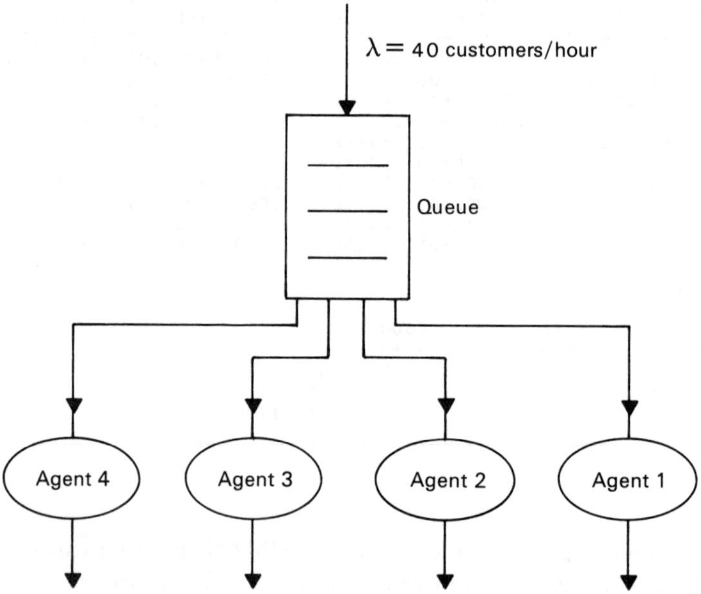

a. Simulate the system assuming two identical agents, given an exponentially distributed service rate with a mean of 20 customers per hour. Compute the average customer waiting time.
b. Simulate the system assuming two agents, given exponentially distributed service rates with means of 10 and 30 customers per hour. Compute the average customer waiting time. Compare your results with those in part (a).
c. Repeat part (a), assuming four identical agents having exponentially distributed service rates with means of 10 customers per hour.

6. Many banks use multiple queues with multiple tellers. Simulate the system with four queues and four tellers. Assume that arrivals are Poisson distributed with means of 40 customers per hour and that the tellers have exponentially distributed service rates with means of 10 customers per hour. Assume that 10%, 20%, 30%, and 40%, respectively, of the customers proceed to the queues. Simulate the system and compute the mean time from arrival to departure for each customer. What is the utilization factor of each teller?
7. The computer communication network of Figure 6.21 has Poisson arrival rates of 40 and 5 messages per second. The message lengths are exponentially distributed with a mean of 256 bytes per message. Compute the average message delay in the network. Define message delay as the time it takes a message to travel from its source to its destination. The line capacities are 56 kilobytes per second (kbps). Assume infinite buffer.

FIGURE 6.21 Problem 7

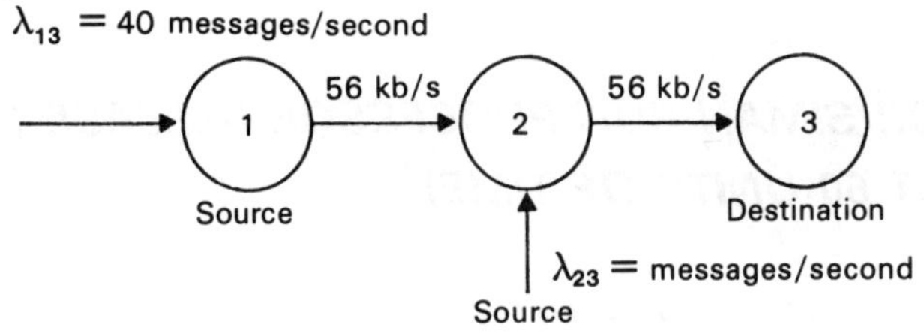

FIGURE 6.22 Problem 8

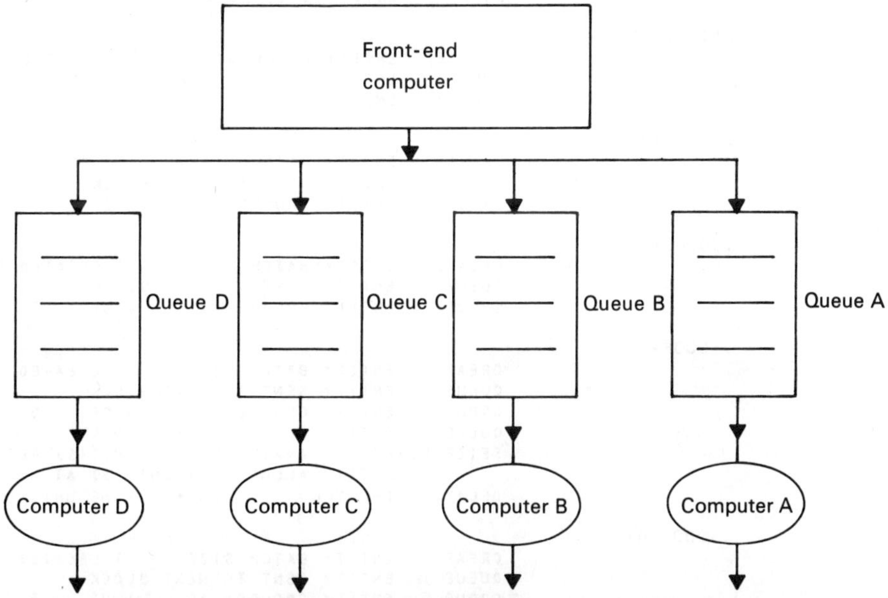

8. The distributed computing system shown in Figure 6.22 is used by a computer services firm. Assume that jobs are submitted to a front end, which assigns each job to one of the parallel computers. Using simulation, determine which of the following strategies will provide the higher throughput. Define throughput as the number of jobs completed per hour.

   Assign a job to each computer in round-robin fashion.
   Assign jobs to the computer with the smallest queue length. In case of a tie, choose randomly among the computers.

   All processes have a deterministic service rate of 10 mega instructions per second (MIPS). Assume that the number of instructions in a job are exponentially distributed with a mean of one MIPS. The arrival rate of jobs is Poisson distributed with a mean of 10 jobs per second.

6.11 APPENDIX: SIMAN RUN PROCESSOR RELEASE (TRACE FOR FIRST 60 UNITS OF TIME)[1]

```
                    SIMAN SYSTEM TRACE BEGINNING AT TIME          .000

   TIME        EVENT OR BLOCK              SYSTEM STATUS CHANGE
-----------------------------------------------------------------------

 .0000E+00
                    CREATE    ENTITY BATCH SIZE      1 CREATED
                    QUEUE     ENTITY SENT TO NEXT BLOCK
                    GROUP     ENTITY GROUPED AS      1 OF     5

 .1000E+01
                    CREATE    ENTITY BATCH SIZE      1 CREATED
                    QUEUE     ENTITY SENT TO NEXT BLOCK
                    GROUP     ENTITY GROUPED AS      2 OF     5

 .2000E+01
                    CREATE    ENTITY BATCH SIZE      1 CREATED
                    QUEUE     ENTITY SENT TO NEXT BLOCK
                    GROUP     ENTITY GROUPED AS      3 OF     5

 .3000E+01
                    CREATE    ENTITY BATCH SIZE      1 CREATED
                    QUEUE     ENTITY SENT TO NEXT BLOCK
                    GROUP     ENTITY GROUPED AS      4 OF     5

 .4000E+01
                    CREATE    ENTITY BATCH SIZE      1 CREATED
                    QUEUE     ENTITY SENT TO NEXT BLOCK
                    GROUP     ENTITY GROUPED AS      5 OF     5
                    QUEUE     ENTITY SENT TO NEXT BLOCK
                    SEIZE MIXER (1) AVAIL IS 3, 1 UNIT(S) REQ'
                              ENTITY ALLOC      1 UNIT(S) AT BLOC
                    DELAY     DELAY BY     .1000E+01 TIME UNIT(S)

 .5000E+01
                    CREATE    ENTITY BATCH SIZE      1 CREATED
                    QUEUE     ENTITY SENT TO NEXT BLOCK
                    GROUP     ENTITY GROUPED AS      1 OF     5
```

```
                 SIMAN SYSTEM TRACE BEGINNING AT TIME        .000

  TIME        EVENT OR BLOCK             SYSTEM STATUS CHANGE
--------------------------------------------------------------------

 .5000E+01
                      RELEASE MIXER (1) AVAIL INCREASED BY 1 TO 3
                      SPLIT     ENTITY SPLIT INTO    5 SUBSET(S)
                      ASSIGN    A( 1) SET TO    .1000E+01
                      BRANCH    SELECTION MADE FROM    1 OF    3 BRANCHES
                                CONDITIONAL,  ENTITY SENT TO LARGE
              LARGE   QUEUE     ENTITY SENT TO NEXT BLOCK
                      COMBINE   ENTITY COMBINED AS     1 OF     5

 .5000E+01
                      ASSIGN    A( 1) SET TO    .1000E+01
                      BRANCH    SELECTION MADE FROM    1 OF    3 BRANCHES
                                CONDITIONAL,  ENTITY SENT TO LARGE
              LARGE   QUEUE     ENTITY SENT TO NEXT BLOCK
                      COMBINE   ENTITY COMBINED AS     2 OF     5

 .5000E+01
                      ASSIGN    A( 1) SET TO    .1000E+01
                      BRANCH    SELECTION MADE FROM    1 OF    3 BRANCHES
                                CONDITIONAL,  ENTITY SENT TO LARGE
              LARGE   QUEUE     ENTITY SENT TO NEXT BLOCK
                      COMBINE   ENTITY COMBINED AS     3 OF     5

 .5000E+01
                      ASSIGN    A( 1) SET TO    .1000E+01
                      BRANCH    SELECTION MADE FROM    1 OF    3 BRANCHES
                                CONDITIONAL,  ENTITY SENT TO LARGE
              LARGE   QUEUE     ENTITY SENT TO NEXT BLOCK
                      COMBINE   ENTITY COMBINED AS     4 OF     5

 .5000E+01
                      ASSIGN    A( 1) SET TO    .2000E+01
                      BRANCH    SELECTION MADE FROM    1 OF    3 BRANCHES
                                CONDITIONAL,  BRANCH NOT SELECTED
                                CONDITIONAL,  ENTITY SENT TO MEDIUM
              MEDIUM  QUEUE     ENTITY SENT TO NEXT BLOCK
                      COMBINE   ENTITY COMBINED AS     1 OF     3

 .6000E+01
                      CREATE    ENTITY BATCH SIZE      1 CREATED
                      QUEUE     ENTITY SENT TO NEXT BLOCK
                      GROUP     ENTITY GROUPED AS    2 OF     5

 .7000E+01
                      CREATE    ENTITY BATCH SIZE      1 CREATED
                      QUEUE     ENTITY SENT TO NEXT BLOCK
                      GROUP     ENTITY GROUPED AS    3 OF     5

 .8000E+01
                      CREATE    ENTITY BATCH SIZE      1 CREATED
                      QUEUE     ENTITY SENT TO NEXT BLOCK
                      GROUP     ENTITY GROUPED AS    4 OF     5

 .9000E+01
                      CREATE    ENTITY BATCH SIZE      1 CREATED
                      QUEUE     ENTITY SENT TO NEXT BLOCK
                      GROUP     ENTITY GROUPED AS    5 OF     5
                      QUEUE     ENTITY SENT TO NEXT BLOCK
                      SEIZE MIXER (1) AVAIL IS 3, 1 UNIT(S) REQ'D
                                ENTITY ALLOC     1 UNIT(S) AT BLOCK
                      DELAY     DELAY BY    .1000E+01 TIME UNIT(S)
```

continues

```
               SIMAN SYSTEM TRACE BEGINNING AT TIME          .000

  TIME        EVENT OR BLOCK              SYSTEM STATUS CHANGE
----------------------------------------------------------------------

.1000E+02
                       CREATE    ENTITY BATCH SIZE      1 CREATED
                       QUEUE     ENTITY SENT TO NEXT BLOCK
                       GROUP     ENTITY GROUPED AS     1 OF     5

.1000E+02
                       RELEASE MIXER (1) AVAIL INCREASED BY 1 TO 3
                       SPLIT     ENTITY SPLIT INTO     5 SUBSET(S)
                       ASSIGN    A(  1) SET TO    .1000E+01
                       BRANCH    SELECTION MADE FROM    1 OF    3 BRANCHES
                                 CONDITIONAL,  ENTITY SENT TO LARGE
             LARGE     QUEUE     ENTITY SENT TO NEXT BLOCK
                       COMBINE   ENTITY COMBINED AS     5 OF     5
             L1        QUEUE     ENTITY SENT TO NEXT BLOCK
                       SEIZE EXTRUD1 (1) AVAIL IS 10, 1 UNIT(S) REQ'D
                                 ENTITY ALLOC     1 UNIT(S) AT BLOCK
                       DELAY     DELAY BY    .1000E+01 TIME UNIT(S)

.1000E+02
                       ASSIGN    A(  1) SET TO    .1000E+01
                       BRANCH    SELECTION MADE FROM    1 OF    3 BRANCHES
                                 CONDITIONAL,  ENTITY SENT TO LARGE
             LARGE     QUEUE     ENTITY SENT TO NEXT BLOCK
                       COMBINE   ENTITY COMBINED AS     1 OF     5

.1000E+02
                       ASSIGN    A(  1) SET TO    .3000E+01
                       BRANCH    SELECTION MADE FROM    1 OF    3 BRANCHES
                                 CONDITIONAL,  BRANCH NOT SELECTED
                                 CONDITIONAL,  BRANCH NOT SELECTED
                                 DETERMINISTIC,ENTITY SENT TO SMALL
             SMALL     QUEUE     ENTITY SENT TO NEXT BLOCK
                       COMBINE   ENTITY COMBINED AS     1 OF     2

.1000E+02
                       ASSIGN    A(  1) SET TO    .1000E+01
                       BRANCH    SELECTION MADE FROM    1 OF    3 BRANCHES
                                 CONDITIONAL,  ENTITY SENT TO LARGE
             LARGE     QUEUE     ENTITY SENT TO NEXT BLOCK
                       COMBINE   ENTITY COMBINED AS     2 OF     5

.1000E+02
                       ASSIGN    A(  1) SET TO    .1000E+01
                       BRANCH    SELECTION MADE FROM    1 OF    3 BRANCHES
                                 CONDITIONAL,  ENTITY SENT TO LARGE
             LARGE     QUEUE     ENTITY SENT TO NEXT BLOCK
                       COMBINE   ENTITY COMBINED AS     3 OF     5

.1100E+02
                       CREATE    ENTITY BATCH SIZE      1 CREATED
                       QUEUE     ENTITY SENT TO NEXT BLOCK
                       GROUP     ENTITY GROUPED AS     2 OF     5

.1100E+02
                       RELEASE EXTRUD1 (1) AVAIL INCREASED BY 1 TO 10
                       QUEUE     ENTITY SENT TO NEXT BLOCK
                       SEIZE WELD1 (1) AVAIL IS 10, 1 UNIT(S) REQ'D
                                 ENTITY ALLOC     1 UNIT(S) AT BLOCK
                       BRANCH    SELECTION MADE FROM    1 OF    2 BRANCHES
                                 PROBABILISTIC,BRANCH NOT SELECTED
                                 DETERMINISTIC,ENTITY SENT TO RUBBER1
             RUBBER1   DELAY     DELAY BY    .1000E+01 TIME UNIT(S)
```

```
                SIMAN SYSTEM TRACE BEGINNING AT TIME        .000

   TIME        EVENT OR BLOCK            SYSTEM STATUS CHANGE
-----------------------------------------------------------------------

.1200E+02
                       CREATE    ENTITY BATCH SIZE      1 CREATED
                       QUEUE     ENTITY SENT TO NEXT BLOCK
                       GROUP     ENTITY GROUPED AS     3 OF     5

.1200E+02
             END1      RELEASE WELD1 (1) AVAIL INCREASED BY 1 TO 10
                       COUNT COUNTER 1 INCREMENTED BY 1 TO 1
             PASS      BRANCH    SELECTION MADE FROM    1 OF   2 BRANCHES
                                 PROBABILISTIC,BRANCH NOT SELECTED
                                 DETERMINISTIC,ENTITY SENT TO HOLD
             HOLD      DELAY     DELAY BY    .1000E+01 TIME UNIT(S)

.1300E+02
                       CREATE    ENTITY BATCH SIZE      1 CREATED
                       QUEUE     ENTITY SENT TO NEXT BLOCK
                       GROUP     ENTITY GROUPED AS     4 OF     5

.1300E+02
                       BRANCH    SELECTION MADE FROM    1 OF   2 BRANCHES
                                 PROBABILISTIC,BRANCH NOT SELECTED
                                 DETERMINISTIC,ENTITY SENT TO EXIT
             EXIT      TALLY     OBS   .8000E+01 RECORDED FOR TALLY 1

.1400E+02
                       CREATE    ENTITY BATCH SIZE      1 CREATED
                       QUEUE     ENTITY SENT TO NEXT BLOCK
                       GROUP     ENTITY GROUPED AS     5 OF     5
                       QUEUE     ENTITY SENT TO NEXT BLOCK
                       SEIZE MIXER (1) AVAIL IS 3, 1 UNIT(S) REQ'D
                                 ENTITY ALLOC     1 UNIT(S) AT BLOCK
                       DELAY     DELAY BY    .1000E+01 TIME UNIT(S)

.1500E+02
                       CREATE    ENTITY BATCH SIZE      1 CREATED
                       QUEUE     ENTITY SENT TO NEXT BLOCK
                       GROUP     ENTITY GROUPED AS     1 OF     5

.1500E+02
                       RELEASE MIXER (1) AVAIL INCREASED BY 1 TO 3
                       SPLIT     ENTITY SPLIT INTO     5 SUBSET(S)
                       ASSIGN    A(  1) SET TO    .1000E+01
                       BRANCH    SELECTION MADE FROM    1 OF   3 BRANCHES
                                 CONDITIONAL,  ENTITY SENT TO LARGE
             LARGE     QUEUE     ENTITY SENT TO NEXT BLOCK
                       COMBINE   ENTITY COMBINED AS     4 OF     5

.1500E+02
                       ASSIGN    A(  1) SET TO    .1000E+01
                       BRANCH    SELECTION MADE FROM    1 OF   3 BRANCHES
                                 CONDITIONAL,  ENTITY SENT TO LARGE
             LARGE     QUEUE     ENTITY SENT TO NEXT BLOCK
                       COMBINE   ENTITY COMBINED AS     5 OF     5
             L1        QUEUE     ENTITY SENT TO NEXT BLOCK
                       SEIZE EXTRUD1 (1) AVAIL IS 10, 1 UNIT(S) REQ'D
                                 ENTITY ALLOC     1 UNIT(S) AT BLOCK
                       DELAY     DELAY BY    .1000E+01 TIME UNIT(S)
```

continues

```
                 SIMAN SYSTEM TRACE BEGINNING AT TIME          .000

   TIME         EVENT OR BLOCK              SYSTEM STATUS CHANGE
-----------------------------------------------------------------

.1500E+02
                   ASSIGN    A( 1) SET TO   .2000E+01
                   BRANCH    SELECTION MADE FROM    1 OF   3 BRANCHES
                             CONDITIONAL, BRANCH NOT SELECTED
                             CONDITIONAL, ENTITY SENT TO MEDIUM
          MEDIUM   QUEUE     ENTITY SENT TO NEXT BLOCK
                   COMBINE   ENTITY COMBINED AS     2 OF    3

.1500E+02
                   ASSIGN    A( 1) SET TO   .1000E+01
                   BRANCH    SELECTION MADE FROM    1 OF   3 BRANCHES
                             CONDITIONAL, ENTITY SENT TO LARGE
          LARGE    QUEUE     ENTITY SENT TO NEXT BLOCK
                   COMBINE   ENTITY COMBINED AS     1 OF    5

.1500E+02
                   ASSIGN    A( 1) SET TO   .1000E+01
                   BRANCH    SELECTION MADE FROM    1 OF   3 BRANCHES
                             CONDITIONAL, ENTITY SENT TO LARGE
          LARGE    QUEUE     ENTITY SENT TO NEXT BLOCK
                   COMBINE   ENTITY COMBINED AS     2 OF    5

.1600E+02
                   CREATE    ENTITY BATCH SIZE      1 CREATED
                   QUEUE     ENTITY SENT TO NEXT BLOCK
                   GROUP     ENTITY GROUPED AS     2 OF    5

.1600E+02
                   RELEASE EXTRUD1 (1) AVAIL INCREASED BY 1 TO 10
                   QUEUE     ENTITY SENT TO NEXT BLOCK
                   SEIZE WELD1 (1) AVAIL IS 10, 1 UNIT(S) REQ'D
                             ENTITY ALLOC     1 UNIT(S) AT BLOCK
                   BRANCH    SELECTION MADE FROM    1 OF   2 BRANCHES
                             PROBABILISTIC,BRANCH NOT SELECTED
                             DETERMINISTIC,ENTITY SENT TO RUBBER1
          RUBBER1  DELAY     DELAY BY   .1000E+01 TIME UNIT(S)

.1700E+02
                   CREATE    ENTITY BATCH SIZE      1 CREATED
                   QUEUE     ENTITY SENT TO NEXT BLOCK
                   GROUP     ENTITY GROUPED AS     3 OF    5

.1700E+02
          END1     RELEASE WELD1 (1) AVAIL INCREASED BY 1 TO 10
                   COUNT COUNTER 1 INCREMENTED BY 1 TO 2
          PASS     BRANCH    SELECTION MADE FROM    1 OF   2 BRANCHES
                             PROBABILISTIC,BRANCH NOT SELECTED
                             DETERMINISTIC,ENTITY SENT TO HOLD
          HOLD     DELAY     DELAY BY   .1000E+01 TIME UNIT(S)

.1800E+02
                   CREATE    ENTITY BATCH SIZE      1 CREATED
                   QUEUE     ENTITY SENT TO NEXT BLOCK
                   GROUP     ENTITY GROUPED AS     4 OF    5

.1800E+02
                   BRANCH    SELECTION MADE FROM    1 OF   2 BRANCHES
                             PROBABILISTIC,BRANCH NOT SELECTED
                             DETERMINISTIC,ENTITY SENT TO EXIT
          EXIT     TALLY     OBS   .7000E+01 RECORDED FOR TALLY 1
```

```
                    SIMAN SYSTEM TRACE BEGINNING AT TIME          .000

   TIME        EVENT OR BLOCK              SYSTEM STATUS CHANGE
---------------------------------------------------------------------

.1900E+02
                        CREATE    ENTITY BATCH SIZE      1 CREATED
                        QUEUE     ENTITY SENT TO NEXT BLOCK
                        GROUP     ENTITY GROUPED AS      5 OF     5
                        QUEUE     ENTITY SENT TO NEXT BLOCK
                        SEIZE MIXER (1) AVAIL IS 3, 1 UNIT(S) REQ'D
                                  ENTITY ALLOC      1 UNIT(S) AT BLOCK
                        DELAY     DELAY BY    .1000E+01 TIME UNIT(S)

.2000E+02
                        CREATE    ENTITY BATCH SIZE      1 CREATED
                        QUEUE     ENTITY SENT TO NEXT BLOCK
                        GROUP     ENTITY GROUPED AS      1 OF     5

.2000E+02
                        RELEASE MIXER (1) AVAIL INCREASED BY 1 TO 3
                        SPLIT     ENTITY SPLIT INTO     5 SUBSET(S)
                        ASSIGN    A(  1) SET TO    .3000E+01
                        BRANCH    SELECTION MADE FROM    1 OF    3 BRANCHES
                                  CONDITIONAL,  BRANCH NOT SELECTED
                                  CONDITIONAL,  BRANCH NOT SELECTED
                                  DETERMINISTIC,ENTITY SENT TO SMALL
              SMALL     QUEUE     ENTITY SENT TO NEXT BLOCK
                        COMBINE   ENTITY COMBINED AS     2 OF     2
              S1        QUEUE     ENTITY SENT TO NEXT BLOCK
                        SEIZE EXTRUD3 (1) AVAIL IS 2, 1 UNIT(S) REQ'D
                                  ENTITY ALLOC      1 UNIT(S) AT BLOCK
                        DELAY     DELAY BY    .1000E+01 TIME UNIT(S)

.2000E+02
                        ASSIGN    A(  1) SET TO    .2000E+01
                        BRANCH    SELECTION MADE FROM    1 OF    3 BRANCHES
                                  CONDITIONAL,  BRANCH NOT SELECTED
                                  CONDITIONAL,  ENTITY SENT TO MEDIUM
              MEDIUM    QUEUE     ENTITY SENT TO NEXT BLOCK
                        COMBINE   ENTITY COMBINED AS     3 OF     3
              M1        QUEUE     ENTITY SENT TO NEXT BLOCK
                        SEIZE EXTRUD2 (1) AVAIL IS 10, 1 UNIT(S) REQ'D
                                  ENTITY ALLOC      1 UNIT(S) AT BLOCK
                        DELAY     DELAY BY    .1000E+01 TIME UNIT(S)

.2000E+02
                        ASSIGN    A(  1) SET TO    .1000E+01
                        BRANCH    SELECTION MADE FROM    1 OF    3 BRANCHES
                                  CONDITIONAL,  ENTITY SENT TO LARGE
              LARGE     QUEUE     ENTITY SENT TO NEXT BLOCK
                        COMBINE   ENTITY COMBINED AS     3 OF     5

.2000E+02
                        ASSIGN    A(  1) SET TO    .3000E+01
                        BRANCH    SELECTION MADE FROM    1 OF    3 BRANCHES
                                  CONDITIONAL,  BRANCH NOT SELECTED
                                  CONDITIONAL,  BRANCH NOT SELECTED
                                  DETERMINISTIC,ENTITY SENT TO SMALL
              SMALL     QUEUE     ENTITY SENT TO NEXT BLOCK
                        COMBINE   ENTITY COMBINED AS     1 OF     2

.2000E+02
                        ASSIGN    A(  1) SET TO    .1000E+01
                        BRANCH    SELECTION MADE FROM    1 OF    3 BRANCHES
                                  CONDITIONAL,  ENTITY SENT TO LARGE
              LARGE     QUEUE     ENTITY SENT TO NEXT BLOCK
                        COMBINE   ENTITY COMBINED AS     4 OF     5
```

continues

```
                    SIMAN SYSTEM TRACE BEGINNING AT TIME        .000

  TIME        EVENT OR BLOCK            SYSTEM STATUS CHANGE
-------------------------------------------------------------------

.2100E+02
                    CREATE   ENTITY BATCH SIZE      1 CREATED
                    QUEUE    ENTITY SENT TO NEXT BLOCK
                    GROUP    ENTITY GROUPED AS     2 OF     5
.2100E+02
                    RELEASE EXTRUD3 (1) AVAIL INCREASED BY 1 TO 2
                    QUEUE    ENTITY SENT TO NEXT BLOCK
                    SEIZE WELD3 (1) AVAIL IS 2, 1 UNIT(S) REQ'D
                             ENTITY ALLOC     1 UNIT(S) AT BLOCK
                    DELAY    DELAY BY   .1000E+01 TIME UNIT(S)
.2100E+02
                    RELEASE EXTRUD2 (1) AVAIL INCREASED BY 1 TO 10
                    QUEUE    ENTITY SENT TO NEXT BLOCK
                    SEIZE WELD2 (1) AVAIL IS 10, 1 UNIT(S) REQ'D
                             ENTITY ALLOC     1 UNIT(S) AT BLOCK
                    BRANCH   SELECTION MADE FROM    1 OF    2 BRANCHES
                             PROBABILISTIC,ENTITY SENT TO SOLV2
          SOLV2     DELAY    DELAY BY   .1000E+01 TIME UNIT(S)
.2200E+02
                    CREATE   ENTITY BATCH SIZE      1 CREATED
                    QUEUE    ENTITY SENT TO NEXT BLOCK
                    GROUP    ENTITY GROUPED AS     3 OF     5
.2200E+02
                    RELEASE WELD3 (1) AVAIL INCREASED BY 1 TO 2
                    COUNT    COUNTER     3 INCREMENTED BY    1 TO 1
          PASS      BRANCH   SELECTION MADE FROM    1 OF    2 BRANCHES
                             PROBABILISTIC,BRANCH NOT SELECTED
                             DETERMINISTIC,ENTITY SENT TO HOLD
          HOLD      DELAY    DELAY BY   .1000E+01 TIME UNIT(S)
.2200E+02
          END2      RELEASE WELD2 (1) AVAIL INCREASED BY 1 TO 10
                    COUNT    COUNTER     2 INCREMENTED BY    1 TO 1
          PASS      BRANCH   SELECTION MADE FROM    1 OF    2 BRANCHES
                             PROBABILISTIC,BRANCH NOT SELECTED
                             DETERMINISTIC,ENTITY SENT TO HOLD
          HOLD      DELAY    DELAY BY   .1000E+01 TIME UNIT(S)
.2300E+02
                    CREATE   ENTITY BATCH SIZE      1 CREATED
                    QUEUE    ENTITY SENT TO NEXT BLOCK
                    GROUP    ENTITY GROUPED AS     4 OF     5
.2300E+02
                    BRANCH   SELECTION MADE FROM    1 OF    2 BRANCHES
                             PROBABILISTIC,BRANCH NOT SELECTED
                             DETERMINISTIC,ENTITY SENT TO EXIT
          EXIT      TALLY    OBS   .8000E+01 RECORDED FOR TALLY 3
.2300E+02
                    BRANCH   SELECTION MADE FROM    1 OF    2 BRANCHES
                             PROBABILISTIC,BRANCH NOT SELECTED
                             DETERMINISTIC,ENTITY SENT TO EXIT
          EXIT      TALLY    OBS   .7000E+01 RECORDED FOR TALLY 2
.2400E+02
                    CREATE   ENTITY BATCH SIZE      1 CREATED
                    QUEUE    ENTITY SENT TO NEXT BLOCK
                    GROUP    ENTITY GROUPED AS     5 OF     5
                    QUEUE    ENTITY SENT TO NEXT BLOCK
                    SEIZE MIXER (1) AVAIL IS 3, 1 UNIT(S) REQ'D
                             ENTITY ALLOC     1 UNIT(S) AT BLOCK
                    DELAY    DELAY BY   .1000E+01 TIME UNIT(S)
```

```
                 SIMAN SYSTEM TRACE BEGINNING AT TIME        .000

   TIME        EVENT OR BLOCK              SYSTEM STATUS CHANGE
------------------------------------------------------------------

.2500E+02
                      CREATE    ENTITY BATCH SIZE      1 CREATED
                      QUEUE     ENTITY SENT TO NEXT BLOCK
                      GROUP     ENTITY GROUPED AS     1 OF     5

.2500E+02
                      RELEASE MIXER (1) AVAIL INCREASED BY 1 TO 3
                      SPLIT     ENTITY SPLIT INTO     5 SUBSET(S)
                      ASSIGN    A( 1) SET TO     .1000E+01
                      BRANCH    SELECTION MADE FROM     1 OF    3 BRANCHES
                                CONDITIONAL, ENTITY SENT TO LARGE
            LARGE     QUEUE     ENTITY SENT TO NEXT BLOCK
                      COMBINE   ENTITY COMBINED AS     5 OF     5
            L1        QUEUE     ENTITY SENT TO NEXT BLOCK
                      SEIZE EXTRUD1 (1) AVAIL IS 10, 1 UNIT(S) REQ'D
                                ENTITY ALLOC     1 UNIT(S) AT BLOCK
                      DELAY     DELAY BY     .1000E+01 TIME UNIT(S)

.2500E+02
                      ASSIGN    A( 1) SET TO     .1000E+01
                      BRANCH    SELECTION MADE FROM     1 OF    3 BRANCHES
                                CONDITIONAL, ENTITY SENT TO LARGE
            LARGE     QUEUE     ENTITY SENT TO NEXT BLOCK
                      COMBINE   ENTITY COMBINED AS     1 OF     5

.2500E+02
                      ASSIGN    A( 1) SET TO     .1000E+01
                      BRANCH    SELECTION MADE FROM     1 OF    3 BRANCHES
                                CONDITIONAL, ENTITY SENT TO LARGE
            LARGE     QUEUE     ENTITY SENT TO NEXT BLOCK
                      COMBINE   ENTITY COMBINED AS     2 OF     5

.2500E+02
                      ASSIGN    A( 1) SET TO     .1000E+01
                      BRANCH    SELECTION MADE FROM     1 OF    3 BRANCHES
                                CONDITIONAL, ENTITY SENT TO LARGE
            LARGE     QUEUE     ENTITY SENT TO NEXT BLOCK
                      COMBINE   ENTITY COMBINED AS     3 OF     5

.2500E+02
                      ASSIGN    A( 1) SET TO     .1000E+01
                      BRANCH    SELECTION MADE FROM     1 OF    3 BRANCHES
                                CONDITIONAL, ENTITY SENT TO LARGE
            LARGE     QUEUE     ENTITY SENT TO NEXT BLOCK
                      COMBINE   ENTITY COMBINED AS     4 OF     5

.2600E+02
                      CREATE    ENTITY BATCH SIZE      1 CREATED
                      QUEUE     ENTITY SENT TO NEXT BLOCK
                      GROUP     ENTITY GROUPED AS     2 OF     5

.2600E+02
                      RELEASE EXTRUD1 (1) AVAIL INCREASED BY 1 TO 10
                      QUEUE     ENTITY SENT TO NEXT BLOCK
                      SEIZE WELD1 (1) AVAIL IS 10, 1 UNIT(S) REQ'D
                                ENTITY ALLOC     1 UNIT(S) AT BLOCK
                      BRANCH    SELECTION MADE FROM     1 OF    2 BRANCHES
                                PROBABILISTIC,BRANCH NOT SELECTED
                                DETERMINISTIC,ENTITY SENT TO RUBBER1
            RUBBER1   DELAY     DELAY BY     .1000E+01 TIME UNIT(S)

.2700E+02
                      CREATE    ENTITY BATCH SIZE      1 CREATED
                      QUEUE     ENTITY SENT TO NEXT BLOCK
                      GROUP     ENTITY GROUPED AS     3 OF     5
```

continues

```
                SIMAN SYSTEM TRACE BEGINNING AT TIME         .000

  TIME        EVENT OR BLOCK               SYSTEM STATUS CHANGE
---------------------------------------------------------------------

.2700E+02
            END1      RELEASE WELD1 (1) AVAIL INCREASED BY 1 TO 10
                      COUNT COUNTER 1 INCREMENTED BY 1 TO 3
            PASS      BRANCH    SELECTION MADE FROM    1 OF    2 BRANCHES
                                PROBABILISTIC,BRANCH NOT SELECTED
                                DETERMINISTIC,ENTITY SENT TO HOLD
            HOLD      DELAY     DELAY BY    .1000E+01 TIME UNIT(S)

.2800E+02
                      CREATE    ENTITY BATCH SIZE       1 CREATED
                      QUEUE     ENTITY SENT TO NEXT BLOCK
                      GROUP     ENTITY GROUPED AS     4 OF     5

.2800E+02
                      BRANCH    SELECTION MADE FROM    1 OF    2 BRANCHES
                                PROBABILISTIC,BRANCH NOT SELECTED
                                DETERMINISTIC,ENTITY SENT TO EXIT
            EXIT      TALLY     OBS    .8000E+01 RECORDED FOR TALLY 1

.2900E+02
                      CREATE    ENTITY BATCH SIZE       1 CREATED
                      QUEUE     ENTITY SENT TO NEXT BLOCK
                      GROUP     ENTITY GROUPED AS     5 OF     5
                      QUEUE     ENTITY SENT TO NEXT BLOCK
                      SEIZE MIXER (1) AVAIL IS 3, 1 UNIT(S) REQ'D
                                ENTITY ALLOC      1 UNIT(S) AT BLOCK
                      DELAY     DELAY BY    .1000E+01 TIME UNIT(S)

.3000E+02
                      CREATE    ENTITY BATCH SIZE       1 CREATED
                      QUEUE     ENTITY SENT TO NEXT BLOCK
                      GROUP     ENTITY GROUPED AS     1 OF     5

.3000E+02
                      RELEASE MIXER (1) AVAIL INCREASED BY 1 TO 3
                      SPLIT     ENTITY SPLIT INTO      5 SUBSET(S)
                      ASSIGN    A(  1) SET TO     .2000E+01
                      BRANCH    SELECTION MADE FROM    1 OF    3 BRANCHES
                                CONDITIONAL,  BRANCH NOT SELECTED
                                CONDITIONAL,  ENTITY SENT TO MEDIUM
            MEDIUM    QUEUE     ENTITY SENT TO NEXT BLOCK
                      COMBINE   ENTITY COMBINED AS     1 OF     3

.3000E+02
                      ASSIGN    A(  1) SET TO     .1000E+01
                      BRANCH    SELECTION MADE FROM    1 OF    3 BRANCHES
                                CONDITIONAL,  ENTITY SENT TO LARGE
            LARGE     QUEUE     ENTITY SENT TO NEXT BLOCK
                      COMBINE   ENTITY COMBINED AS     5 OF     5
            L1        QUEUE     ENTITY SENT TO NEXT BLOCK
                      SEIZE EXTRUD1 (1) AVAIL IS 10, 1 UNIT(S) REQ'D
                                ENTITY ALLOC      1 UNIT(S) AT BLOCK
                      DELAY     DELAY BY    .1000E+01 TIME UNIT(S)

.3000E+02
                      ASSIGN    A(  1) SET TO     .2000E+01
                      BRANCH    SELECTION MADE FROM    1 OF    3 BRANCHES
                                CONDITIONAL,  BRANCH NOT SELECTED
                                CONDITIONAL,  ENTITY SENT TO MEDIUM
            MEDIUM    QUEUE     ENTITY SENT TO NEXT BLOCK
                      COMBINE   ENTITY COMBINED AS     2 OF     3
```

```
              SIMAN SYSTEM TRACE BEGINNING AT TIME        .000

  TIME        EVENT OR BLOCK             SYSTEM STATUS CHANGE
----------------------------------------------------------------

.3000E+02
                   ASSIGN   A(  1) SET TO    .1000E+01
                   BRANCH   SELECTION MADE FROM    1 OF   3 BRANCHES
                            CONDITIONAL,  ENTITY SENT TO LARGE
          LARGE    QUEUE    ENTITY SENT TO NEXT BLOCK
                   COMBINE  ENTITY COMBINED AS     1 OF     5

.3000E+02
                   ASSIGN   A(  1) SET TO    .1000E+01
                   BRANCH   SELECTION MADE FROM    1 OF   3 BRANCHES
                            CONDITIONAL,  ENTITY SENT TO LARGE
          LARGE    QUEUE    ENTITY SENT TO NEXT BLOCK
                   COMBINE  ENTITY COMBINED AS     2 OF     5

.3100E+02
                   CREATE   ENTITY BATCH SIZE      1 CREATED
                   QUEUE    ENTITY SENT TO NEXT BLOCK
                   GROUP    ENTITY GROUPED AS     2 OF     5

.3100E+02
                   RELEASE EXTRUD1 (1) AVAIL INCREASED BY 1 TO 10
                   QUEUE    ENTITY SENT TO NEXT BLOCK
                   SEIZE WELD1 (1) AVAIL IS 10, 1 UNIT(S) REQ'D
                            ENTITY ALLOC      1 UNIT(S) AT BLOCK
                   BRANCH   SELECTION MADE FROM    1 OF   2 BRANCHES
                            PROBABILISTIC,BRANCH NOT SELECTED
                            DETERMINISTIC,ENTITY SENT TO RUBBER1
          RUBBER1  DELAY    DELAY BY    .1000E+01 TIME UNIT(S)

.3200E+02
                   CREATE   ENTITY BATCH SIZE      1 CREATED
                   QUEUE    ENTITY SENT TO NEXT BLOCK
                   GROUP    ENTITY GROUPED AS     3 OF     5

.3200E+02
          END1     RELEASE WELD1 (1) AVAIL INCREASED BY 1 TO 10
                   COUNT COUNTER 1 INCREMENTED BY 1 TO 4
          PASS     BRANCH   SELECTION MADE FROM    1 OF   2 BRANCHES
                            PROBABILISTIC,BRANCH NOT SELECTED
                            DETERMINISTIC,ENTITY SENT TO HOLD
          HOLD     DELAY    DELAY BY    .1000E+01 TIME UNIT(S)

.3300E+02
                   CREATE   ENTITY BATCH SIZE      1 CREATED
                   QUEUE    ENTITY SENT TO NEXT BLOCK
                   GROUP    ENTITY GROUPED AS     4 OF     5

.3300E+02
                   BRANCH   SELECTION MADE FROM    1 OF   2 BRANCHES
                            PROBABILISTIC,BRANCH NOT SELECTED
                            DETERMINISTIC,ENTITY SENT TO EXIT
          EXIT     TALLY    OBS   .7000E+01 RECORDED FOR TALLY 1

.3400E+02
                   CREATE   ENTITY BATCH SIZE      1 CREATED
                   QUEUE    ENTITY SENT TO NEXT BLOCK
                   GROUP    ENTITY GROUPED AS     5 OF     5
                   QUEUE    ENTITY SENT TO NEXT BLOCK
                   SEIZE MIXER (1) AVAIL IS 3, 1 UNIT(S) REQ'D
                            ENTITY ALLOC     1 UNIT(S) AT BLOCK
                   DELAY    DELAY BY    .1000E+01 TIME UNIT(S)
```

continues

```
                 SIMAN SYSTEM TRACE BEGINNING AT TIME          .000

   TIME        EVENT OR BLOCK            SYSTEM STATUS CHANGE
-------------------------------------------------------------------

.3500E+02
                      CREATE    ENTITY BATCH SIZE      1 CREATED
                      QUEUE     ENTITY SENT TO NEXT BLOCK
                      GROUP     ENTITY GROUPED AS     1 OF     5

.3500E+02
                      RELEASE MIXER (1) AVAIL INCREASED BY 1 TO 3
                      SPLIT     ENTITY SPLIT INTO     5 SUBSET(S)
                      ASSIGN    A(  1) SET TO    .1000E+01
                      BRANCH    SELECTION MADE FROM    1 OF    3 BRANCHES
                                CONDITIONAL,  ENTITY SENT TO LARGE
            LARGE     QUEUE     ENTITY SENT TO NEXT BLOCK
                      COMBINE   ENTITY COMBINED AS     3 OF     5

.3500E+02
                      ASSIGN    A(  1) SET TO    .1000E+01
                      BRANCH    SELECTION MADE FROM    1 OF    3 BRANCHES
                                CONDITIONAL,  ENTITY SENT TO LARGE
            LARGE     QUEUE     ENTITY SENT TO NEXT BLOCK
                      COMBINE   ENTITY COMBINED AS     4 OF     5

.3500E+02
                      ASSIGN    A(  1) SET TO    .2000E+01
                      BRANCH    SELECTION MADE FROM    1 OF    3 BRANCHES
                                CONDITIONAL,  BRANCH NOT SELECTED
                                CONDITIONAL,  ENTITY SENT TO MEDIUM
            MEDIUM    QUEUE     ENTITY SENT TO NEXT BLOCK
                      COMBINE   ENTITY COMBINED AS     3 OF     3
            M1        QUEUE     ENTITY SENT TO NEXT BLOCK
                      SEIZE EXTRUD2 (1) AVAIL IS 10, 1 UNIT(S) REQ'D
                                ENTITY ALLOC     1 UNIT(S) AT BLOCK
                      DELAY     DELAY BY   .1000E+01 TIME UNIT(S)

.3500E+02
                      ASSIGN    A(  1) SET TO    .1000E+01
                      BRANCH    SELECTION MADE FROM    1 OF    3 BRANCHES
                                CONDITIONAL,  ENTITY SENT TO LARGE
            LARGE     QUEUE     ENTITY SENT TO NEXT BLOCK
                      COMBINE   ENTITY COMBINED AS     5 OF     5
            L1        QUEUE     ENTITY SENT TO NEXT BLOCK
                      SEIZE EXTRUD1 (1) AVAIL IS 10, 1 UNIT(S) REQ'D
                                ENTITY ALLOC     1 UNIT(S) AT BLOCK
                      DELAY     DELAY BY   .1000E+01 TIME UNIT(S)

.3500E+02
                      ASSIGN    A(  1) SET TO    .1000E+01
                      BRANCH    SELECTION MADE FROM    1 OF    3 BRANCHES
                                CONDITIONAL,  ENTITY SENT TO LARGE
            LARGE     QUEUE     ENTITY SENT TO NEXT BLOCK
                      COMBINE   ENTITY COMBINED AS     1 OF     5

.3600E+02
                      CREATE    ENTITY BATCH SIZE      1 CREATED
                      QUEUE     ENTITY SENT TO NEXT BLOCK
                      GROUP     ENTITY GROUPED AS     2 OF     5

.3600E+02
                      RELEASE EXTRUD2 (1) AVAIL INCREASED BY 1 TO 10
                      QUEUE     ENTITY SENT TO NEXT BLOCK
                      SEIZE WELD2 (1) AVAIL IS 10, 1 UNIT(S) REQ'D
                                ENTITY ALLOC     1 UNIT(S) AT BLOCK
                      BRANCH    SELECTION MADE FROM    1 OF    2 BRANCHES
                                PROBABILISTIC,BRANCH NOT SELECTED
                                DETERMINISTIC,ENTITY SENT TO RUBBER2
            RUBBER2   DELAY     DELAY BY   .1000E+01 TIME UNIT(S)
```

```
                    SIMAN SYSTEM TRACE BEGINNING AT TIME        .000

   TIME        EVENT OR BLOCK          SYSTEM STATUS CHANGE
---------------------------------------------------------------------

.3600E+02
                         RELEASE EXTRUD1 (1) AVAIL INCREASED BY 1 TO 10
                         QUEUE     ENTITY SENT TO NEXT BLOCK
                         SEIZE WELD1 (1) AVAIL IS 10, 1 UNIT(S) REQ'D
                                   ENTITY ALLOC     1 UNIT(S) AT BLOCK
                         BRANCH    SELECTION MADE FROM    1 OF    2 BRANCHES
                                   PROBABILISTIC,ENTITY SENT TO SOLV1
               SOLV1     DELAY     DELAY BY    .1000E+01 TIME UNIT(S)

.3700E+02
                         CREATE    ENTITY BATCH SIZE       1 CREATED
                         QUEUE     ENTITY SENT TO NEXT BLOCK
                         GROUP     ENTITY GROUPED AS     3 OF     5

.3700E+02
               END2      RELEASE WELD2 (1) AVAIL INCREASED BY 1 TO 10
                         COUNT COUNTER 2 INCREMENTED BY 1 TO 2
               PASS      BRANCH    SELECTION MADE FROM    1 OF    2 BRANCHES
                                   PROBABILISTIC,BRANCH NOT SELECTED
                                   DETERMINISTIC,ENTITY SENT TO HOLD
               HOLD      DELAY     DELAY BY    .1000E+01 TIME UNIT(S)

.3700E+02
               END1      RELEASE WELD1 (1) AVAIL INCREASED BY 1 TO 10
                         COUNT     COUNTER     1 INCREMENTED BY    1 TO 5
               PASS      BRANCH    SELECTION MADE FROM    1 OF    2 BRANCHES
                                   PROBABILISTIC,BRANCH NOT SELECTED
                                   DETERMINISTIC,ENTITY SENT TO HOLD
               HOLD      DELAY     DELAY BY    .1000E+01 TIME UNIT(S)

.3800E+02
                         CREATE    ENTITY BATCH SIZE       1 CREATED
                         QUEUE     ENTITY SENT TO NEXT BLOCK
                         GROUP     ENTITY GROUPED AS     4 OF     5

.3800E+02
                         BRANCH    SELECTION MADE FROM    1 OF    2 BRANCHES
                                   PROBABILISTIC,BRANCH NOT SELECTED
                                   DETERMINISTIC,ENTITY SENT TO EXIT
               EXIT      TALLY     OBS    .6000E+01 RECORDED FOR TALLY 2

.3800E+02
                         BRANCH    SELECTION MADE FROM    1 OF    2 BRANCHES
                                   PROBABILISTIC,BRANCH NOT SELECTED
                                   DETERMINISTIC,ENTITY SENT TO EXIT
               EXIT      TALLY     OBS    .5000E+01 RECORDED FOR TALLY 1

.3900E+02
                         CREATE    ENTITY BATCH SIZE       1 CREATED
                         QUEUE     ENTITY SENT TO NEXT BLOCK
                         GROUP     ENTITY GROUPED AS     5 OF     5
                         QUEUE     ENTITY SENT TO NEXT BLOCK
                         SEIZE MIXER (1) AVAIL IS 3, 1 UNIT(S) REQ'D
                                   ENTITY ALLOC     1 UNIT(S) AT BLOCK
                         DELAY     DELAY BY    .1000E+01 TIME UNIT(S)

.4000E+02
                         CREATE    ENTITY BATCH SIZE       1 CREATED
                         QUEUE     ENTITY SENT TO NEXT BLOCK
                         GROUP     ENTITY GROUPED AS     1 OF     5
```

continues

```
                 SIMAN SYSTEM TRACE BEGINNING AT TIME        .000

  TIME        EVENT OR BLOCK             SYSTEM STATUS CHANGE
--------------------------------------------------------------------

.4000E+02
                         RELEASE MIXER (1) AVAIL INCREASED BY 1 TO 3
                         SPLIT    ENTITY SPLIT INTO    5 SUBSET(S)
                         ASSIGN   A(  1) SET TO    .1000E+01
                         BRANCH   SELECTION MADE FROM    1 OF   3 BRANCHES
                                  CONDITIONAL, ENTITY SENT TO LARGE
              LARGE      QUEUE    ENTITY SENT TO NEXT BLOCK
                         COMBINE  ENTITY COMBINED AS    2 OF    5

.4000E+02
                         ASSIGN   A(  1) SET TO    .2000E+01
                         BRANCH   SELECTION MADE FROM    1 OF   3 BRANCHES
                                  CONDITIONAL, BRANCH NOT SELECTED
                                  CONDITIONAL, ENTITY SENT TO MEDIUM
              MEDIUM     QUEUE    ENTITY SENT TO NEXT BLOCK
                         COMBINE  ENTITY COMBINED AS    1 OF    3

.4000E+02
                         ASSIGN   A(  1) SET TO    .1000E+01
                         BRANCH   SELECTION MADE FROM    1 OF   3 BRANCHES
                                  CONDITIONAL, ENTITY SENT TO LARGE
              LARGE      QUEUE    ENTITY SENT TO NEXT BLOCK
                         COMBINE  ENTITY COMBINED AS    3 OF    5

.4000E+02
                         ASSIGN   A(  1) SET TO    .1000E+01
                         BRANCH   SELECTION MADE FROM    1 OF   3 BRANCHES
                                  CONDITIONAL, ENTITY SENT TO LARGE
              LARGE      QUEUE    ENTITY SENT TO NEXT BLOCK
                         COMBINE  ENTITY COMBINED AS    4 OF    5

.4000E+02
                         ASSIGN   A(  1) SET TO    .2000E+01
                         BRANCH   SELECTION MADE FROM    1 OF   3 BRANCHES
                                  CONDITIONAL, BRANCH NOT SELECTED
                                  CONDITIONAL, ENTITY SENT TO MEDIUM
              MEDIUM     QUEUE    ENTITY SENT TO NEXT BLOCK
                         COMBINE  ENTITY COMBINED AS    2 OF    3

.4100E+02
                         CREATE   ENTITY BATCH SIZE     1 CREATED
                         QUEUE    ENTITY SENT TO NEXT BLOCK
                         GROUP    ENTITY GROUPED AS    2 OF    5

.4200E+02
                         CREATE   ENTITY BATCH SIZE     1 CREATED
                         QUEUE    ENTITY SENT TO NEXT BLOCK
                         GROUP    ENTITY GROUPED AS    3 OF    5

.4300E+02
                         CREATE   ENTITY BATCH SIZE     1 CREATED
                         QUEUE    ENTITY SENT TO NEXT BLOCK
                         GROUP    ENTITY GROUPED AS    4 OF    5

.4400E+02
                         CREATE   ENTITY BATCH SIZE     1 CREATED
                         QUEUE    ENTITY SENT TO NEXT BLOCK
                         GROUP    ENTITY GROUPED AS    5 OF    5
                         QUEUE    ENTITY SENT TO NEXT BLOCK
                         SEIZE MIXER (1) AVAIL IS 3, 1 UNIT(S) REQ'D
                                  ENTITY ALLOC     1 UNIT(S) AT BLOCK
                         DELAY    DELAY BY   .1000E+01 TIME UNIT(S)
```

```
                  SIMAN SYSTEM TRACE BEGINNING AT TIME        .000

  TIME        EVENT OR BLOCK               SYSTEM STATUS CHANGE
---------------------------------------------------------------------

.4500E+02
                      CREATE    ENTITY BATCH SIZE      1 CREATED
                      QUEUE     ENTITY SENT TO NEXT BLOCK
                      GROUP     ENTITY GROUPED AS     1 OF     5
.4500E+02
                      RELEASE MIXER (1) AVAIL INCREASED BY 1 TO 3
                      SPLIT     ENTITY SPLIT INTO     5 SUBSET(S)
                      ASSIGN    A(  1) SET TO    .1000E+01
                      BRANCH    SELECTION MADE FROM    1 OF    3 BRANCHES
                                CONDITIONAL,  ENTITY SENT TO LARGE
              LARGE   QUEUE     ENTITY SENT TO NEXT BLOCK
                      COMBINE   ENTITY COMBINED AS     5 OF     5
              L1      QUEUE     ENTITY SENT TO NEXT BLOCK
                      SEIZE EXTRUD1 (1) AVAIL IS 10, 1 UNIT(S) REQ'D
                                ENTITY ALLOC     1 UNIT(S) AT BLOCK
                      DELAY     DELAY BY    .1000E+01 TIME UNIT(S)
.4500E+02
                      ASSIGN    A(  1) SET TO    .3000E+01
                      BRANCH    SELECTION MADE FROM    1 OF    3 BRANCHES
                                CONDITIONAL,  BRANCH NOT SELECTED
                                CONDITIONAL,  BRANCH NOT SELECTED
                                DETERMINISTIC,ENTITY SENT TO SMALL
              SMALL   QUEUE     ENTITY SENT TO NEXT BLOCK
                      COMBINE   ENTITY COMBINED AS     2 OF     2
              S1      QUEUE     ENTITY SENT TO NEXT BLOCK
                      SEIZE EXTRUD3 (1) AVAIL IS 2, 1 UNIT(S) REQ'D
                                ENTITY ALLOC     1 UNIT(S) AT BLOCK
                      DELAY     DELAY BY    .1000E+01 TIME UNIT(S)
.4500E+02
                      ASSIGN    A(  1) SET TO    .2000E+01
                      BRANCH    SELECTION MADE FROM    1 OF    3 BRANCHES
                                CONDITIONAL,  BRANCH NOT SELECTED
                                CONDITIONAL,  ENTITY SENT TO MEDIUM
              MEDIUM  QUEUE     ENTITY SENT TO NEXT BLOCK
                      COMBINE   ENTITY COMBINED AS     3 OF     3
              M1      QUEUE     ENTITY SENT TO NEXT BLOCK
                      SEIZE EXTRUD2 (1) AVAIL IS 10, 1 UNIT(S) REQ'D
                                ENTITY ALLOC     1 UNIT(S) AT BLOCK
                      DELAY     DELAY BY    .1000E+01 TIME UNIT(S)
.4500E+02
                      ASSIGN    A(  1) SET TO    .2000E+01
                      BRANCH    SELECTION MADE FROM    1 OF    3 BRANCHES
                                CONDITIONAL,  BRANCH NOT SELECTED
                                CONDITIONAL,  ENTITY SENT TO MEDIUM
              MEDIUM  QUEUE     ENTITY SENT TO NEXT BLOCK
                      COMBINE   ENTITY COMBINED AS     1 OF     3
.4500E+02
                      ASSIGN    A(  1) SET TO    .2000E+01
                      BRANCH    SELECTION MADE FROM    1 OF    3 BRANCHES
                                CONDITIONAL,  BRANCH NOT SELECTED
                                CONDITIONAL,  ENTITY SENT TO MEDIUM
              MEDIUM  QUEUE     ENTITY SENT TO NEXT BLOCK
                      COMBINE   ENTITY COMBINED AS     2 OF     3
.4600E+02
                      CREATE    ENTITY BATCH SIZE      1 CREATED
                      QUEUE     ENTITY SENT TO NEXT BLOCK
                      GROUP     ENTITY GROUPED AS     2 OF     5
```

continues

```
                SIMAN SYSTEM TRACE BEGINNING AT TIME          .000

  TIME       EVENT OR BLOCK              SYSTEM STATUS CHANGE
---------------------------------------------------------------------

.4600E+02
                      RELEASE EXTRUD1 (1) AVAIL INCREASED BY 1 TO 10
                      QUEUE    ENTITY SENT TO NEXT BLOCK
                      SEIZE WELD1 (1) AVAIL IS 10, 1 UNIT(S) REQ'D
                               ENTITY ALLOC    1 UNIT(S) AT BLOCK
                      BRANCH   SELECTION MADE FROM   1 OF   2 BRANCHES
                               PROBABILISTIC,BRANCH NOT SELECTED
                               DETERMINISTIC,ENTITY SENT TO RUBBER1
            RUBBER1   DELAY    DELAY BY   .1000E+01 TIME UNIT(S)

.4600E+02
                      RELEASE EXTRUD3 (1) AVAIL INCREASED BY 1 TO 2
                      QUEUE    ENTITY SENT TO NEXT BLOCK
                      SEIZE WELD3 (1) AVAIL IS 2, 1 UNIT(S) REQ'D
                               ENTITY ALLOC    1 UNIT(S) AT BLOCK
                      DELAY    DELAY BY   .1000E+01 TIME UNIT(S)

.4600E+02
                      RELEASE EXTRUD2 (1) AVAIL INCREASED BY 1 TO 10
                      QUEUE    ENTITY SENT TO NEXT BLOCK
                      SEIZE WELD2 (1) AVAIL IS 10, 1 UNIT(S) REQ'D
                               ENTITY ALLOC    1 UNIT(S) AT BLOCK
                      BRANCH   SELECTION MADE FROM   1 OF   2 BRANCHES
                               PROBABILISTIC,BRANCH NOT SELECTED
                               DETERMINISTIC,ENTITY SENT TO RUBBER2
            RUBBER2   DELAY    DELAY BY   .1000E+01 TIME UNIT(S)

.4700E+02
                      CREATE   ENTITY BATCH SIZE     1 CREATED
                      QUEUE    ENTITY SENT TO NEXT BLOCK
                      GROUP    ENTITY GROUPED AS   3 OF    5

.4700E+02
            END1      RELEASE WELD1 (1) AVAIL INCREASED BY 1 TO 10
                      COUNT COUNTER 1 INCREMENTED BY 1 TO 6
            PASS      BRANCH   SELECTION MADE FROM   1 OF   2 BRANCHES
                               PROBABILISTIC,BRANCH NOT SELECTED
                               DETERMINISTIC,ENTITY SENT TO HOLD
            HOLD      DELAY    DELAY BY   .1000E+01 TIME UNIT(S)

.4700E+02
                      RELEASE WELD3 (1) AVAIL INCREASED BY 1 TO 2
                      COUNT COUNTER 3 INCREMENTED BY 1 TO 2
            PASS      BRANCH   SELECTION MADE FROM   1 OF   2 BRANCHES
                               PROBABILISTIC,ENTITY SENT TO CRUSH
            CRUSH     QUEUE    ENTITY SENT TO NEXT BLOCK
                      SEIZE CRUSHER (1) AVAIL IS 1, 1 UNIT(S) REQ'D
                               ENTITY ALLOC    1 UNIT(S) AT BLOCK
                      DELAY    DELAY BY   .1000E+01 TIME UNIT(S)

.4700E+02
            END2      RELEASE WELD2 (1) AVAIL INCREASED BY 1 TO 10
                      COUNT COUNTER 2 INCREMENTED BY 1 TO 3
            PASS      BRANCH   SELECTION MADE FROM   1 OF   2 BRANCHES
                               PROBABILISTIC,BRANCH NOT SELECTED
                               DETERMINISTIC,ENTITY SENT TO HOLD
            HOLD      DELAY    DELAY BY   .1000E+01 TIME UNIT(S)

.4800E+02
                      CREATE   ENTITY BATCH SIZE     1 CREATED
                      QUEUE    ENTITY SENT TO NEXT BLOCK
                      GROUP    ENTITY GROUPED AS   4 OF    5
```

```
                SIMAN SYSTEM TRACE BEGINNING AT TIME        .000

  TIME       EVENT OR BLOCK              SYSTEM STATUS CHANGE
----------------------------------------------------------------------

.4800E+02
                        BRANCH    SELECTION MADE FROM    1 OF    2 BRANCHES
                                  PROBABILISTIC,BRANCH NOT SELECTED
                                  DETERMINISTIC,ENTITY SENT TO EXIT
              EXIT      TALLY     OBS    .8000E+01 RECORDED FOR TALLY 1

.4800E+02
                        RELEASE CRUSHER (1) AVAIL INCREASED BY 1 TO 1
                        BRANCH    SELECTION MADE FROM    1 OF    3 BRANCHES
                                  CONDITIONAL,  BRANCH NOT SELECTED
                                  CONDITIONAL,  BRANCH NOT SELECTED
                                  DETERMINISTIC,ENTITY SENT TO S1
              S1        QUEUE     ENTITY SENT TO NEXT BLOCK
                        SEIZE EXTRUD3 (1) AVAIL IS 2, 1 UNIT(S) REQ'D
                                  ENTITY ALLOC     1 UNIT(S) AT BLOCK
                        DELAY     DELAY BY    .1000E+01 TIME UNIT(S)

.4800E+02
                        BRANCH    SELECTION MADE FROM    1 OF    2 BRANCHES
                                  PROBABILISTIC,BRANCH NOT SELECTED
                                  DETERMINISTIC,ENTITY SENT TO EXIT
              EXIT      TALLY     OBS    .6000E+01 RECORDED FOR TALLY 2

.4900E+02
                        CREATE    ENTITY BATCH SIZE      1 CREATED
                        QUEUE     ENTITY SENT TO NEXT BLOCK
                        GROUP     ENTITY GROUPED AS     5 OF     5
                        QUEUE     ENTITY SENT TO NEXT BLOCK
                        SEIZE MIXER (1) AVAIL IS 3, 1 UNIT(S) REQ'D
                                  ENTITY ALLOC     1 UNIT(S) AT BLOCK
                        DELAY     DELAY BY    .1000E+01 TIME UNIT(S)

.4900E+02
                        RELEASE EXTRUD3 (1) AVAIL INCREASED BY 1 TO 2
                        QUEUE     ENTITY SENT TO NEXT BLOCK
                        SEIZE WELD3 (1) AVAIL IS 2, 1 UNIT(S) REQ'D
                                  ENTITY ALLOC     1 UNIT(S) AT BLOCK
                        DELAY     DELAY BY    .1000E+01 TIME UNIT(S)

.5000E+02
                        CREATE    ENTITY BATCH SIZE      1 CREATED
                        QUEUE     ENTITY SENT TO NEXT BLOCK
                        GROUP     ENTITY GROUPED AS     1 OF     5

.5000E+02
                        RELEASE MIXER (1) AVAIL INCREASED BY 1 TO 3
                        SPLIT     ENTITY SPLIT INTO      5 SUBSET(S)
                        ASSIGN    A(  1) SET TO    .1000E+01
                        BRANCH    SELECTION MADE FROM    1 OF    3 BRANCHES
                                  CONDITIONAL,  ENTITY SENT TO LARGE
              LARGE     QUEUE     ENTITY SENT TO NEXT BLOCK
                        COMBINE   ENTITY COMBINED AS     1 OF     5

.5000E+02
                        ASSIGN    A(  1) SET TO    .1000E+01
                        BRANCH    SELECTION MADE FROM    1 OF    3 BRANCHES
                                  CONDITIONAL,  ENTITY SENT TO LARGE
              LARGE     QUEUE     ENTITY SENT TO NEXT BLOCK
                        COMBINE   ENTITY COMBINED AS     2 OF     5
```

continues

```
                    SIMAN SYSTEM TRACE BEGINNING AT TIME        .000

   TIME        EVENT OR BLOCK               SYSTEM STATUS CHANGE
---------------------------------------------------------------------

.5000E+02
                       ASSIGN    A( 1) SET TO    .1000E+01
                       BRANCH    SELECTION MADE FROM   1 OF   3 BRANCHES
                                 CONDITIONAL, ENTITY SENT TO LARGE
                LARGE  QUEUE     ENTITY SENT TO NEXT BLOCK
                       COMBINE   ENTITY COMBINED AS    3 OF    5

.5000E+02
                       ASSIGN    A( 1) SET TO    .3000E+01
                       BRANCH    SELECTION MADE FROM   1 OF   3 BRANCHES
                                 CONDITIONAL, BRANCH NOT SELECTED
                                 CONDITIONAL, BRANCH NOT SELECTED
                                 DETERMINISTIC,ENTITY SENT TO SMALL
                SMALL  QUEUE     ENTITY SENT TO NEXT BLOCK
                       COMBINE   ENTITY COMBINED AS    1 OF    2

.5000E+02
                       ASSIGN    A( 1) SET TO    .2000E+01
                       BRANCH    SELECTION MADE FROM   1 OF   3 BRANCHES
                                 CONDITIONAL, BRANCH NOT SELECTED
                                 CONDITIONAL, ENTITY SENT TO MEDIUM
                MEDIUM QUEUE     ENTITY SENT TO NEXT BLOCK
                       COMBINE   ENTITY COMBINED AS    3 OF    3
                M1     QUEUE     ENTITY SENT TO NEXT BLOCK
                       SEIZE EXTRUD2 (1) AVAIL IS 10, 1 UNIT(S) REQ'D
                                 ENTITY ALLOC    1 UNIT(S) AT BLOCK
                       DELAY     DELAY BY    .1000E+01 TIME UNIT(S)

.5000E+02
                       RELEASE WELD3 (1) AVAIL INCREASED BY 1 TO 2
                       COUNT COUNTER 3 INCREMENTED BY 1 TO 3
                PASS   BRANCH    SELECTION MADE FROM   1 OF   2 BRANCHES
                                 PROBABILISTIC,BRANCH NOT SELECTED
                                 DETERMINISTIC,ENTITY SENT TO HOLD
                HOLD   DELAY     DELAY BY    .1000E+01 TIME UNIT(S)

.5100E+02
                       CREATE    ENTITY BATCH SIZE     1 CREATED
                       QUEUE     ENTITY SENT TO NEXT BLOCK
                       GROUP     ENTITY GROUPED AS    2 OF    5

.5100E+02
                       RELEASE EXTRUD2 (1) AVAIL INCREASED BY 1 TO 10
                       QUEUE     ENTITY SENT TO NEXT BLOCK
                       SEIZE WELD2 (1) AVAIL IS 10, 1 UNIT(S) REQ'D
                                 ENTITY ALLOC    1 UNIT(S) AT BLOCK
                       BRANCH    SELECTION MADE FROM   1 OF   2 BRANCHES
                                 PROBABILISTIC,BRANCH NOT SELECTED
                                 DETERMINISTIC,ENTITY SENT TO RUBBER2
                RUBBER2 DELAY    DELAY BY    .1000E+01 TIME UNIT(S)

.5100E+02
                       BRANCH    SELECTION MADE FROM   1 OF   2 BRANCHES
                                 PROBABILISTIC,BRANCH NOT SELECTED
                                 DETERMINISTIC,ENTITY SENT TO EXIT
                EXIT   TALLY     OBS    .1000E+02 RECORDED FOR TALLY 3

.5200E+02
                       CREATE    ENTITY BATCH SIZE     1 CREATED
                       QUEUE     ENTITY SENT TO NEXT BLOCK
                       GROUP     ENTITY GROUPED AS    3 OF    5
```

```
                 SIMAN SYSTEM TRACE BEGINNING AT TIME          .000

   TIME        EVENT OR BLOCK              SYSTEM STATUS CHANGE
-------------------------------------------------------------------------

.5200E+02
             END2       RELEASE WELD2 (1) AVAIL INCREASED BY 1 TO 10
                        COUNT COUNTER 2 INCREMENTED BY 1 TO 4
             PASS       BRANCH    SELECTION MADE FROM    1 OF    2 BRANCHES
                                  PROBABILISTIC,BRANCH NOT SELECTED
                                  DETERMINISTIC,ENTITY SENT TO HOLD
             HOLD       DELAY     DELAY BY    .1000E+01 TIME UNIT(S)

.5300E+02
                        CREATE    ENTITY BATCH SIZE       1 CREATED
                        QUEUE     ENTITY SENT TO NEXT BLOCK
                        GROUP     ENTITY GROUPED AS     4 OF      5
.5300E+02
                        BRANCH    SELECTION MADE FROM    1 OF    2 BRANCHES
                                  PROBABILISTIC,BRANCH NOT SELECTED
                                  DETERMINISTIC,ENTITY SENT TO EXIT
             EXIT       TALLY     OBS    .4000E+01 RECORDED FOR TALLY 2

.5400E+02
                        CREATE    ENTITY BATCH SIZE       1 CREATED
                        QUEUE     ENTITY SENT TO NEXT BLOCK
                        GROUP     ENTITY GROUPED AS     5 OF      5
                        QUEUE     ENTITY SENT TO NEXT BLOCK
                        SEIZE MIXER (1) AVAIL IS 3, 1 UNIT(S) REQ'D
                                  ENTITY ALLOC     1 UNIT(S) AT BLOCK
                        DELAY     DELAY BY    .1000E+01 TIME UNIT(S)
.5500E+02
                        CREATE    ENTITY BATCH SIZE       1 CREATED
                        QUEUE     ENTITY SENT TO NEXT BLOCK
                        GROUP     ENTITY GROUPED AS     1 OF      5
.5500E+02
                        RELEASE MIXER (1) AVAIL INCREASED BY 1 TO 3
                        SPLIT     ENTITY SPLIT INTO      5 SUBSET(S)
                        ASSIGN    A(  1) SET TO    .3000E+01
                        BRANCH    SELECTION MADE FROM    1 OF    3 BRANCHES
                                  CONDITIONAL,  BRANCH NOT SELECTED
                                  CONDITIONAL,  BRANCH NOT SELECTED
                                  DETERMINISTIC,ENTITY SENT TO SMALL
             SMALL      QUEUE     ENTITY SENT TO NEXT BLOCK
                        COMBINE   ENTITY COMBINED AS     2 OF     2
             S1         QUEUE     ENTITY SENT TO NEXT BLOCK
                        SEIZE EXTRUD3 (1) AVAIL IS 2, 1 UNIT(S) REQ'D
                                  ENTITY ALLOC     1 UNIT(S) AT BLOCK
                        DELAY     DELAY BY    .1000E+01 TIME UNIT(S)
.5500E+02
                        ASSIGN    A(  1) SET TO    .1000E+01
                        BRANCH    SELECTION MADE FROM    1 OF    3 BRANCHES
                                  CONDITIONAL,  ENTITY SENT TO LARGE
             LARGE      QUEUE     ENTITY SENT TO NEXT BLOCK
                        COMBINE   ENTITY COMBINED AS     4 OF     5

.5500E+02
                        ASSIGN    A(  1) SET TO    .1000E+01
                        BRANCH    SELECTION MADE FROM    1 OF    3 BRANCHES
                                  CONDITIONAL,  ENTITY SENT TO LARGE
             LARGE      QUEUE     ENTITY SENT TO NEXT BLOCK
                        COMBINE   ENTITY COMBINED AS     5 OF     5
             L1         QUEUE     ENTITY SENT TO NEXT BLOCK
                        SEIZE EXTRUD1 (1) AVAIL IS 10, 1 UNIT(S) REQ'D
                                  ENTITY ALLOC     1 UNIT(S) AT BLOCK
                        DELAY     DELAY BY    .1000E+01 TIME UNIT(S)
```

continues

```
                SIMAN SYSTEM TRACE BEGINNING AT TIME          .000

  TIME        EVENT OR BLOCK              SYSTEM STATUS CHANGE
----------------------------------------------------------------------

.5500E+02
                     ASSIGN    A(  1) SET TO    .3000E+01
                     BRANCH    SELECTION MADE FROM    1 OF   3 BRANCHES
                               CONDITIONAL,  BRANCH NOT SELECTED
                               CONDITIONAL,  BRANCH NOT SELECTED
                               DETERMINISTIC,ENTITY SENT TO SMALL
          SMALL      QUEUE     ENTITY SENT TO NEXT BLOCK
                     COMBINE   ENTITY COMBINED AS     1 OF     2

.5500E+02
                     ASSIGN    A(  1) SET TO    .1000E+01
                     BRANCH    SELECTION MADE FROM    1 OF   3 BRANCHES
                               CONDITIONAL,  ENTITY SENT TO LARGE
          LARGE      QUEUE     ENTITY SENT TO NEXT BLOCK
                     COMBINE   ENTITY COMBINED AS     1 OF     5

.5600E+02
                     CREATE    ENTITY BATCH SIZE       1 CREATED
                     QUEUE     ENTITY SENT TO NEXT BLOCK
                     GROUP     ENTITY GROUPED AS     2 OF     5

.5600E+02
                     RELEASE EXTRUD3 (1) AVAIL INCREASED BY 1 TO 2
                     QUEUE     ENTITY SENT TO NEXT BLOCK
                     SEIZE WELD3 (1) AVAIL IS 2, 1 UNIT(S) REQ'D
                               ENTITY ALLOC     1 UNIT(S) AT BLOCK
                     DELAY     DELAY BY    .1000E+01 TIME UNIT(S)

.5600E+02
                     RELEASE EXTRUD1 (1) AVAIL INCREASED BY 1 TO 10
                     QUEUE     ENTITY SENT TO NEXT BLOCK
                     SEIZE WELD1 (1) AVAIL IS 10, 1 UNIT(S) REQ'D
                               ENTITY ALLOC     1 UNIT(S) AT BLOCK
                     BRANCH    SELECTION MADE FROM    1 OF   2 BRANCHES
                               PROBABILISTIC,BRANCH NOT SELECTED
                               DETERMINISTIC,ENTITY SENT TO RUBBER1
          RUBBER1    DELAY     DELAY BY    .1000E+01 TIME UNIT(S)

.5700E+02
                     CREATE    ENTITY BATCH SIZE       1 CREATED
                     QUEUE     ENTITY SENT TO NEXT BLOCK
                     GROUP     ENTITY GROUPED AS     3 OF     5

.5700E+02
                     RELEASE WELD3 (1) AVAIL INCREASED BY 1 TO 2
                     COUNT COUNTER 3 INCREMENTED BY 1 TO 4
          PASS       BRANCH    SELECTION MADE FROM    1 OF   2 BRANCHES
                               PROBABILISTIC,BRANCH NOT SELECTED
                               DETERMINISTIC,ENTITY SENT TO HOLD
          HOLD       DELAY     DELAY BY    .1000E+01 TIME UNIT(S)

.5700E+02
          END1       RELEASE WELD1 (1) AVAIL INCREASED BY 1 TO 10
                     COUNT COUNTER 1 INCREMENTED BY 1 TO 7
          PASS       BRANCH    SELECTION MADE FROM    1 OF   2 BRANCHES
                               PROBABILISTIC,BRANCH NOT SELECTED
                               DETERMINISTIC,ENTITY SENT TO HOLD
          HOLD       DELAY     DELAY BY    .1000E+01 TIME UNIT(S)

.5800E+02
                     CREATE    ENTITY BATCH SIZE       1 CREATED
                     QUEUE     ENTITY SENT TO NEXT BLOCK
                     GROUP     ENTITY GROUPED AS     4 OF     5
```

```
                      SIMAN SYSTEM TRACE BEGINNING AT TIME        .000

   TIME       EVENT OR BLOCK               SYSTEM STATUS CHANGE
-------------------------------------------------------------------------

.5800E+02
                        BRANCH    SELECTION MADE FROM     1 OF    2 BRANCHES
                                  PROBABILISTIC,BRANCH NOT SELECTED
                                  DETERMINISTIC,ENTITY SENT TO EXIT
              EXIT      TALLY     OBS   .8000E+01 RECORDED FOR TALLY 3

.5800E+02
                        BRANCH    SELECTION MADE FROM     1 OF    2 BRANCHES
                                  PROBABILISTIC,BRANCH NOT SELECTED
                                  DETERMINISTIC,ENTITY SENT TO EXIT
              EXIT      TALLY     OBS   .6000E+01 RECORDED FOR TALLY 1

.5900E+02
                        CREATE    ENTITY BATCH SIZE      1 CREATED
                        QUEUE     ENTITY SENT TO NEXT BLOCK
                        GROUP     ENTITY GROUPED AS      5 OF     5
                        QUEUE     ENTITY SENT TO NEXT BLOCK
                        SEIZE MIXER (1) AVAIL IS 3, 1 UNIT(S) REQ'D
                                  ENTITY ALLOC     1 UNIT(S) AT BLOCK
                        DELAY     DELAY BY    .1000E+01 TIME UNIT(S)

.6000E+02
                        CREATE    ENTITY BATCH SIZE      1 CREATED
                        QUEUE     ENTITY SENT TO NEXT BLOCK
                        GROUP     ENTITY GROUPED AS      1 OF     5

.6000E+02
                        RELEASE MIXER (1) AVAIL INCREASED BY 1 TO 3
                        SPLIT     ENTITY SPLIT INTO      5 SUBSET(S)
                        ASSIGN    A( 1) SET TO    .2000E+01
                        BRANCH    SELECTION MADE FROM     1 OF    3 BRANCHES
                                  CONDITIONAL,  BRANCH NOT SELECTED
                                  CONDITIONAL,  ENTITY SENT TO MEDIUM
              MEDIUM    QUEUE     ENTITY SENT TO NEXT BLOCK
                        COMBINE   ENTITY COMBINED AS      1 OF     3
.6000E+02
                        ASSIGN    A( 1) SET TO    .1000E+01
                        BRANCH    SELECTION MADE FROM     1 OF    3 BRANCHES
                                  CONDITIONAL,  ENTITY SENT TO LARGE
              LARGE     QUEUE     ENTITY SENT TO NEXT BLOCK
                        COMBINE   ENTITY COMBINED AS      2 OF     5
.6000E+02
                        ASSIGN    A(  1) SET TO    .2000E+01
                        BRANCH    SELECTION MADE FROM     1 OF    3 BRANCHES
                                  CONDITIONAL,  BRANCH NOT SELECTED
                                  CONDITIONAL,  ENTITY SENT TO MEDIUM
              MEDIUM    QUEUE     ENTITY SENT TO NEXT BLOCK
                        COMBINE   ENTITY COMBINED AS      2 OF     3

.6000E+02
                        ASSIGN    A(  1) SET TO    .3000E+01
                        BRANCH    SELECTION MADE FROM     1 OF    3 BRANCHES
                                  CONDITIONAL,  BRANCH NOT SELECTED
                                  CONDITIONAL,  BRANCH NOT SELECTED
                                  DETERMINISTIC,ENTITY SENT TO SMALL
              SMALL     QUEUE     ENTITY SENT TO NEXT BLOCK
                        COMBINE   ENTITY COMBINED AS      2 OF     2
              S1        QUEUE     ENTITY SENT TO NEXT BLOCK
                        SEIZE EXTRUD3 (1) AVAIL IS 2, 1 UNIT(S) REQ'D
                                  ENTITY ALLOC     1 UNIT(S) AT BLOCK
                        DELAY     DELAY BY  .1000E+01 TIME UNIT(S)
```

continues

```
                SIMAN SYSTEM TRACE BEGINNING AT TIME          .000

   TIME       EVENT OR BLOCK             SYSTEM STATUS CHANGE
------------------------------------------------------------------

.6000E+02
                        ASSIGN    A(  1) SET TO    .2000E+01
                        BRANCH    SELECTION MADE FROM    1 OF    3 BRANCHES
                                  CONDITIONAL,  BRANCH NOT SELECTED
                                  CONDITIONAL,  ENTITY SENT TO MEDIUM
              MEDIUM    QUEUE     ENTITY SENT TO NEXT BLOCK
                        COMBINE   ENTITY COMBINED AS     3 OF     3
              M1        QUEUE     ENTITY SENT TO NEXT BLOCK
                        SEIZE EXTRUD2 (1) AVAIL IS 10, 1 UNIT(S) REQ'D
                                  ENTITY ALLOC     1 UNIT(S) AT BLOCK
                        DELAY     DELAY BY    .1000E+01 TIME UNIT(S)


                              ***    ***    ***
```

6.12 REFERENCES

1. Al-Aggad, T., "Simulation of an Extrusion Process Using SIMAN" (Master's essay, Bucknell University, April 1985). Excerpts reprinted by permission.
2. Banks, J., and Carson, J.S. *Discrete-Event System Simulation* (Englewood Cliffs, NJ: Prentice-Hall, 1983).
3. Birtwistle, G.M., *Discrete-Event Modelling in SIMULA* (London: Macmillan, 1979).
4. Bryant, R.M., *SIMPAS 5.0 User Manual*, Technical Report 456, Computer Science Department, University of Wisconsin, Madison, WI, 1981.
5. Burroughs Corporation, *Burroughs General Purpose Simulation System (GPSS): Quick Reference Manual* (Detroit: Burroughs Corporation, 1978).
6. CACI, *PC SIMSCRIPT II.5: Introduction and User's Manual* (Los Angeles: CACI, 1985).
7. Control Data Corporation, *General Purpose Simulation System V (GPSS) V/6000 General Information Manual* (Minneapolis: Control Data Corporation, 1974).
8. Dahl, O.J., and Nygard, K.E., "SIMULA—An Algol-based Simulation Language," *CACM* 9, no. 9 (1966): 671–78.
9. Garcia, Ricardo F.; Garzio, Mario R.; and Zeigler, Bernard P., "Discrete-Event Simulation," *IEEE Spectrum* 23, no. 12 (December 1986): 32–36.
10. Gordon, G., *The Application of GPSS V to Discrete-Event Simulation* (Englewood Cliffs, NJ: Prentice-Hall, 1975).
11. Gordon, Geoffrey, "GPSS: Sound Recording," recorded at the History of Programming Languages Conference, ACM Sig Plan, Los Angeles, June 1978.
12. Gordon, Geoffrey, *System Simulation* (Englewood Cliffs, NJ: Prentice-Hall, 1978).
13. Gordon, G., "The Design of the GPSS Language," in *Current Issues in Computer Simulation*, edited by N.R. Adams and A. Dogramaci (New York: Wiley, 1979), pp. 15–26.
14. IBM (U.K.), *SIMPL II Operations Guide*, SH19-5038-0, IBM Corporation, White Plains, NY, 1972.

15. IBM, *General Purpose Simulation System V User's Manual*, SH20-0851-2, IBM Corporation, White Plains, NY, 1977.

16. Katzke, J., and Reitman, J., "Approaching a Universal GPSS," in *Progress in Simulation*, edited by K.W. Gohring et al. (New York: Gordon & Breach, 1972), pp. 43–58.

17. Kay, M.I.; Swain, N.D.; and Sandler, R.L., "Digital Discrete Simulation Languages: A Discussion and an Inventory," in *Progress in Simulation*, edited by K.W. Gohring et al. (New York: Gordon & Breach, 1972), pp. 115–42.

18. Kiviat, P.J.; Markowitz, H.M.; and Villanueva, R., *SIMSCRIPT II.5 Programming Language* (Los Angeles: CACI, 1983).

19. Kiviat, P.J.; Villanueva, R.; and Markowitz, H.M., *The SIMSCRIPT II Programming Language* (Englewood Cliffs, NJ: Prentice-Hall, 1965).

20. Kreutzer, Wolfgang, *System Simulation*: *Programming Styles and Languages* (Reading, MA: Addison-Wesley, 1986).

21. Law, Averill M., and Larmey, Christophes, *An Introduction to Simulation Using SIMSCRIPT II.5* (*Los Angeles*: *CACI*, 1981).

22. Papazoglou, M.P.; Georgiadis, P.I.; and Maritsas, D.G., "An Outline of the Programming Language SIMULA," *Computer Languages* 9, no. 2 (April 1984): 107–31.

23. Pegden, C.D., *Introduction to SIMAN* (State College, PA: Systems Modeling Corporation, 1982).

24. Pritsker, A.A.B., and Pegden, P., *Introduction to Simulation and SLAM*, 2nd ed. (New York: Wiley, Halstead Press, 1984).

25. Pritsker, A. Alan B., and Young, R.E., *Simulation with GASP: APL/Based Continuous/Discrete Simulation Language* (New York: Wiley, 1975).

26. Ripley, Brian D., *Stochastic Simulation* (New York: Wiley, 1987).

27. Russell, Edward C., *Building Simulation Models with SIMSCRIPT II.5* (Los Angeles: CACI, 1983).

28. Schriber, T.J., *Simulation Using GPSS* (New York: Wiley, 1974).

29. Speckhart, H., and Green, W.H., *A Guide to Using CSMP* (Englewood Cliffs, NJ: Prentice-Hall, 1976).

30. Thomas, G., and Aburdene, M.F., "Simulation of a Five Node Computer Network," in *Modeling and Simulation*, edited by William G. Vogt and Marlin H. Mickle, vol. 14 (Pittsburgh: Instrument Society of America, 1983), pp. 253–57.

31. Thomas, G., and Aburdene, M.F., "Simulation of Time Division Multi-Access Systems Using SLAM," in *Modeling and Simulation*, edited by William G. Vogt and Marlin H. Mickle, vol. 14 (Pittsburgh: Instrument Society of America, 1983), pp. 259–64.

32. Tocker, K.D., "Review of Simulation Languages," *Operations Research Quarterly* 16, no. 2 (1965): 189–217.

7 Discrete Systems and Markov Processes

7.1 INTRODUCTION

The theory of stochastic (random) processes plays an important role in the study of physical, engineering, and social systems. The concept of "states" or "conditions" of the system occurs frequently in the models representing these systems, and we are often interested in the transition from one state to another. The states of a system can be either discrete or continuous.

The Markov process is a special type of stochastic process often used to simulate social and physical systems. It is one in which the probability that the system will be in a given state at any given time t_2, may be obtained from knowledge of its state at any earlier time, t_1, but does not depend on how the process reached the state at time t_1. If the states are discrete and the transitions occur at discrete points in time, the process is called a Markov chain. The analysis and simulation of Markov chains is relatively simple and, for this reason, Markov chains are used to approximate more complex physical systems, even though the transition to the next state depends on more than just the present state.

In this chapter, we will provide a mathematical representation of Markov processes, along with several examples of Markov chains and a method to simulate them using uniform random number generators.

7.2 MATHEMATICAL REPRESENTATION OF MARKOV PROCESSES

A stochastic process M(t) is called a Markov process if, for any set of n time points (continuous or discrete) $t_1 < t_2 < \cdots < t_n$, the following is true:

$$p(\mathrm{M}(t_n) = m_n \mid \mathrm{M}(t_{n-1}) = m_{n-1}, \ldots, \mathrm{M}(t_1) = m_1)$$
$$= p(\mathrm{M}(t_n) = m_n \mid \mathrm{M}(t_{n-1}) = m_{n-1}) \qquad \textbf{(7.1)}$$

where m_i is the state of the process at time t_i ($t_1 \le t_i \le t_n$). The set of possible states of the Markov process $\{m_i\colon\ i = 1, 2, \ldots, q\ (q \le n)\}$ is called the state space of the process.[11] Equation (7.1) tells us that the probability that the Markov process is in state m_n at t_n, given that the Markov process was in state m_{n-1} at t_{n-1}, and the probability of being in state m_{n-2} at time t_{n-2}, and so on, is equal to the probability of finding the system in state m_n at t_n, given that the system was in state m_{n-1} at time t_{n-1}. In short, the future state of the system depends on the present state of the system and does not depend on how the system arrived at the present state.

Markov processes are classified into four basic types, as shown in Table 7.1. The classification depends on

The nature of the parameter of the process (t)

The nature of the states of the process (m_i); when the states are discrete, the Markov process is called a Markov chain.

A Markov process is fully described by its probability transition matrix:

$$[P] = [p_{ij}] = \begin{bmatrix} p_{11} & p_{12} & \cdots & p_{1q} \\ p_{21} & p_{22} & \cdots & p_{2q} \\ \vdots & & & \\ p_{q1} & p_{q2} & \cdots & p_{qq} \end{bmatrix} \qquad (7.2)$$

where p_{ij} is the probability of transition from state i to state j. The initial probability vector is

$$\mathbf{p}(0) = \begin{bmatrix} p_1(0) \\ p_2(0) \\ \vdots \\ p_q(0) \end{bmatrix} \qquad (7.3)$$

subject to

$$\sum_{j=1}^{q} p_{ij} = 1 \qquad \text{(for all } i\text{)} \qquad (7.4)$$

TABLE 7.1 Classification of Markov Processes

Parameter of Process (t)	States of Process (m_i): Discrete	States of Process (m_i): Continuous
Discrete	Discrete-parameter Markov chain	Discrete-parameter Markov process
Continuous	Continuous-parameter Markov chain	Continuous-parameter Markov process

and $$\sum_{i=1}^{q} p_i(0) = 1 \tag{7.5}$$

where $p_1(0)$ = the probability of the system being in state 1 at $t = 0$
$p_i(0)$ = the probability of the system being in state i at $t = 0$

The probability of the system being in state 1 at $t = t_1$ is given by

$$p_1(t = t_1) = p_{11}p_1(0) + p_{21}p_2(0) + \cdots + p_{q1}p_q(0) \tag{7.6}$$

which can be written

$$p_1(t = t_1) = [p_{11}p_{21} \cdots p_{q1}] \begin{bmatrix} p_1(0) \\ p_2(0) \\ \vdots \\ p_q(0) \end{bmatrix} \tag{7.7}$$

In general, we can write

$$\begin{bmatrix} p_1(t = t_1) \\ p_2(t = t_1) \\ \vdots \\ p_q(t = t_1) \end{bmatrix} = \begin{bmatrix} p_{11} & p_{12} & \cdots & p_{1q} \\ p_{21} & p_{22} & \cdots & p_{2q} \\ \vdots & & \cdots & \\ & & \cdots & \\ p_{q1} & p_{q2} & \cdots & p_{qq} \end{bmatrix}^T \begin{bmatrix} p_1(0) \\ p_2(0) \\ \vdots \\ p_q(0) \end{bmatrix} \tag{7.8}$$

(where T indicates the transpose of the matrix), and we can easily show that

$$\mathbf{p}(t = t_n) = [P^{\mathrm{T}}]^n \mathbf{p}(0)$$

This is often called the nth step transition because it forms a matrix of probabilities answering the generic question, If the process is in state i at time $t = t_a$, what is the probability that the process is in state j at time $t = t_{a+n}$? After we look at some examples, we will return to describe how we can obtain the probability of transition matrix.

A finite Markov chain is a Markov chain with a finite number of states. As we stated earlier, finite Markov chains occur frequently in sociology, psychology, physical systems, and other fields of study. The model for the "naive subject" experiment first performed by Asch is a discrete-parameter Markov chain with four states; the model for "opinion change during repeated balloting" is a five-state discrete-parameter chain; the "survival after treatment of cancer" model, developed by Zahl, is a four-state continuous-parameter Markov chain; and the ALOHA packet radio communication system can be modelled using a Markov chain. We will now discuss each of these models in some detail.

7.3 EXAMPLES OF MARKOV PROCESSES

The examples presented here relate to some interesting social, behavioral, and engineering experiments. They also demonstrate the application of Markov chains for modelling highly complex social, behavioral, and engineering systems found in real field studies. We will walk through these examples to gain insight into the types of systems described by the Markov process. Once we are familiar with these systems, we will move to Section 7.4, where we will learn the methods for constructing probability transition matrices.

A Markov Chain Model in Sociology[2,6]

This model comes from an experiment first performed by Asch.[2] A naive subject answers a question aloud after seven pretrained subjects answered the same question unanimously, though incorrectly. The study attempts to determine whether the naive subject answers according to his own belief or tries to conform to the unanimous choice of the group. The model was proposed by B. Cohen.[6] It assumes that the subject can be in one of four mental states:

Nonconforming
Temporary nonconforming
Temporary conforming
Conforming

The changes of state were assumed to form a Markov chain with the following transition matrix:

$$[P] = \begin{bmatrix} 1.00 & 0.00 & 0.00 & 0.00 \\ 0.04 & 0.63 & 0.33 & 0.00 \\ 0.00 & 0.46 & 0.49 & 0.05 \\ 0.00 & 0.00 & 0.00 & 1.00 \end{bmatrix}$$

This matrix was obtained from experimental data and computed as shown in Section 7.4. The model assumes that the subject starts in state 2 (temporary nonconforming). Therefore, $\mathbf{p}(0) = [0100]^{\mathrm{T}}$. Figure 7.1 is a graphical representation of the Markov process. We will now illustrate the computation of the probabilities of being in state 1, 2, 3, and 4 at time $t = 1$, given that the system was in state 2 at $t = 0$. Using equation (7.8), we find

$$\begin{bmatrix} p_1(1) \\ p_2(1) \\ p_3(1) \\ p_4(1) \end{bmatrix} = \begin{bmatrix} 1.00 & 0.00 & 0.00 & 0.00 \\ 0.04 & 0.63 & 0.33 & 0.00 \\ 0.00 & 0.46 & 0.49 & 0.05 \\ 0.00 & 0.00 & 0.00 & 1.00 \end{bmatrix}^T \begin{bmatrix} 0.00 \\ 1.00 \\ 0.00 \\ 0.00 \end{bmatrix}$$

FIGURE 7.1 State Transition Matrix for the Naive Subject Model

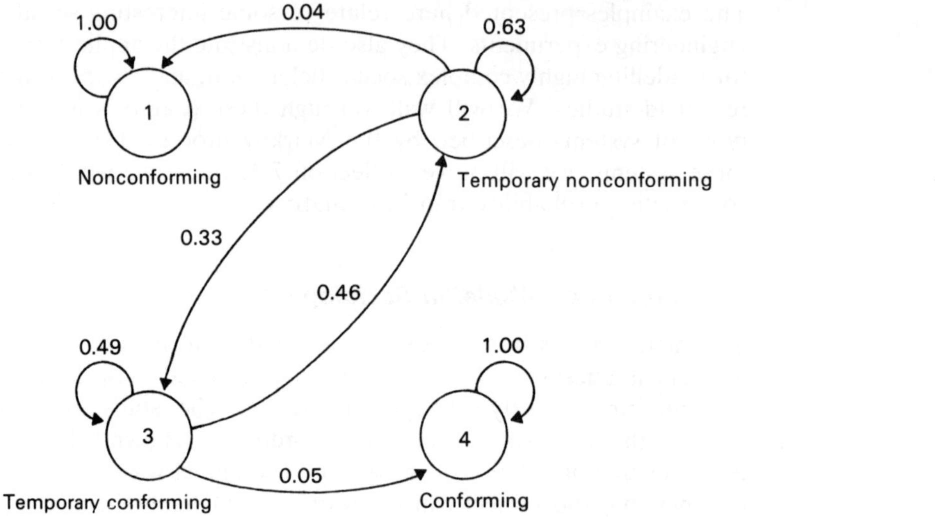

which leads to

$$\mathbf{p}(1) = \begin{bmatrix} p_1(1) \\ p_2(1) \\ p_3(1) \\ p_4(1) \end{bmatrix} = \begin{bmatrix} 0.04 \\ 0.63 \\ 0.33 \\ 0.00 \end{bmatrix}$$

Model for Opinion Change during Repeated Balloting[10]

This model is an application of Markov chains to the study of the change of opinion expressed by a group (e.g., jury, board of trustees, city council) on a given topic. The group has v voters who are asked to express a choice between two options; thus, there are $v + 1$ conceivable numerical states. These states are the number of ways the group can vote. Usually, the aim of such a study is to find a "majority" (two-thirds majority or qualified majority). This model is designed to track the formation of a majority through successive ballots. We assume that some voters change their vote from one ballot to another. More specifically, we hypothesize that voters belong to one of two categories: decided or undecided. Letting u signify the undecided, $v - u$ represents the decided. The decided never change their vote. The behavior of the undecided, however, is stochastic. The probability is assumed to be proportional to the total number of votes the two options received on the preceding ballot. For example, suppose that there are seven voters; option 1 had five votes and option 2 had two votes. This balloting state is designated as state (5, 2). Then each of the undecided voters is assumed to vote in favor of option 1 with probability $\frac{5}{7}$. However, if the states are defined to be the number of undecided voters leaning toward option 1, then the sequence of ballots forms a Markov process with $u+1$ states. In this example, let $u=4$, and the number of voters decided for option 1 be two, and the number of voters decided for option 2 be one. Therefore, the five states are defined as

State 1 = state (2,5)
State 2 = state (3,4)
State 3 = state (4,3)
State 4 = state (5,2)
State 5 = state (6,1)

The probability transition matrix can be calculated to be

$$[P] = \begin{bmatrix} 0.2603 & 0.4165 & 0.2499 & 0.0666 & 0.0067 \\ 0.1066 & 0.3199 & 0.3599 & 0.1799 & 0.0337 \\ 0.0337 & 0.1799 & 0.3599 & 0.3199 & 0.1066 \\ 0.0067 & 0.0666 & 0.2499 & 0.4165 & 0.2603 \\ 0.0004 & 0.0100 & 0.0900 & 0.3599 & 0.5397 \end{bmatrix}$$

To attain a majority, we must have a two-thirds majority. This corresponds to states 1, 4, and 5. Problem 11 at the end of the chapter will give you the opportunity to derive this matrix once you have completed Section 7.4.

Model for Social Mobility[3,12]

This model describes the changes in social class. The assumptions are that the process is a Markov chain, the transition probabilities are independent of time, and the family lines are such that each father has exactly one son. The transition probabilities are for England and Wales and were estimated by Glass and Hall.[8]

Seven social classes (states) were assumed to exist:

Professional and higher administrative
Managerial and executive
Higher-grade supervisory and nonmanual
Lower-grade supervisory and nonmanual
Skilled manual and routine nonmanual
Semiskilled
Unskilled

The transition matrix is given by[8,12]

$$[P] = \begin{bmatrix} 0.388 & 0.146 & 0.202 & 0.062 & 0.410 & 0.047 & 0.015 \\ 0.107 & 0.267 & 0.227 & 0.120 & 0.206 & 0.053 & 0.020 \\ 0.035 & 0.101 & 0.188 & 0.191 & 0.357 & 0.067 & 0.061 \\ 0.021 & 0.039 & 0.112 & 0.212 & 0.430 & 0.124 & 0.062 \\ 0.009 & 0.024 & 0.075 & 0.123 & 0.473 & 0.171 & 0.125 \\ 0.000 & 0.013 & 0.041 & 0.088 & 0.391 & 0.312 & 0.155 \\ 0.000 & 0.008 & 0.036 & 0.083 & 0.364 & 0.235 & 0.274 \end{bmatrix}$$

Problem 5 at the end of the chapter requires the use of this transition matrix.

Model for Survival after Treatment of Cancer[3,14]

In this example, the parameter time is continuous, unlike in the previous three examples, where we were considering processes at discrete points in time. In this type of problem, a subject has been treated for cancer and, some time later, may be in one of a number of states. The model defines four states:

Suffering from active cancer

Dead from cancer

Cured or with the cancer in remission

Dead from other causes

In this case, since changes can occur at any time, the *infinitesimal transition matrix*[3,14] is used instead of the probability transition matrix:

$$[P] = \begin{bmatrix} 0.6930 & 0.0840 & 0.2210 & 0.0020 \\ 0.0000 & 1.0000 & 0.0000 & 0.0000 \\ 0.0160 & 0.0010 & 0.9630 & 0.0200 \\ 0.0000 & 0.0000 & 0.0000 & 1.0000 \end{bmatrix}$$

Here, as well as in the other examples, we have assumed that the transition matrix does not change with time.

Slotted ALOHA Model[5]

As mentioned, the slotted ALOHA communication system can be modelled as a Markov chain. The system consists of a single communication channel resource to be shared by many users. Time is divided into discrete intervals called slots, and users contend for the channel at the beginning of a slot. Users can be in one of two states: blocked or unblocked. For example, when two or more users begin to transmit their packets (small messages) during the same slot, a "collision" occurs and the users become blocked. These users cannot transmit new packets until the blocked packet is retransmitted successfully. Blocked users contend for the

FIGURE 7.2 State Transition Diagram for a Two-User Slotted ALOHA System

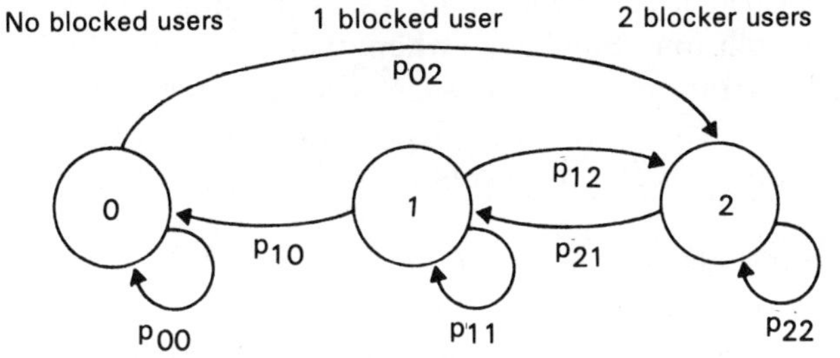

channel in future slots with all users who have a packet to transmit. If only one user transmits during a slot, the packet is transmitted and the user remains in the unblocked state. If we assume that the system has N users, we can define the states to correspond to the number of blocked users, and therefore the system will have $N + 1$ states (no blocked users, one blocked user, . . . , N blocked users). To complete our model, we can assume that unblocked users have probability p of transmission at the beginning of a slot, and blocked users have probability of retransmission α. We can develop the state transition diagram for a two-user system and see that the system moves among a set of discrete states, as illustrated in Figure 7.2.

7.4 DETERMINING THE PROBABILITY TRANSITION MATRIX

Now that we have defined and given examples of Markov chains, our next task in simulating a system we judge to be Markov-describable is to determine the states of the system. To simplify our model without losing accuracy, we can reduce the number of states we consider by grouping rarely visited states into one state. We must be certain, however, that all information needed to predict future behavior is included within these "grouped" states.

Once the states are determined, we are ready to construct the probability transition matrix. The fewer the number of states, the fewer the number of transition probabilities we need to compute or determine experimentally. Earlier, we presented examples of Markov chains where states were already determined for us. In this section, we will present methods for determining probability transition matrices. The following example illustrates an analytical method.

Slotted ALOHA System Model

Let us determine the probability transition matrix for the slotted ALOHA system of Section 7.3. Our objective is to determine p_{00}, p_{02}, p_{10}, p_{11}, p_{12}, p_{21}, and p_{22}. We know from the assumptions about the system that p_{01} and p_{20} are zero.

Probability p_{01} is zero because when we have no blocked users, the only way we can have a blocked user is when both users transmit and both become blocked because of a collision. Similarly, when two users are blocked, both cannot simultaneously retransmit their message without experiencing a collision.

Let us first compute p_{00}. The system will remain in the same state if one of the following three conditions occurs:

1. User 1 transmits and user 2 remains silent. No collisions occur, and the probability of this condition is $p(1-p)$.
2. User 1 is silent and user 2 transmits. The probability of this condition is $(1-p)p$, by symmetry with condition 1.
3. User 1 and user 2 remain silent. The probability of this occurring is $(1-p)(1-p)$.

Thus,

$$p_{00} = p(1-p) + (1-p)p + (1-p)^2 = 2p(1-p) + (1-p)^2 = 1 - p^2$$

We can compute p_{02} in two ways:

$$p_{02} = 1 - p_{00} = 1 - (1-p^2) = p^2$$

or $$p_{02} = \text{probability that user 1 transmits and user 2 transmits} = p \cdot p = p^2$$

We can compute p_{10} because we know that this condition occurs when the blocked user transmits and the unblocked user remains silent. The probability of this condition is $\alpha(1-p)$. Thus,

$$p_{10} = \alpha(1-p)$$

Computing p_{11} is a little more difficult. The system remains in the same state if either of the following conditions occurs:

1. The blocked user does not retransmit and the unblocked user does not transmit. This occurs with probability $(1-\alpha)(1-p)$.
2. The blocked user does not retransmit and the unblocked user transmits. The probability of this condition is $(1-\alpha)p$.

Therefore,

$$p_{11} = (1-\alpha)(1-p) + (1-\alpha)p = (1-\alpha)$$

Probability p_{12} is easily computed to be

$$p_{12} = 1 - (p_{10} + p_{11}) = \alpha p$$

We can compute p_{22} because we know that when we have two blocked users, they will remain blocked if one of the following conditions occurs:

1. Both users retransmit. This occurs with probability $\alpha \cdot \alpha = \alpha^2$.
2. Both users do not retransmit. This occurs with probability $(1-\alpha)(1-\alpha) = (1-\alpha)^2$.

Therefore,

$$p_{22} = \alpha^2 + (1-\alpha)^2 = 1 - 2\alpha + 2\alpha^2$$

It follows that

$$p_{21} = 1 - p_{22} = 2\alpha - 2\alpha^2 = 2\alpha(1-\alpha)$$

Thus, the state transition matrix is

$$[P] = \begin{bmatrix} 1-p^2 & 0 & p^2 \\ \alpha(1-p) & 1-\alpha & \alpha p \\ 0 & 2\alpha(1-\alpha) & 1-2\alpha+2\alpha^2 \end{bmatrix}$$

Now we will illustrate an experimental method for determining a probability transition matrix.

Soft Drink Model

SDC, a soft drink company, is interested in determining the long-term trend of its market share. This year, its market share was 50% and its competitor's share was 50%. SDC management decides to randomly survey 2000 consumers as they exit grocery stores across the country. Two questions are asked:

> Which brand of soft drink did you purchase today?
>
> Which brand of soft drink did you purchase on your last trip to the grocery store?

The data are tabulated as follows:

	Today's Purchase		
Last Purchase	*SDC*	*Competition*	*Totals*
SDC	400	140	540
Competition	460	1000	1460
			2000

TABLE 7.2

State	1	2	...	N	Sum of Row
1	n_{11}	n_{12}	...	n_{1N}	S_1
2	n_{21}	n_{22}	...	n_{2N}	S_2
⋮	⋮				
N	n_{N1}	n_{N2}	...	n_{NN}	S_N

We can now develop a two-state model, where state 1 denotes the purchase of the SDC soft drink and state 2 denotes the purchase of the competitor's soft drink. To compute the probability of transition from state 1 to state 1, we note that 400 out of 540, or 74% of consumers, purchased the SDC soft drink. The transition from state 1 to state 2 is 140 out of 540, or 26% of consumers. Similarly, the transition from state 2 to state 1 is 460 out of 1460, or 32%, and the transition from state 2 to state 2 is 1000 out of 1460, or 68%.

Our probability transition matrix is then given by

$$[P] = \begin{bmatrix} 0.74 & 0.26 \\ 0.32 & 0.68 \end{bmatrix}$$

To determine the long-term effects, SDC can simulate or calculate consumer preference after n units of time. The simulation can be done using the algorithm of Section 7.5. To calculate consumer preference, SDC can use equation (7.8), which leads to

$$\begin{bmatrix} p_1(n) \\ p_2(n) \end{bmatrix} = \begin{bmatrix} 0.74 & 0.32 \\ 0.26 & 0.68 \end{bmatrix}^n \begin{bmatrix} p_1(0) \\ p_2(0) \end{bmatrix}$$

with $p_1(0) = 0.5$ and $p_2(0) = 0.5$. The initial probabilities $p_1(0)$ and $p_2(0)$ are taken as this year's market share.

This example has shown us that to compute the probability matrix for a Markov process with N states, we can compile a table similar to Table 7.2. Each entry n_{ij} in the table refers to the number of times a transition has occurred from state i to state j. The probability transition matrix is formed by dividing each element in every row by the sum of each row. Because of the nature of these calculations, this type of data is usually expensive to obtain and we must be careful in our choice of sampling techniques.

7.5 SAMPLE MARKOV SIMULATION ALGORITHM

Once the probability transition matrix is determined, the simulation of a Markov process with N states requires information about the starting state. If we do not know the starting state but know an initial probability vector, we determine a starting state as follows, using uniform random generators, presented in Chapter 5:

1. Divide the interval [0,1] into N intervals of lengths equal to initial probabilities of being in each of the N states. The length of the first interval should be $p_1(0)$; the length of the second interval should be $p_2(0)$; and so on.
2. Generate a uniformly distributed random number in the range [0,1].
3. Determine in what interval the random number falls. The number of this interval corresponds to the starting state, because the probability of the random number falling in the first interval is equal to $p_1(0)$, the probability of it falling in the second interval is equal to $p_2(0)$, and so on.

Our program should determine the transition from one state (Last) to the next state (New). The first step involves setting the State (Last) equal to the initial state. This locates the row of the probability transition matrix. As we did earlier when we determined the initial state, we execute steps 1 through 3, with the row probabilities replacing the initial probabilities in step 1. The number of the interval in step 3 is the next state (New). This sequence of steps is then repeated to find each one of the other transitions, with State (Last) set to State (New). Since we are also interested in the number of transitions (TotNoTr) from every state to every other state for comparison to the transition probability matrix, the program tallies each transition n_{ij}, as shown in Table 7.2. Our program should also determine when the process reaches an absorbing state (Absorbing), from which no transition to another state can occur. For example, in the survival after treatment of cancer process, states 2 and 4 are absorbing. A simulation algorithm written in Pascal is shown in Algorithm 7.1. We can make a slight modification to this program to print the state at each instant, thus providing us with a history of the states of the system. Figure 7.3 shows the output results of executing the Markov program for the ALOHA-type system of the examples in Sections 7.3 and 7.4, with $\alpha = 0.6$ and $p = 0.5$. We can also use the Markov algorithm to simulate any stochastic phenomena where the new state depends on the probability transition matrix which, in turn, depends on how the system reached the present state.

ALGORITHM 7.1 Markov Simulation Program (Continued)

```
PROGRAM MarkovProcess (INPUT,OUTPUT);

            {A program to simulate Markov processes.}

CONST {constants for the main program}

   Maxstates=10;
   MaxTrans=5000;
   MaxTrials=500;


TYPE {for the main program}

   ARRAY1=ARRAY[1..MaxStates,1..MaxStates] OF REAL;
   ARRAY2=ARRAY[1..MaxStates] OF REAL;
   ARRAY3=ARRAY[1..MaxStates,1..MaxStates] OF INTEGER;
```

continues

ALGORITHM 7.1 Markov Simulation Program (Continued)

```
VAR {for the main program}

   MarkovArray:ARRAY1;
   InitialProbVector:ARRAY2;
   NTrials,NStates,I,K,J,N,Last,New:INTEGER;
   DT,TMax,Time:REAL;
   Absorbing:BOOLEAN;
   NoOfTransitions:ARRAY3;
   TotNoTr:ARRAY3;
PROCEDURE Calculate(VAR New:INTEGER;IPV:ARRAY2);

   {This procedure calculates the initial state of the system}
               {from the initial probability matrix}


VAR {for the procedure Calculate}


   I:INTEGER;
   PR,RndHolder:REAL;


BEGIN {Begin the procedure Calculate}


   RndHolder:=RANDOM;
   PR:=0;
   I:=1;
   WHILE (RndHolder>PR+IPV[I]) DO
      BEGIN
         PR:=IPV[I]+PR;
         I:=I+1;
      END;
   New:=I;
END;
PROCEDURE OutputInfo(VAR NoOfT:ARRAY3;I,N,NS,New:INTEGER);

   {This procedure outputs the transitions matrix for each}
                  {simulation run}


VAR {of the procedure OutputInfo}

   J,K:INTEGER;


BEGIN {Begin the procedure OutputInfo}


   WRITELN('For trial ',I,' the number of transitions was ',N);
   WRITELN('The final state of the process was state ',New);
   WRITELN('and the transitions matrix was:');
   WRITELN;
   WRITELN;
   WRITE('      ');
   FOR J:=1 TO NStates-1 DO
      WRITE(J,'    ');
   WRITELN(NStates);
   WRITELN;
   FOR J:=1 TO NStates DO
```

ALGORITHM 7.1 Markov Simulation Program (Continued)

```
           BEGIN
              WRITE(J,'   ');
              FOR K:=1 TO NStates-1 DO
                  WRITE(NoOfT[J,K],'   ');
              WRITELN(NoOfT[J,NStates]);
           END;
        WRITELN;
        WRITELN;
        WRITELN;
     END;
     PROCEDURE InputInfo(VAR IPV:ARRAY2;VAR PTM:ARRAY1;
                         VAR NTrials,NStates:INTEGER;VAR DT,TMax:REAL);

        {This procedure prompts the user for input and does}
       {some simple tests on the input to check its validity}

     VAR {for the procedure InputInfo}

        Ch:CHAR;
        Sum:REAL;
        Int,I,J:INTEGER;

     BEGIN {Begin the procedure InputInfo}

        WRITELN('Please enter the length of time that you wish each simulation');
        WRITELN('run to last and the interval between events that you want');
        REPEAT
            READLN(TMax,DT);
            IF TMax/DT>MaxTrans THEN
               BEGIN
                  WRITELN('The maximum number of events per trial is');
                  WRITELN(MaxTrans,' Please input different numbers.');
               END;
        UNTIL TMax/DT<MaxTrans;
        WRITELN('How many simulation trials do you want');
        REPEAT
            READLN(NTrials);
            IF NTrials>MaxTrials THEN
               BEGIN
                  WRITELN('Maximum number of trials allowed is ',MaxTrials);
                  WRITELN('Try again');
               END
        UNTIL MaxTrials>NTrials;
        WRITELN('Is the initial state of the process known?');
        WRITELN('                   (Y/N)                  ');
        REPEAT
            READLN(Ch);
            IF Ch='Y' THEN
               BEGIN
                  WRITELN('Now input the initial state');
                  REPEAT
                      READLN(Int);
                      IF (Int>NStates) OR (Int<0) THEN
                          WRITELN('Incorrect Input');
                  UNTIL (Int<=NStates) AND (Int>0);
                  FOR I:=1 TO NStates DO
                      IPV[I]:=0;
                  IPV[Int]:=1;
               END
            ELSE
               IF Ch='N' THEN
                  BEGIN
                      WRITELN('Then input the initial state probability');
                      WRITELN('vector.');
```

continues

ALGORITHM 7.1 Markov Simulation Program (Continued)

```
                    REPEAT
                        FOR I:=1 TO NStates-1 DO
                            READ(IPV[I]);
                        READLN(IPV[NStates]);
                        Sum:=0;
                        FOR I:=1 TO NStates DO
                            Sum:=Sum+IPV[I];
                        IF (Sum>1.000001) OR (Sum<0.999999) THEN
                            BEGIN
                                WRITELN('Sum of Probabilities must be 1
                                WRITELN('Try Again.');
                            END;
                    UNTIL    (Sum<1.000001) AND (Sum>0.999999);
                END
            ELSE
                WRITELN('Incorrect Input');
        UNTIL (Ch='Y') OR (Ch='N');
        WRITELN('Now input the probability matrix');
        FOR I:=1 TO NStates DO
            BEGIN
                WRITELN('Input the probabilities of transition from');
                WRITELN('state ',I,' to each state');
                REPEAT
                    FOR J:=1 TO NStates-1 DO
                        READ(PTM[I,J]);
                    READLN(PTM[I,NStates]);
                    Sum:=0;
                    FOR J:=1 TO NStates DO
                        Sum:=Sum+PTM[I,J];
                    IF (Sum>1.000001) OR (Sum<0.999999) THEN
                        BEGIN
                            WRITELN('Sum of probabilities must be 1.');
                            WRITELN('Try again.');
                        END;
                UNTIL (Sum<1.000001) AND (Sum>0.999999);
            END;
END;
PROCEDURE Markov(VAR Ab:BOOLEAN;PTM:ARRAY1;
                VAR NoTr:ARRAY3;VAR Last,New:INTEGER);


{This procedure calculates the next state and checks if it}
            {is an absorbing state or not}



VAR {for the procedure Markov}

   I:INTEGER;
   PR,RndHolder:REAL;


BEGIN {Begin the procedure Markov}

   RndHolder:=RANDOM;
   Last:=New;
   I:=1;
   PR:=0;
   WHILE (RndHolder>PR+PTM[Last,I]) DO
       BEGIN
           PR:=PTM[Last,I]+PR;
           I:=I+1;
       END;
```

ALGORITHM 7.1 Markov Simulation Program (Continued)

```
    New:=I;
    NoTr[Last,New]:=NoTr[Last,New]+1;
    IF PTM[New,New]=1 THEN
        Ab:=TRUE;
END;

BEGIN {Begin the main program MarkovProcess}

    WRITELN('Welcome to the Markov Process Generator');
    WRITELN;
    WRITELN('Please enter the number of states in your process');
    REPEAT
        READLN(NStates);
        IF NStates > MaxStates THEN
            WRITELN('Maximum number of states is ',MaxStates);
    UNTIL NStates<=MaxStates;
    InputInfo(InitialProbVector,MarkovArray,NTrials,NStates,DT,TMax);
    FOR K:=1 TO NStates DO
        BEGIN
            FOR J:=1 TO NStates DO
                TotNoTr[K,J]:=0;
        END;
    FOR I:=1 TO NTrials DO
        BEGIN
            FOR K:=1 TO NStates DO
                BEGIN
                    FOR J:=1 TO NStates DO
                        NoOfTransitions[K,J]:=0;
                END;
            Calculate(New,InitialProbVector);
            Time:=0;
            N:=0;
            Absorbing:=FALSE;
            REPEAT
                Markov(Absorbing,MarkovArray,NoOfTransitions,Last,New);
                Time:=Time+DT;
                N:=N+1;
            UNTIL (Time>=TMax) OR (Absorbing);
            FOR K:=1 TO NStates DO
                BEGIN
                    FOR J:=1 TO NStates DO
                        TotNoTr[K,J]:=TotNoTr[K,J]+NoOfTransitions[K,J];
                END;
            OutputInfo(NoOfTransitions,I,N,NStates,New)
        END;
    WRITELN;
    WRITELN('For the entire experiment the transitions matrix was:');
    WRITELN;
    WRITELN;
    WRITE('       ');
    FOR J:=1 TO NStates-1 DO
        WRITE(J,'      ');
    WRITELN(NStates);
    WRITELN;
    FOR J:=1 TO NStates DO
        BEGIN
            WRITE(J,'      ');
            FOR K:=1 TO NStates-1 DO
                WRITE(TotNoTr[J,K],'    ');
            WRITELN(TotNoTr[J,NStates]);
        END;
END.
```

FIGURE 7.3 Output Results

```
Welcome to the Markov Process Generator

Please enter the number of states in your process
*3
Please enter the length of time that you wish each simulation
run to last and the interval between events that you want
*100 1
How many simulation trials do you want
*2
Is the initial state of the process known?
                (Y/N)
*Y
Now input the initial state
*1
Now input the probability matrix
Input the probabilities of transition from
state 1 to each state
*0.75 0.00 0.25
Input the probabilities of transition from
state 2 to each state
*0.30 0.40 0.30
Input the probabilities of transition from
state 3 to each state
*0.00 0.48 0.52
For trial 1 the number of transitions was 100
The final state of the process was state 3
and the transitions matrix was:

      1     2     3

1     39    0     9
2     8     8     7
3     0     15    14


For trial 2 the number of transitions was 100
The final state of the process was state 2
and the transitions matrix was:

      1     2     3

1     31    0     10
2     9     14    5
3     0     15    16


For the entire experiment the transitions matrix was:

      1     2     3

1     70    0     19
2     17    22    12
3     0     30    30
```

7.6 SUMMARY

A Markov process is a system whose future probabilistic behavior depends only on the present state and not on how it arrived at its present state. Knowing the system's history does not change the probability of any future transitions. A Markov system or model is completely determined by the number of states and its probability transition matrix. Simulations of Markov processes are often carried out to determine the time history of the states of the system, that is, the transition from state to state over time. In this chapter, we presented several examples of Markov chains and demonstrated the principles of simulating these systems. In addition, a simulation algorithm in Pascal was presented for student use.

7.7 REVIEW PROBLEMS

1. Describe a Markov process.
2. What are the classifications of Markov processes?
3. Simulate the naive subject example. Assume that the subject starts in state 2. What is the probability that he becomes nonconforming? What is the probability that he becomes conforming?
4. Simulate the repeated balloting process. In this case, define states 1 and 5 as majority states. What is the expected number of ballots needed to reach a majority state? What is the probability of reaching state 1? Assume that the starting state is state 2. Repeat assuming the starting state is state 3.
5. Use the data of the social mobility system. What is the expected number of transitions for a person in the unskilled social class to reach the professional and higher administrative class?
6. Define the infinitesimal probability transition matrix.
7. Simulate the survival after treatment of cancer system, assuming the starting state is 3. What is the probability of the person reaching state 1? What is the expected number of transitions to reach state 1?
8. Can we use a Markov model to predict the standing of sports teams? Why or why not?
9. Can foreign policy decisions be looked at as a Markov chain? Discuss your reasoning.
10. List some other examples of Markov processes.
11. Derive the probability transition matrix of the repeated balloting example.
12. Derive the probability transition matrix for a three-user slotted ALOHA system with $p = 0.1$ and $\alpha = 0.2$. By simulation, determine the mean number of blocked users at equilibrium. Define equilibrium as the time after which the probabilities of being in a particular state do not change with time.
13. Consider the computer network shown in Figure 7.4. The probability that a message in computer A is routed to computer B is p and to node G is $1 - p$. In general, when a message is at a computer it is forwarded clockwise with probability p and counterclockwise with probability $1 - p$.

FIGURE 7.4 Problem 13

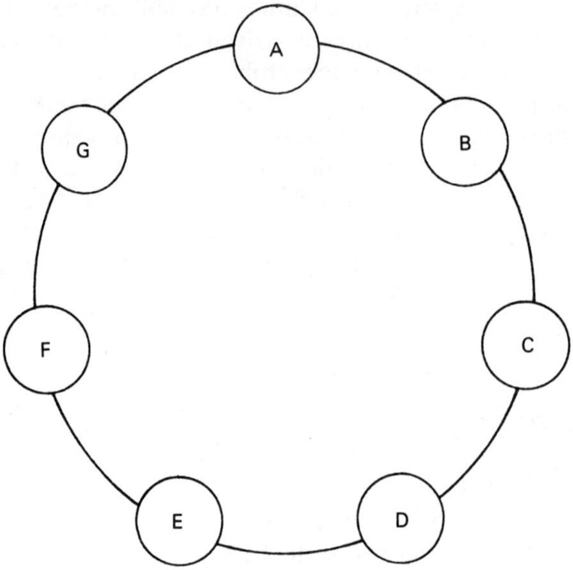

a. Can you model this as a Markov chain? If yes, define the states and the probability transition matrix. If no, specify why not. Assume that when a message reaches its destination, it is removed from the network.

b. Using simulation, determine the average number of hops (links) a message travels before it reaches its destination. Assume that all destinations are equally likely for any message entering the network.

7.8 REFERENCES

1. Aburdene, M.F., and Kochenburger, R.J., "MARKOV: A General Purpose Simulation Program," 1974 Winter Simulation Conference, Jan. 1974, pp. 729–30.
2. Asch, S.E., *Social Psychology* (Englewood Cliffs, NJ: Prentice-Hall, 1952).
3. Bartholomew, D.J., *Stochastic Models for Social Processes*, 3rd ed. (New York: Wiley, 1982).
4. Bharacha-Reid, A.T., *Elements of the Theory of Markov Processes and Their Applications* (New York: McGraw-Hill, 1960).
5. Carleil, A.B., and Hellman, M.E., "Bistable Behavior of ALOHA-Type Systems," *IEEE Transactions on Communication*, vol. COM-23, no. 4 (April 1975): 401–10.
6. Cohen, B.P., "A Probability Model for Conformity,"*Sociometry* 21, no. 1 (January 1958): 69–81.
7. Daellenbach, H.G.; George, J.A.; and McNickle, D.C., *Introduction to Operations Research Techniques*, 2nd ed. (Boston: Allyn & Bacon, 1983).

8. Glass, D.V., and Hall, J.R., "Social Mobility in Great Britain: A Study of Inter-Generation Changes in Status," in *Social Mobility in Britain*, edited by D.V. Glass (London: Routledge & Kegan Paul, 1954), pp. 177–254.

9. Kemeny, J.G., and Snell, L.J., *Mathematical Models in the Social Sciences* (Lexington, MA: Ginn & Co., 1962).

10. Kreweras, G., "A Model for Opinion Change during Repeated Balloting," in *Readings in Mathematical Social Science*, edited by P.F. Lazarsfeld and N.W. Henry (Chicago: Science Research Associates, 1966).

11. Parzen, E., *Stochastic Processes* (San Francisco: Holden-Day, 1967).

12. Paris, S.J., "Measuring Social Mobility," *Journal of the Royal Statistical Society* A 118 (1955): 56–66.

13. Render, B., and Stair, R.M., Jr., *Quantitative Analysis for Management*, 2nd ed. (Boston: Allyn & Bacon, 1985).

14. Zahl, S., "A Markov Process Model for Follow-up Studies," *Human Biology* 27, no. 2 (May 1955): 90–120.

8 System Dynamics and Integration

8.1 INTRODUCTION

System dynamics, originally called industrial dynamics, is a method for simulating complex dynamic systems. Any engineering or social system of interest to us, as simulation analysts, is bound to be complex. Accordingly, to simulate a system successfully, we must have a systematic approach to the simulation. The basic procedure presented in Chapter 1 involved six steps:

1. Define the system boundary.
2. Quantify the interactions among the system's components.
3. Formulate a computer simulation for the lumped model.
4. Run the computer simulation program.
5. Verify the simulation program.
6. Validate the simulation model.

The validity of any simulation depends on careful completion of these steps. Defining a system's boundary is usually straightforward. Quantifying the interactions within the system, however, is often the most difficult step. In this chapter, we will cover several concepts and tools that are necessary to simulate a system, focusing on steps 3 and 4. As mentioned earlier, simulation is defined as model construction and model use.

Feedback loops are fundamental parameters of dynamic systems. Understanding them is important because they determine the dynamic behavior of the system. The system dynamics approach concentrates on two variables to express the model of the system: the *level* and the *rate*. Level variables represent accumulations over time of whatever is being measured. The volume of ocean water is a level. System levels describe the conditions or states of the system at any point in time. Levels are called stocks in economics and state variables in engineering.

Rates describe the inflow to and outflow from a system, changing the system's levels. The melting rate of the polar ice cap would be a rate affecting the ocean water level. A rate is the derivative of the state, referred to as $\dot{x}$ in Chapter 3. Numerical integration techniques are used to compute levels from rates.

The system dynamics approach applies only to continuous systems. Often, however, we are interested in the aggregate behavior of discrete systems and we treat these systems as continuous. For example, students' arrivals at a professor's office are discrete events, but when we view the system as continuous, we aggregate the events and represent the flow of students as a rate.

This chapter will present methods that are implemented in commercially available simulation languages. Through examples, we will gain a comprehensive understanding of the properties of those languages, which will allow us to use them most effectively. In addition, we will study the results of the simulation of dynamic systems, which illustrate the various types of behavior of linear and nonlinear second-order systems.

8.2 INTEGRATION AND EXPONENTIATION

The mathematical process that relates a level to inflows and outflows over time is called *integration*. The concept is basic to the simulation of continuous dynamic systems. We use integration because it reflects processes that occur frequently in natural and social phenomena.

Suppose an individual decides to put his money in a savings account. The bank's interest rate is r; interest is compounded periodically (e.g., daily, weekly, monthly, quarterly). If we let

$m(t)$ = the amount of money in the savings account at any time t (a level)

r = the interest rate (a rate)

Δt = the compounding period

$m(t + \Delta t)$ = the amount of money in the savings account at time $t + \Delta t$

then

$$m(t + \Delta t) = m(t) + m(t)r\Delta t \tag{8.1}$$

The term $m(t)r\Delta t$ is the amount of increase in money in the period Δt due to interest. Quantity r is assumed to be fixed, but it could be a function of time, that is, $r(t)$. Furthermore,

$$m(t + 2\Delta t) = m(t + \Delta t) + m(t + \Delta t)r\Delta t \tag{8.2}$$

The recursive process can be repeated to find the amount of money in the savings account at any desired point in time. Quantity $\Delta m(t)$ is called the *first forward difference* of $m(t)$, relative to Δt, and is defined by

$$\Delta m(t) = m(t + \Delta t) - m(t) \tag{8.3}$$

From equation (8.1),

$$\Delta m(t) = m(t)r\Delta t$$

By iteration, the *second forward difference* is defined as

$$\begin{aligned}\Delta^2 m(t) &= \Delta(\Delta m(t)) \\ &= \Delta(m(t+\Delta t) - m(t)) \\ &= \Delta m(t+\Delta t) - \Delta m(t) \\ &= m(t+2\Delta t) - m(t+\Delta t) - m(t+\Delta t) + m(t)\end{aligned}$$

$$\Delta^2 m(t) = m(t+2\Delta t) - 2m(t+\Delta t) + m(t) \quad \textbf{(8.4)}$$

And so on.

Because they relate the difference of $m(t)$ at different times, equations (8.3) and (8.4) are called *difference equations*. The difference between the largest and smallest coefficients of Δt when the equations are written in the form of equation (8.3) is called the *order* of the equations. Equation (8.3) is of order 1, equation (8.4) of order 2. A linear difference equation involves no products or nonlinear functions of $m(t)$ and its differences. These difference equations define the behavior of a discrete dynamic system. In most applications, $\Delta m(t)$ can be readily obtained. From this information, we can calculate $m(t + k\Delta t)$, where $k = 0, 1, 2, \ldots$. For example, $m(t + k\Delta t)$ can be shown graphically as in Figure 8.1 with $t = 0$ and $\Delta t = 1$.

Customarily, the amount of money in a savings account is calculated only at regular intervals. Given these discrete values, we want to know the amount of money in the account between interest accrual dates. Graphically, the question is how to connect the calculated data points. Two possible schemes are shown in Figure 8.2.

One scheme, indicated by the dashed line, adds interest at the end of each period. This is normally the way in which the increasing principal is recorded by

FIGURE 8.1 Graphical Representation of Money in a Savings Account

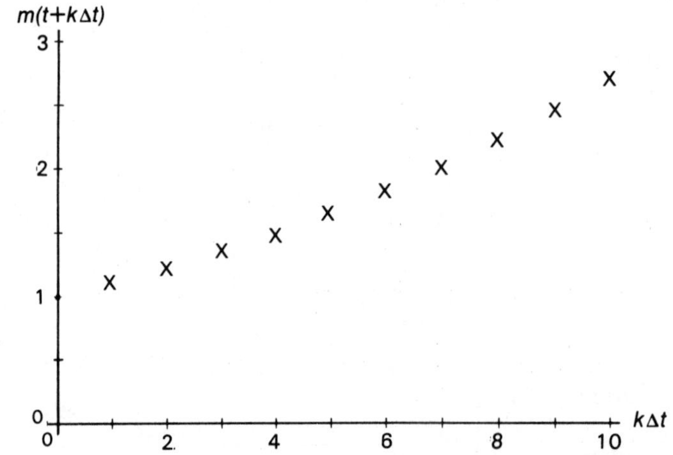

FIGURE 8.2 Two Schemes for Showing $m(t + k\Delta t)$

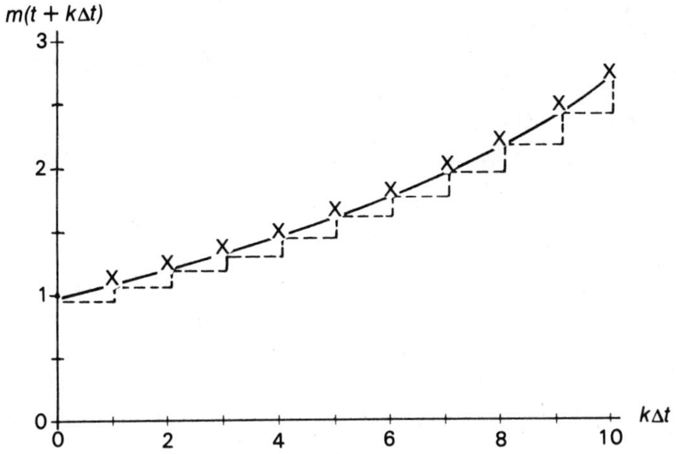

the bank. The interest at point $t + k\Delta t$ is given by $m(t + (k-1)\Delta t)r\Delta t$, where $k \geq 1$. An alternative is a smooth curve, which can be understood by a closer inspection of the equation $m(t + \Delta t) = m(t)r\Delta t$. By taking Δt to be small, and solving the equation recursively, we obtain a smooth exponential curve (see Section 8.18). The time rate of increase of money in the savings account, $\Delta m(t)/\Delta t$, can be obtained by dividing both sides of equation (8.3) by Δt, resulting in

$$\frac{\Delta m(t)}{\Delta t} = \frac{m(t + \Delta t) - m(t)}{\Delta t} \tag{8.5}$$

If Δt is allowed to approach zero, equation (8.5) can be expressed mathematically as

$$\lim_{\Delta t \to 0} \frac{\Delta m(t)}{\Delta t} = \frac{dm(t)}{dt} = \dot{m}(t) \tag{8.6}$$

Quantity $\dot{m}(t)$, the derivative of $m(t)$ with respect to t, is the rate of the level $m(t)$. Graphically, $m(t)$ evaluated for a specific value of t is the slope of the line in Figure 8.2 at t.

From equations (8.6) and (8.3),

$$\dot{m}(t) = \frac{dm(t)}{dt} = m(t)r \tag{8.7}$$

Equations of the form of equation (8.7) are called first-order linear *differential* equations. As we stated in Chapter 3, they describe the behavior of continuous dynamic systems. In the bank interest example, this is called continuous compounding.

Again, in most cases $\dot{m}(t)$ can be determined for all time from the initial value of $m(t)$ and r. In simulation, we want to know the value of $m(t)$ for all t within some interval of time.

Equation (8.7) defines a positive feedback loop, where $m(t)$ is the level and r is the rate. In this example, the amount of money in the savings account is used to calculate the interest for the next time instant. Thus, the interest is feeding back positively, increasing the associated level, that is, the principal.

Rectangular integration, also called Euler integration, is a simple algorithm for computing $m(t)$. The process is well suited to computer implementation. For finite Δt's, we can calculate $m(t)$ from knowledge of $\dot{m}(t)$ by the following process. If $m(t_0)$ is the initial deposit in the savings account, then $\dot{m}(t_0)$ is known and given by equation (8.7). Table 8.1 shows the sequence of calculations for approximating $m(t)$.

Table 8.1 demonstrates an integration or accumulation process. The mathematical notation for the area under the curve from t_0 to t is

$$m(t) = \int_{t_0}^{t} \dot{m}(t)\, dt \qquad \textbf{(8.8)}$$

Quantity $\dot{m}(t)$ is called the integrand.

Looking at Figure 8.3, the integration process can be described graphically as the summation of the area under the graph of $\dot{m}(t)$ if Δt is very small ($\Delta t \rightarrow 0$). The value of $m(t_1)$ is equal to the value of $m(t_0)$ plus the value of the shaded area A_1. The unshaded area B_1 is the error in calculating $m(t)$ up to the time t_1. The value of $m(t_2)$ is equal to the value of $m(t_1)$ plus the shaded are A_2, and so on. The smaller the Δt, the more accurate the estimate of $m(t)$ will be, but at the expense of more calculations that consume more computer time. The error in computing $m(t)$ is due to the assumption that $\dot{m}(t)$ is constant over the interval Δt, as was discussed in Chapter 2.

This type of integration, the rectangular or Euler method, is one of many computational techniques of numerical integration that are commonly used and implemented in commercially available simulation packages. An algorithm for Euler integration is presented in the next section.

TABLE 8.1 Sequence of Integration Scheme

t	$\dot{m}(t)$	$m(t)$
t_0	$\dot{m}(t_0)$	$m(t_0)$
t_1	$\dot{m}(t_1)$	$m(t_1) = m(t_0) + \dot{m}(t_0)\Delta t$
t_2	$\dot{m}(t_2)$	$m(t_2) = m(t_1) + \dot{m}(t_1)\Delta t$
.	.	.
.	.	.
.	.	.

FIGURE 8.3 Graphical Integration

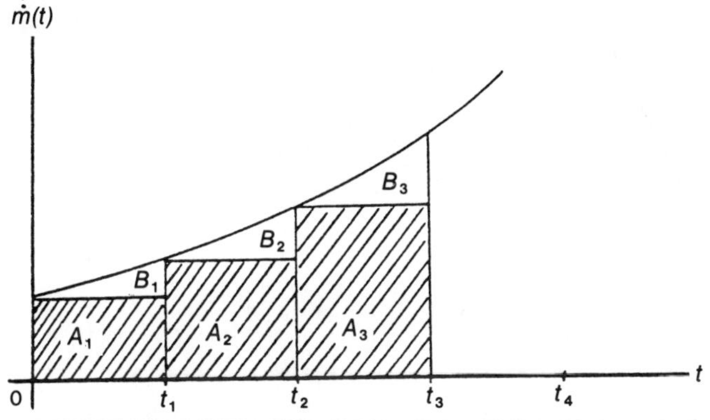

8.3 EULER INTEGRATION ALGORITHM

The Euler integration algorithm can be used for the simulation of continuous systems on a digital computer. Algorithm 8.1, presented in Pascal, is capable of integrating integrands, denoted by Input. The integrals (levels or state variables) are denoted by the variable Out, and the independent variable of integration T (usually time) is incremented by DT (Δt) units.

A main program calls the Euler subprogram by providing the values of N and DT, and initial values of T, Input, and Out. On subsequent calls, the Euler algorithm is provided with the new values of Input and calculates the new values of T and Out. The Euler method is implemented in many simulation packages.

ALGORITHM 8.1 Euler Method

```
PROCEDURE EulerIntegration(Input:VectorType; VAR Out:VectorType;
 {Input and Out are one dimensional arrays.  In has the rates you wish to
 integrate and Out will have the levels corresponding to those rates.}
                           N:INTEGER; {the number of levels}
                           DT:REAL; {the integration increment}
                           VAR T:REAL);


VAR
    I:INTEGER;


BEGIN
    T:=T+DT;
    FOR I:=1 TO N DO
        Out[I]:=Out[I]+Input[I]*DT;
END;
```

FIGURE 8.4 Graph of $f(x)$

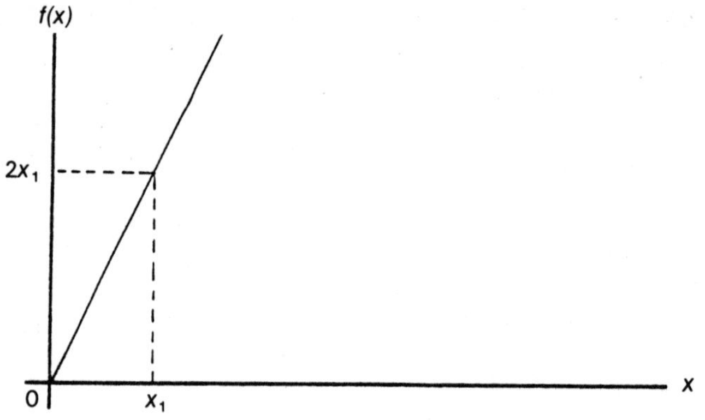

EXAMPLE 8.1 Integrate the function $f(x) = 2x$ using the Euler method. The graph of $f(x)$ is shown in Figure 8.4.

Solution From elementary geometry, we know that the area under the graph is given by $\frac{1}{2}x_1 f(x_1) = x_1^2$. The computed values of the area under $f(x)$ for several values of Δx (increment of integration) are shown in the table below. The percent error is defined as

$$\% \text{ error} = \frac{\text{theoretical value} - \text{computed value}}{\text{theoretical value}} \times 100$$

x_1	Δx	$\int_0^{x_1} f(x)\,dx$	% Error
1.000	0.001	1.001	−0.10
1.000	0.010	1.010	−1.00
1.000	0.025	1.025	−2.50
1.000	0.050	1.050	−5.00
1.000	0.100	1.100	−10.00

The percent error in this example, as in most cases, is a linear function of Δx. A method for choosing the integration increment is presented in Section 8.13. Section 8.20 covers another popular integration method, the Runge-Kutta.

8.4 DEFENSE EXPENDITURES MODEL

In this and the following sections, we will show how one man's convictions were served by the use of simulation. Through this example, we will demonstrate model development procedure and present the solutions of a dynamic system described by second-order differential equations.

Lewis Fry Richardson, a Quaker, was opposed to war. He felt, however, that it would be easier to stop wars if he had an understanding of their origins. His approach was to study the defense budgets of Russia, France, Germany, and the Austro-Hungarian empire for the years before World War I. Table 8.2 shows the defense budgets for the pre–World War I years. The plot of the rate of increase of defense budgets versus the total defense budgets is linear.

The regression equation using the rate of increase of defense spending as the dependent variable, $\Delta x(t)/\Delta t$, and total defense spending as the independent variable, $x(t)$, is

$$\frac{\Delta x(t)}{\Delta t} = 0.73(x(t) - 194) \tag{8.9}$$

Thus, Richardson hypothesized that the rate of increase in defense spending is proportional to the total defense spending. That is, there is positive feedback in the military funding process. In general, this can be expressed as

$$\frac{\Delta x(t)}{\Delta t} = kx(t)$$

given the boundary condition $x(t_0) = x_0$. The total defense spending at time t can be expressed as

$$x(t + \Delta t) = x(t) + \Delta x(t) = x(t) + kx(t)\Delta t$$

This is a first-order system, where $x(t)$ is the level and $kx(t)$ is the rate, and is of the same form as equation (8.1).

The simulation of this process exhibits exponential growth. Total defense spending increases by a factor of e (approximately 2.718) every $1/k$ time units. This is a basic characteristic of exponential growth, and is illustrated in Table 8.3. This model assumes a never-ending source of funds for defense expenditures and is therefore not entirely realistic.

TABLE 8.2 Defense Budgets in £10^6 Sterling

	1909	*1910*	*1911*	*1912*	*1913*
Russia	66.7	50.9	57.1	81.8	74.7
France	48.6	68.5	70.7	63.2	92.0
Germany	63.1	62.0	62.5	68.2	95.4
Austria-Hungary	20.8	23.4	24.6	25.2	26.9
	199.2	204.8	214.9	238.4	289.0
Rate of increase		5.7	10.1	23.5	50.6

Source: Lewis F. Richardson, *Arms and Insecurity*, edited by Nicolas Rashevsky and Ernesto Trucco (Pittsburgh: Boxwood Press, 1960), p. 32.

TABLE 8.3

Time	*Total Defense Spending (£10⁶)*
1909	199.2
1910	204.8
1911	214.9
1912	238.7
1913	289.0

8.5 SECOND-ORDER SYSTEM: AN ARMS RACE MODEL[15]

Richardson then proposed an arms race model for a two-nation (group) system as follows. If x_1 is the total defense expenditure for nation 1, and x_2 is the total defense expenditure for nation 2, then the time rate of change of defense expenditures for each nation is proportional to the "threat factor" of the opposing nation. This leads to

$$\dot{x}_1 = \frac{dx_1(t)}{dt} = kx_2(t) \tag{8.10}$$

$$\dot{x}_2 = \frac{dx_2(t)}{dt} = lx_1(t) \tag{8.11}$$

where k and l are called the defense coefficients and are a measure of how seriously each nation views military spending by a possible adversary. The system description is a set of two coupled linear first-order differential equations, containing two feedback loops. The feedback can be positive or negative, depending on the signs of k and l. This system is a linear second-order system. The system states or levels are $x_1(t)$ and $x_2(t)$, and the rates are $kx_2(t)$ and $lx_1(t)$.

The behavior of this second-order system depends on the values of k, l, $x_1(t_0)$, and $x_2(t_0)$. If k and l are both greater than zero, then the system exhibits exponential growth as shown in Figure 8.5 and the analytic solution is given by

$$x_1(t) = x_1(t_0)\left[\left(\frac{1}{2} - \frac{1}{2}\sqrt{\frac{k}{l}}\right)e^{-\sqrt{kl}t} + \left(\frac{1}{2} + \frac{1}{2}\sqrt{\frac{k}{l}}\right)e^{\sqrt{kl}t}\right] \tag{8.12}$$

$$x_2(t) = x_2(t_0)\left[\left(\frac{1}{2} - \frac{1}{2}\sqrt{\frac{l}{k}}\right)e^{-\sqrt{kl}t} + \left(\frac{1}{2} + \frac{1}{2}\sqrt{\frac{l}{k}}\right)e^{\sqrt{kl}t}\right] \tag{8.13}$$

The growth of $x_1(t)$ and $x_2(t)$ cannot continue forever. In the past, growth of defense expenditures has flattened out after declarations of war and when domestic politics forced reconsideration of the trade-off between guns and butter. Thus, the model should include economic burden/society fatigue coefficients α

FIGURE 8.5 Exponential Growth of Defense Expenditures vs. Time

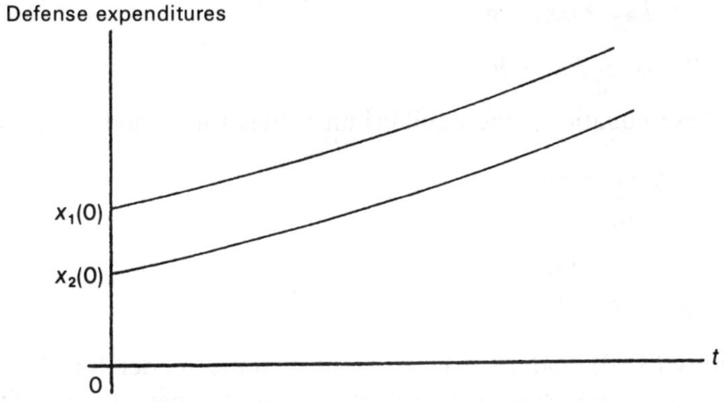

and β. With this inclusion, the system equations become

$$\dot{x}_1 = \frac{dx_1(t)}{dt} = kx_2(t) - \alpha x_1(t) \quad \textbf{(8.14)}$$

$$\dot{x}_2 = \frac{dx_2(t)}{dt} = lx_1(t) - \beta x_2(t) \quad \textbf{(8.15)}$$

To further complete the Richardson model, grievance or goodwill factors are added to the time rate of change of defense expenditures. The system equations then become

$$\dot{x}_1 = \frac{dx_1(t)}{dt} = kx_2(t) - \alpha x_1(t) + g \quad \textbf{(8.16)}$$

$$\dot{x}_2 = \frac{dx_2(t)}{dt} = lx_1(t) - \beta x_2(t) + h \quad \textbf{(8.17)}$$

We will assume g and h to be constant. A positive g or h is called the grievance coefficient since it increases the rate of arms spending regardless of the other nation's defense expenditures. When g or h is negative, it is called the goodwill coefficient since it decreases the rate of arms spending regardless of the other nation's defense expenditures.

8.6 EQUILIBRIUM CONDITIONS

The state of equilibrium is reached when the rates of the system equal zero, as we defined equilibrium in Chapter 2. Using the arms race model, the equilibrium conditions are given by

$$0 = kx_2 - \alpha x_1 + g \tag{8.18}$$

$$0 = lx_1 - \beta x_2 + h \tag{8.19}$$

From these equations, the equilibrium values for x_1 and x_2 are shown to be

$$x_1 = \frac{g\beta + kh}{\alpha\beta - kl} \tag{8.20}$$

$$x_2 = \frac{lg + h\alpha}{\alpha\beta - kl} \tag{8.21}$$

Equations (8.20) and (8.21) represent a static model for the arms race. If g changes, the equilibrium values change. The manner in which they change over time cannot be evaluated. Therefore, as we can see, simulation would not be necessary for static systems. However, equilibrium values will help us verify, and possibly validate, the simulation model.

8.7 SOLUTION OF THE ARMS RACE MODEL

In this section, we will present the solution for the arms race model, demonstrating the dependence of the dynamic behavior upon equations (8.14) and (8.15). The solutions to equations (8.16) and (8.17) have the same form, but the equilibrium conditions, when they exist, are different.

If α, β, k, l, g, and h are allowed to take on any values, the solutions of equations (8.16) and (8.17) fall into one of six categories. Each category represents a characteristic type of system behavior. Within each of these categories, the system behavior can be either free or forced. When g and h are both equal to zero, the system is said to be free. This means that the system's internal feedback rates are proportional only to the levels of the system, and system behavior depends on the initial values of the states. When g and h are nonzero, the system is said to be forced, with g and h being the "driving" or "forcing" functions. In this case, the rates are not only proportional to the system levels, but are also dependent on g and h. The behavior of a forced system before it reaches equilibrium (transient time response) is basically the same as it would be if the system were not forced. The difference is that the system's equilibrium position may be changed as a result of forces g and h.

Category 1: Undamped or Oscillatory System

If $\alpha = -\beta$ and $kl - \alpha\beta < 0$, the system is said to be an undamped system. The behavior of such a system is oscillatory and, accordingly, its simulation should show an oscillatory response for $x_1(t)$ and $x_2(t)$. The time between two successive maxima is called the period and is given by $2\pi/\sqrt{\alpha\beta - kl}$. The values of x_1 and x_2 oscillate around an equilibrium value.

FIGURE 8.6 Oscillatory Response

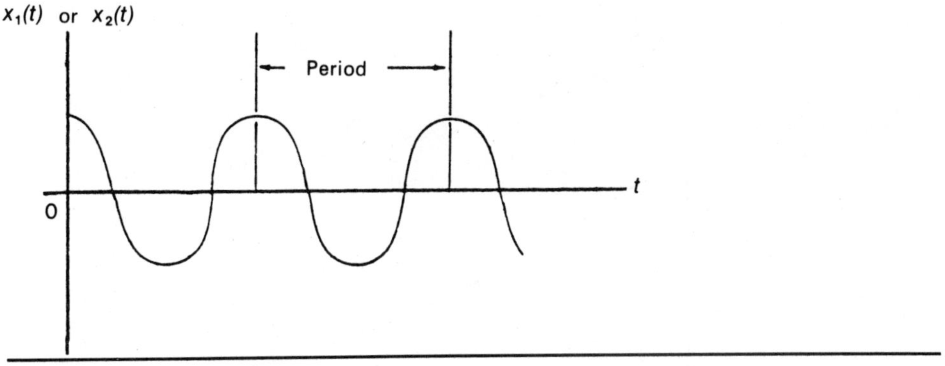

The general plot of a solution is shown in Figure 8.6. This is similar to the simple harmonic motion example of Chapter 1, where air resistance and other friction factors were assumed to be zero. An oscillatory system, theoretically, never achieves equilibrium but continues to oscillate forever. A system with both an ideal capacitor and inductor with an initial voltage on the capacitor exhibits similar time response behavior.

Category 2: Damped Oscillatory System

If $\alpha + \beta > 0$ and $(\alpha + \beta)/2 < \sqrt{\alpha\beta - kl}$, the system is said to be a damped oscillatory system. The response is not truly oscillatory, but the interval between two sucessive maxima or minima is still called the period. The general form of the response is shown in Figure 8.7. The response will eventually reach the equilibrium value computed from equations (8.20) and (8.21). The mass-spring system of Chapter 1 could exhibit this behavior when the effects of air resistance are included in the model. A series RLC circuit can have a similar time response, depending on the values of R, L, and C.

FIGURE 8.7 Damped Oscillatory Response

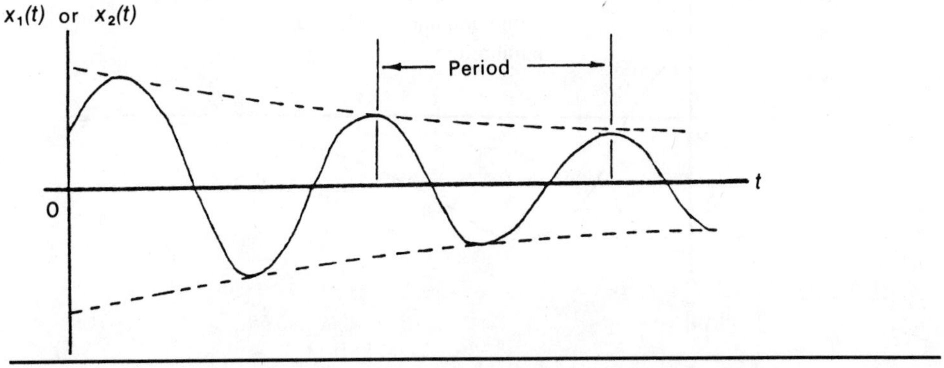

FIGURE 8.8 Critically Damped Response

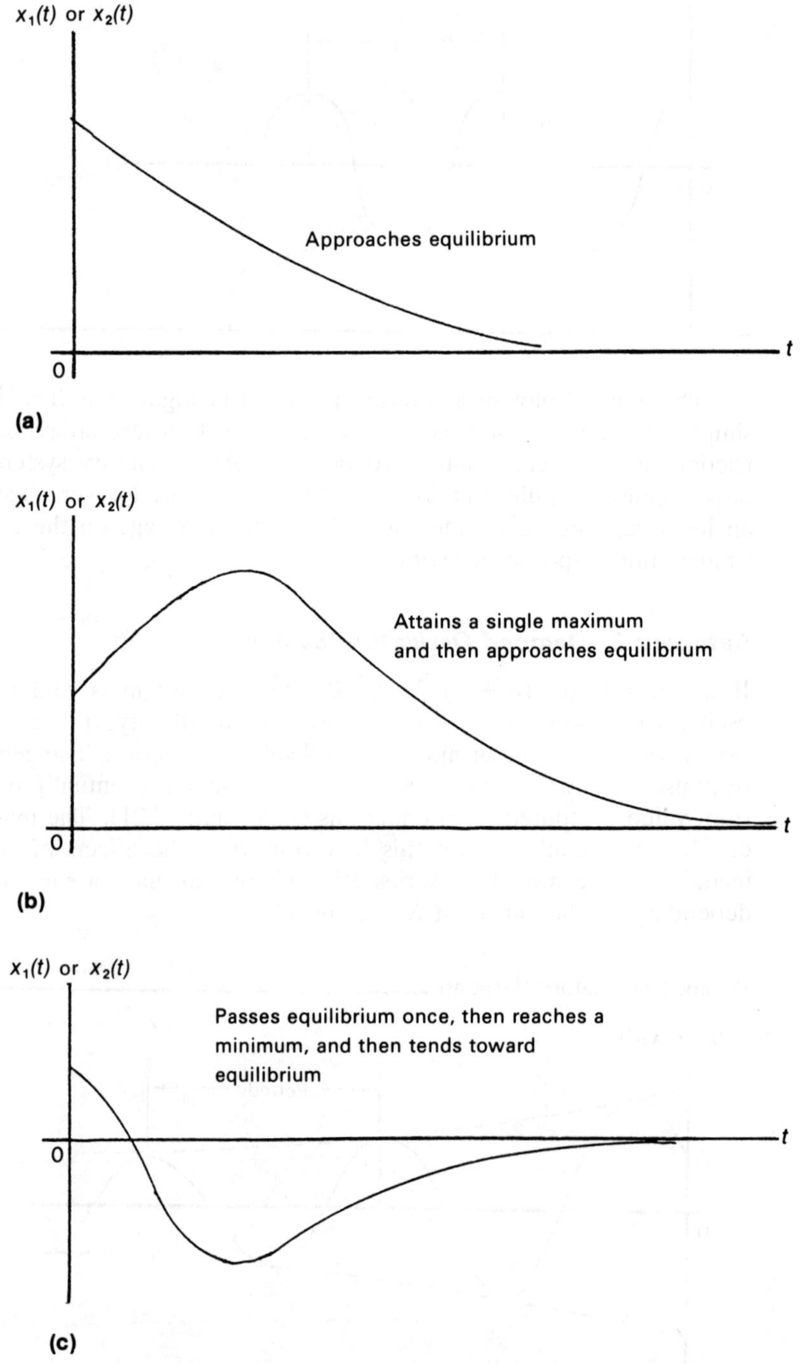

Category 3: Critically Damped System

If $a + \beta > 0$ and $(\alpha + \beta)/2 = \sqrt{\alpha\beta - kl}$, the system is said to be critically damped. The response is no longer oscillatory, but a slight decrease in the value of $\alpha + \beta$ will change the response to that of category 2. This is a borderline case for oscillations. Depending on the initial conditions, the system exhibits one of the three responses shown in Figure 8.8.

Category 4: Overdamped System

If $(\alpha + \beta)/2 > \sqrt{\alpha\beta - kl} > 0$, the system is said to be overdamped. The response of this system will not oscillate even if $\alpha + \beta$ is slightly decreased. The general form of the system response is similar to that of category 3.

Category 5: Explosive Oscillatory System

If $\alpha + \beta < 0$ and $(\alpha + \beta)/2 < \sqrt{\alpha\beta - kl}$, the system is said to be explosive oscillatory. The general form of the system response is shown in Figure 8.9.

Category 6: Explosive System

If $\alpha + \beta < 0$ and $(\alpha + \beta)/2 > \sqrt{\alpha\beta - kl}$, the system is said to be explosive (exhibits exponential growth). The general form of the solution is shown in Figure 8.10. The arms race example, as specified by equations (8.10) and (8.11), is said to have this response.

System dynamics is concerned with describing the behavior of systems. In the arms race example, the behavior of what initially appeared to be a simple system is, in fact, complex and varies greatly from category to category. We often use simulation to answer the question, What if α, β, k, l, g, or h were to change? We often compare the behavior of more complex systems (higher-order systems, i.e., those with more levels or states) with the behavior of second-order systems.

FIGURE 8.9 Explosive Oscillatory Response

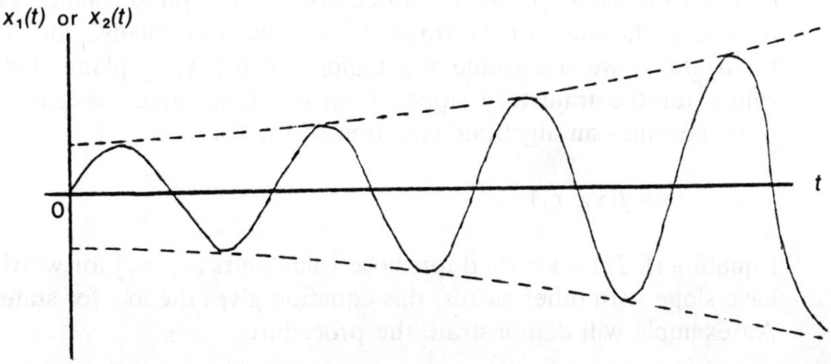

FIGURE 8.10 Explosive Response

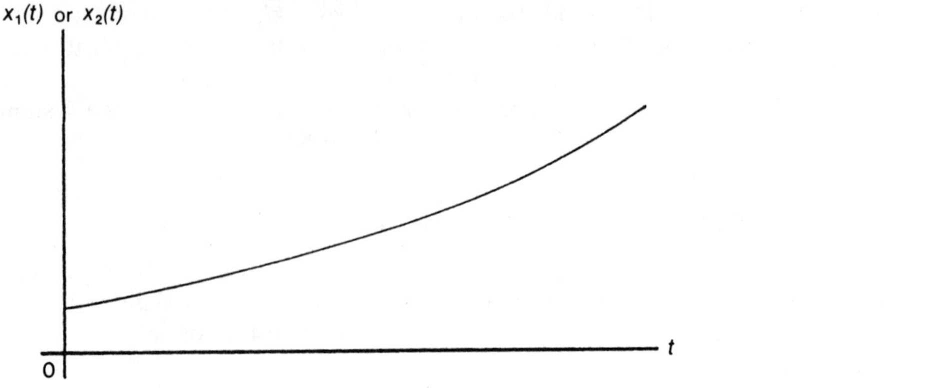

To demonstrate the six possible categories, we suggest the student use Algorithm 8.1 to simulate the behavior of this interesting system. For those who are patient and do not like programming, we will simulate this system in Chapter 9, using commercially available packages.

8.8 SYSTEM TRAJECTORIES

The response of the systems considered thus far has been illustrated by a plot of the levels versus time. Another useful approach is to plot the level variables against each other, with time as a parameter. Such a plot is called the system's trajectory or phase plane plot. In the arms race example, a plot of total defense expenditures by nation 2 (x_2) versus total defense expenditures by nation 1 (x_1) in the x_1-x_2 plane would be the trajectory. The x_1-x_2 plane is called the phase plane. An obvious approach to plotting the trajectory is to obtain the time response of $x_1(t)$ and $x_2(t)$ and to plot $x_2(t)$ versus $x_1(t)$ for all time. The process is repeated for all possible initial conditions of $x_1(0)$ and $x_2(0)$.

A second way to obtain the phase plane plot is by the *iso-cline* method, which is based on the slope of the trajectories in the phase plane. The ratio dx_2/dx_1 represents the slope of the trajectories in the x_1-x_2 plane. Aided by the equation for dx_2/dx_1, we determine the regions of the x_1-x_2 plane that have the same values for the trajectory slope. With constant values assigned to dx_2/dx_1, the slope becomes an algebraic equation of the form

$$s = f(x_1, x_2) \tag{8.22}$$

Equation (8.22) is satisfied for those value pairs (x_1, x_2) for which the trajectories have slope s. In other words, this equation gives the loci for states of equal slope. An example will demonstrate the procedure.

EXAMPLE 8.2 Let

$$\frac{dx_1}{dt} = x_2 \tag{8.23}$$

$$\frac{dx_2}{dt} = -0.25x_1 - x_2 \tag{8.24}$$

The equilibrium conditions are $x_1 = x_2 = 0$. Thus, the trajectories should approach the origin of the x_1-x_2 plane. This can be shown by forming the following equation:

$$\frac{dx_2/dt}{dx_1/dt} = \frac{dx_2}{dx_1} = \frac{x_2}{-0.25x_1 - x_2} \tag{8.25}$$

Letting $s = dx_2/dx_1$ and solving for x_2 results in

$$x_2 = \frac{-0.25x_1}{s+1} \tag{8.26}$$

The slope of the trajectory described by equation (8.26) is equal to s. Some representative iso-clines are shown in Figure 8.11. The short line segments along the iso-cline indicate the slope associated with the iso-cline.

Figure 8.11 also shows the trajectory of x_2 versus x_1 for many possible initial conditions. This type of sketch can be used to qualitatively compare the results either with analytic solutions or with solutions obtained by simulation.

FIGURE 8.11 System Trajectories

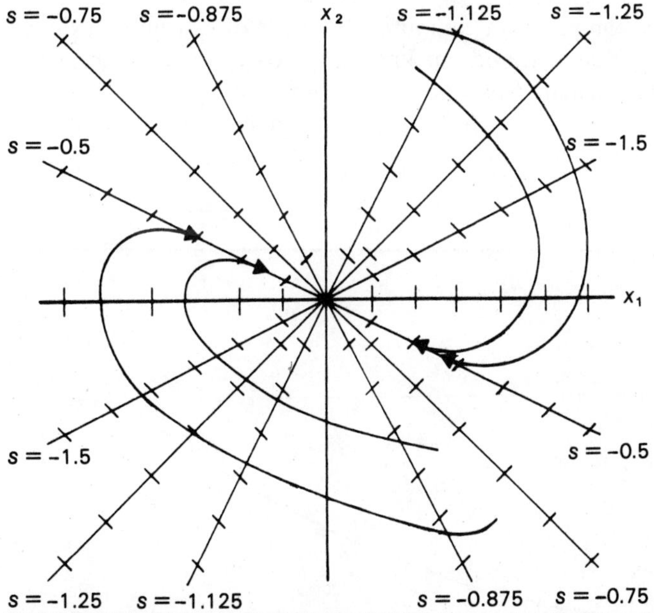

8.9 TIME-VARYING SYSTEMS

In our analysis up to this point, we have been assuming that our system has been time-invariant. That is, the values of k, l, α, and β have been assumed to be constant. If, instead, they are functions of time, then the system is said to be time-varying. Time-varying systems are a good example of systems for which simulation is an ideal tool because of the complexity of analytic solutions.

8.10 NONLINEAR SYSTEMS

If the feedback in a system is a function of the system's states or levels, then the system's behavior is nonlinear. Solutions of nonlinear systems are extremely difficult to obtain, making simulation the best tool for determining system behavior.

For example, suppose that α is a function of x_1, as shown in Figure 8.12. Analytic solutions for this nonlinear system do not generally exist. However, the system's behavior can be readily simulated. To implement Figure 8.12 in simulation, two techniques can be used: the regression technique, as presented in Chapter 4, and a tabular technique, which will be discussed shortly.

The regression technique computes a relationship between α and x_1. In our example, the relationship would be

$$\alpha = 0.001x_1^2$$

The simulation program computes the value of α from knowledge of x_1. From the computed value of α and our knowledge of k, l, x_1, and x_2, the feedback rates $\dot{x}_1$ and $\dot{x}_2$ are calculated for the next Δt interval. The change in x_1 is $\dot{x}_1 \Delta t$, and the change in x_2 is $\dot{x}_2 \Delta t$. The process is repeated as many times as is required to complete the simulation.

FIGURE 8.12 Plot of α vs. x_1

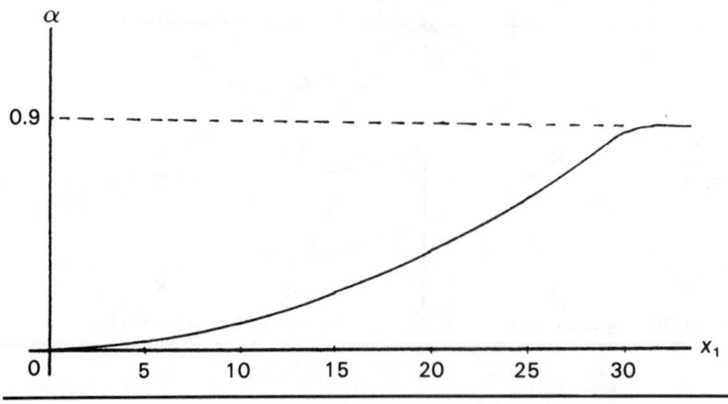

TABLE 8.4

Table Entry	x_1	α
1	0	0.000
2	5	0.025
3	10	0.100
4	15	0.225
5	20	0.400
6	25	0.625
7	30	0.900

The tabular method places the value of x_1 and α in a table, as shown in Table 8.4. This technique is often implemented in simulation languages. From the initial value of x_1, the value of α is calculated by linear interpolation in the table. For example, if x_1 has a value of 14, then α is given by

$$\alpha = 0.1 + \frac{0.225 - 0.1}{15 - 10}(14 - 10) = 0.2$$

Algorithm 8.2, presented in Pascal, can be used for this purpose.

ALGORITHM 8.2 Table Algorithm and Test Algorithm in Pascal

```
PROCEDURE Table(VAR X,Y:ArrayType; {ArrayType must be declared in
                                     your main program as an array with
                                     indices from 0 to some maximum}
                XT:REAL;VAR YT:REAL;DelX,XMax,XMin:REAL;
                TableRead:BOOLEAN);

VAR
    NumberOfEntries,I,Place:INTEGER;

BEGIN
    NumberOfEntries:=TRUNC((XMax-XMin)/Delx);
    IF NOT TableRead THEN
        BEGIN
            WRITELN('Input the pairs of X and Y values in ascending');
            WRITELN('order of X');
            FOR I:=0 TO NumberOfEntries DO
                READLN(X[I],Y[I]);
            YT:=0;
        END
    ELSE
        BEGIN
            IF XT<=XMin THEN
                YT:=Y[0]
            ELSE IF XT>=XMax THEN
                YT:=Y[NumberOfEntries]
            ELSE
                BEGIN
                    Place:=TRUNC((XT-XMin)/DelX)+1;
                    YT:=Y[Place-1]+((Y[Place]-Y[Place-1])/DelX)*
                                    (XT-X[Place-1]);
                END;
        END;
END;
```

continues

ALGORITHM 8.2 Table Algorithm and Test Algorithm in Pascal (Continued)

```
PROGRAM TESTER(INPUT,OUTPUT);
TYPE
   ARRAYTYPE=ARRAY[0..100] OF REAL;
VAR
   XT,YT,DELX,XMAX,XMIN:REAL;
   X,Y:ARRAYTYPE;
PROCEDURE Table(VAR X,Y:ArrayType; {ArrayType must be declared in
                                     your main program as an array with
                                     indices from 0 to some maximum}
                XT:REAL;VAR YT:REAL;DelX,XMax,XMin:REAL;
                TableRead:BOOLEAN);


VAR
    NumberOfEntries,I,Place:INTEGER;


BEGIN
    NumberOfEntries:=TRUNC((XMax-XMin)/Delx);
    IF NOT TableRead THEN
        BEGIN
            WRITELN('Input the pairs of X and Y values in ascending');
            WRITELN('order of X');
            FOR I:=0 TO NumberOfEntries DO
                READLN(X[I],Y[I]);
            YT:=0;
        END
    ELSE
        BEGIN
            IF XT<=XMin THEN
                YT:=Y[0]
            ELSE IF XT>=XMax THEN
                YT:=Y[NumberOfEntries]
            ELSE
                BEGIN
                    Place:=TRUNC((XT-XMin)/DelX)+1;
                    YT:=Y[Place-1]+((Y[Place]-Y[Place-1])/DelX)*
                                 (XT-X[Place-1]);
                END;
        END;
END;
BEGIN
    WRITELN('INPUT XT,XMIN,XMAX,DELX');
    READLN(XT,XMIN,XMAX,DELX);
    TABLE(X,Y,XT,YT,DELX,XMAX,XMIN,FALSE);
    WHILE 2>1 DO
        BEGIN
            READLN(XT);
            TABLE(X,Y,XT,YT,DELX,XMAX,XMIN,TRUE);
            WRITELN(YT);
        END;
END.
```

Now, let us look at both a linear and a nonlinear case of the Richardson arms race model. Let α and $\beta = 0.2$, k and $l = 0.1$, $g = 2.0$, $h = 1.0$, $x_1(0) = 2.0$, and $x_2(0) = 1.0$. Figure 8.13 is a plot of $x_1(t)$ and $x_2(t)$ versus time for the linear case. The simulation of the nonlinear system, using α values from Table 8.4, is shown in Figure 8.14. Note that the nonlinear system reaches equilibrium at a faster rate than the linear system, and that the equilibrium values are different.

FIGURE 8.13 Results of Linear Simulation

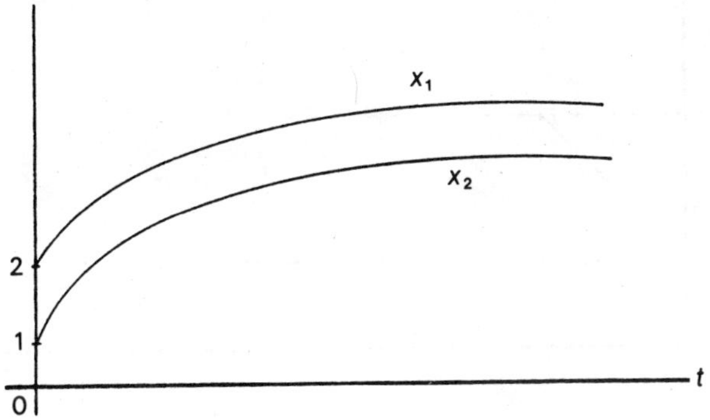

FIGURE 8.14 Results of Nonlinear Simulation

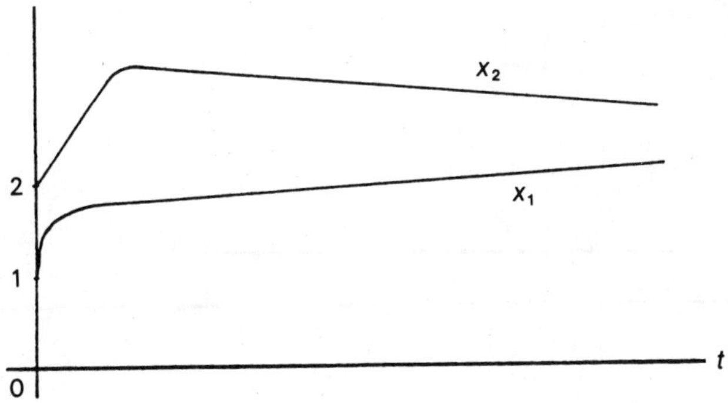

8.11 LIMITING VALUES

Thus far, our analysis of the arms race model has assumed that defense expenditures can be unlimited. In reality, the defense expenditure of nations is limited, and our model must reflect this fact. Let $x_{1(\max)}$ be the maximum allowable defense expenditure for nation 1. This limit introduces nonlinearity into the model. The limiting relationship can be shown by Figure 8.15. This relationship is often implemented in simulation languages and is called the clip or limit function. Engineers call it saturation.

Let us illustrate the behavior of an overdamped system. If $x_{\max} = 10$, the simulation results are plotted as shown in Figure 8.16. Note that the defense expenditures for nation 2 continue to grow for the duration of the simulation instead of reaching an equilibrium as in the case where we placed no limit on nation 1's defense expenditures.

FIGURE 8.15 Limiting Values for x_1

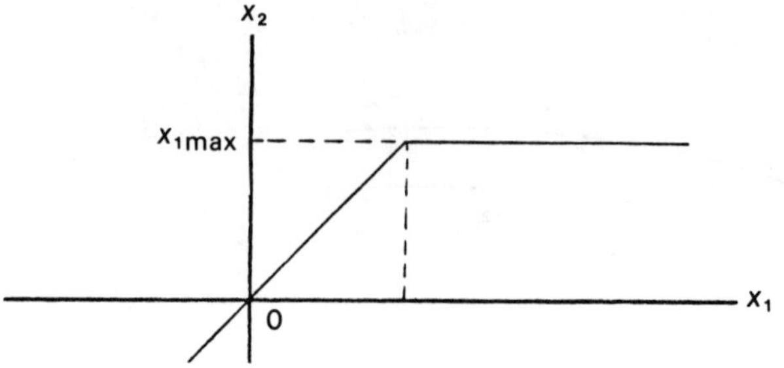

FIGURE 8.16 Simulation Results with Limiting Values

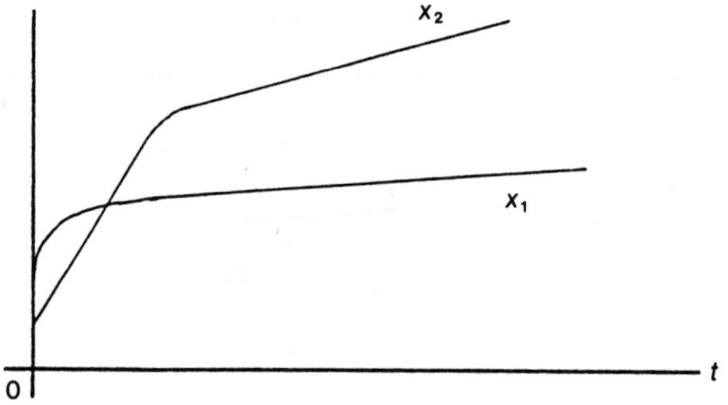

8.12 TIME DELAYS IN CONTINUOUS SYSTEMS

The assumption that an increase in the defense expenditures of nation 2 is immediately known by nation 1 is not necessarily correct. Knowledge depends on the intelligence network of nation 1. The same is true of nation 2 with respect to nation 1. Also, implementation of policy is not instantaneous. In general, there are two types of delay: transportation delay and continuous delay. Transportation delay occurs when there exists an interval of time over which no action occurs. At the end of this interval, the full response is enacted. This phenomenon is shown graphically in Figure 8.17. Algorithm 8.3, presented in Pascal, is used for simulation of transportation delays.

Continuous delay refers to a lag that takes place across an interval, starting at a low level initially and gradually building to its full effect. An illustration of continuous delay is the response of a first-order system. Recall that a first-order system is described by $\dot{y} = ay + bu$. The response of this system is shown graphically in Figure 8.18 for $a < 0$ and $b > 0$. The input is u and the output is y.

FIGURE 8.17 Transportation Delay

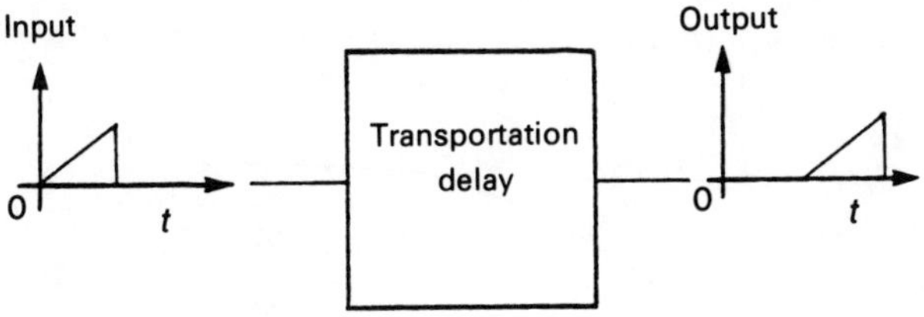

ALGORITHM 8.3 Transportation Delay Algorithm in Pascal

```
PROCEDURE TransportationDelay(Delay,Input:REAL;
                              VAR DelayedInput:REAL;
                              VAR Start:BOOLEAN;
                              VAR Point:INTEGER;
                              VAR S:ArrayType);

{ArrayType must be defined in the main program as an array of
 REALs indexed from 1 to some maximum which should be larger than
                  any expected (Delay/DT)}

VAR
    N,I:INTEGER;

BEGIN
    N:=TRUNC(Delay/DT);
    {DT is assumed to be defined in the main program}
{as the time increment used throughout the main program}

    IF Start THEN
        BEGIN
            Start:=FALSE;
            Point:=1;
            FOR I:=1 TO N DO
                S[I]:=Input
        END
    ELSE
        BEGIN
            DelayedInput:=S[Point];
            S[Point]:=Input;
            Point:=Point+1;
            IF Point>N THEN
                Point:=1;
        END;
END;
```

FIGURE 8.18 Continuous Time Delay

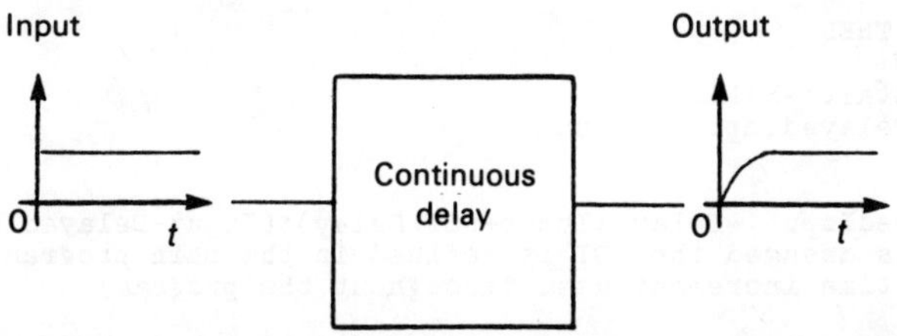

FIGURE 8.19 Response of an Overdamped System When a Continuous Time Delay Is Introduced

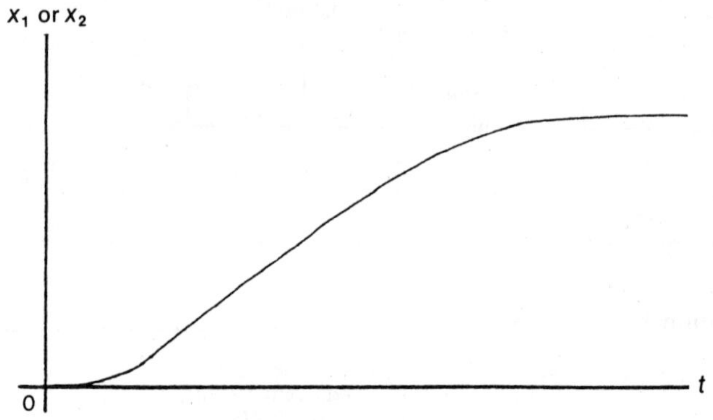

The delay is often taken to be the time it takes the system to reach 50% of its final value.

For the overdamped second-order system of the arms race model, the response of the system is as shown in Figure 8.19 when g and h have positive values. Inspection of Figures 8.18 and 8.19 around the time origin shows that the initial change is less for a second-order system. In general, the higher the order of the system, the greater the delay, until the continuous delay approximates a transportation delay. To implement an nth-order delay, DEL, we can use a cascade of n first-order delays, with each delay equal to DEL/n. Algorithm 8.4 implements a first-order delay in Pascal.

The first-order delay can also be used as a smoothing function, as shown in Figure 8.20. The input change is abrupt at time zero. However, the output appears as a smooth curve and the change is slower depending on the value of the delay.

Now that we have studied the implementation of delays, let us investigate the effects of delays on system behavior. By way of a specific example, the over-

ALGORITHM 8.4 First-Order Delay Algorithm in Pascal

```
PROCEDURE FirstOrderDelay(Input,Delay:REAL;
                          VAR DelayedInput:REAL;
                          VAR Start:BOOLEAN);

BEGIN
    IF Start THEN
        BEGIN
            Start:=FALSE;
            DelayedInput:=Input
        END
    ELSE
        DelayedInput:=DelayedInput+(DT/Delay)*(Input-DelayedInput);
        {It is assumed that DT is defined in the main program as}
        {the time increment used throughout the program}
END;
```

FIGURE 8.20 Use of Delay to Smooth a Function

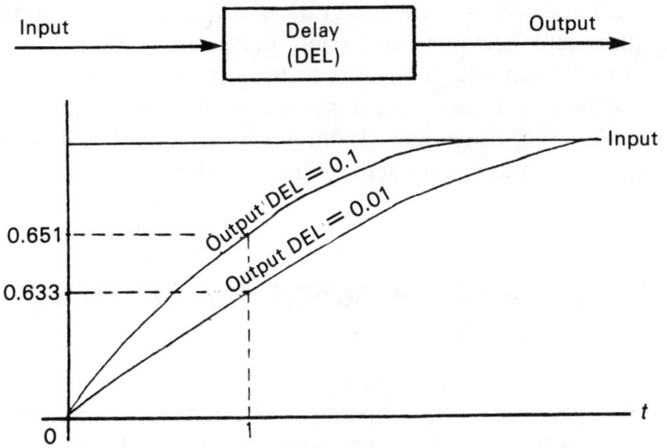

damped arms race model was simulated using a first-order delay of one year. The simulation's form is

$$\frac{dx_1}{dt} = kx_2(t) - \alpha\ \text{DELAY}(x_1(t), 1) + g$$

$$\frac{dx_2}{dt} = l\ \text{DELAY}(x_1(t), 1) - \beta x_2(t) + h$$

The simulation results are shown in Figure 8.21. As can readily be seen, the introduction of delays changed the behavior of the system from an overdamped stable system to an explosive system. In general, the presence of time delays contributes to system oscillation or explosiveness.

FIGURE 8.21 Effects of Delay on an Overdamped Arms Race Model

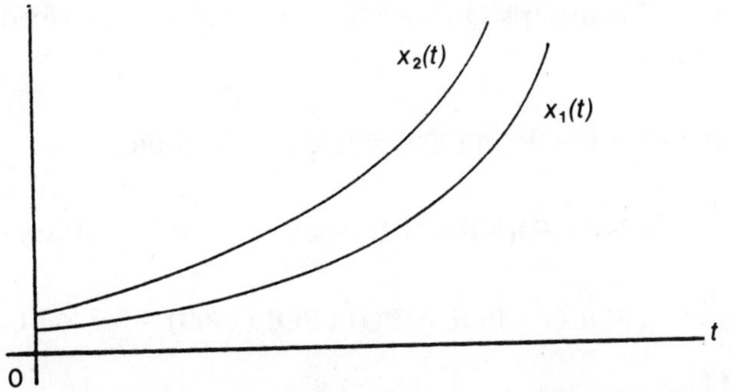

8.13 CHOICE OF INTEGRATION INCREMENT

The choice of the size of the integration increment Δt requires a compromise between numerical accuracy and computer time. A small Δt gives better numerical accuracy than a larger value, at the expense of more computer time. Too large a Δt causes a nonexplosive system to become explosive, as we saw in Chapter 2. As a rule of thumb, Δt should be less than half the smallest delay in the system, where delay is loosely defined as the level divided by the rate.

8.14 SIMULATION OF LARGE-SCALE SOCIAL SYSTEMS

The simulation model of a large-scale social system is a statement of a social theory. Translation of social theory to the computer model usually requires inclusion of a large number of variables and a large amount of data. Policy changes often produce counterintuitive results in the long term. Feedback loops form the basic structural elements of dynamic systems. As was mentioned earlier, levels and rates are the fundamental components of the loops. Rates are statements of policy.

In 1971, Jay W. Forrester published his book *World Dynamics*.[10] The model he presented there has received many critical reviews, both positive and negative, by experts in the social sciences. The social theory used to develop this *World* model is based on three fundamental assumptions:

1. The physical characteristics of the world system are described by exponential growth or death.
2. The world's ecosystem cannot sustain exponential growth.
3. The world society recognizes the problems of overpopulation and shrinking natural resources after a long delay.

Forrester considered the pertinent variables of the world system to be population (P), natural resources (NR), pollution (POL), capital investment (CI), and capital investment fraction in agriculture (CIAF). The model is described by a system of five nonlinear, coupled difference equations. A simplified version of Forrester's DYNAMO model can be described by the following five state equations:

$$\frac{d\mathrm{P}}{dt} = (\mathrm{BR} - \mathrm{DR})\mathrm{P};\ \mathrm{P}(1900) = 1.65 \times 10^9 \text{ people}$$

$$\frac{d\mathrm{NR}}{dt} = -(\mathrm{NRUR})(\mathrm{NR});\ \mathrm{NR}(1900) = 9 \times 10^{11} \text{ arbitrary units}$$

$$\frac{d\mathrm{POL}}{dt} = (\mathrm{POLG} - \mathrm{POLA})\mathrm{POL};\ \mathrm{POL}(1900) = 0.2 \times 10^9 \text{ arbitrary units}$$

$$\frac{d\mathrm{CI}}{dt} = (\mathrm{CIG} - \mathrm{CID})\mathrm{CI};\ \mathrm{CI}(1900) = 0.4 \times 10^9 \text{ arbitrary units}$$

$$\frac{d\text{CIAF}}{dt} = \frac{\text{CFIFR} \times \text{CIQR} - \text{CIAF}}{\text{CIAFT}} \text{CIAF; CIAF(1900)}$$
$$= 0.2 \text{ arbitrary unit}$$

where the auxiliary variables are defined as

BR = birth rate, people/year
DR = death rate, people/year
NRUR = natural resource usage rate, arbitrary units
POLG = pollution generation, units/year
POLA = pollution absorption, units/year
CIG = capital investment generation, units/year
CID = capital investment discard, units/year
CFIFR = capital fraction as determined by food ratio
CIQR = capital investment in agriculture as a function of the ratio of material quality of life to food quality of life
CIAFT = adjustment time for capital investment fraction devoted to agriculture

Table 8.5 shows the five levels and a measure called the "quality of life" (QL), which are plotted versus time in Figure 8.22. Quality of life is based on food availability, crowding, pollution, and the availability of material goods.

Perhaps the other most widely reported large-scale system simulation study is Forrester's *Urban Dynamics* model.[8] Clark and Cole present a detailed survey and comparative study of global simulation models.[4] The better-known national large-scale econometric models are the Brookings model, the Wharton model, and the Federal Reserve model. The paper by Adams and Burmeister describes some econometric models in detail.[1]

TABLE 8.5 Simulation Results of World Model Using Rectangular Integration Method

T	*P*	*NR*	*CI*	*CIAF*	*POL*	*QL*
1900	1.65E + 09	9.00E + 11	4.00E + 08	0.200	2.00E + 08	0.611
1910	1.67E + 09	8.94E + 11	6.16E + 08	0.208	4.07E + 08	0.808
1920	1.86E + 09	8.85E + 11	8.95E + 08	0.210	5.82E + 08	0.927
1930	2.15E + 09	8.73E + 11	1.26E + 09	0.215	8.22E + 08	1.013
1940	2.49E + 09	8.57E + 11	1.73E + 09	0.224	1.14E + 09	1.055
1950	2.87E + 09	8.36E + 11	2.32E + 09	0.238	1.57E + 09	1.055
1960	3.23E + 09	8.09E + 11	3.03E + 09	0.256	2.12E + 09	1.019
1970	3.68E + 09	7.76E + 11	3.84E + 09	0.280	2.93E + 09	0.982
1980	4.14E + 09	7.36E + 11	4.74E + 09	0.295	4.28E + 09	0.943
1990	4.58E + 09	6.91E + 11	5.69E + 09	0.307	6.08E + 09	0.889
2000	4.95E + 09	6.41E + 11	6.65E + 09	0.317	8.47E + 09	0.814

FIGURE 8.22 Simulation Results of Forrester's World Model

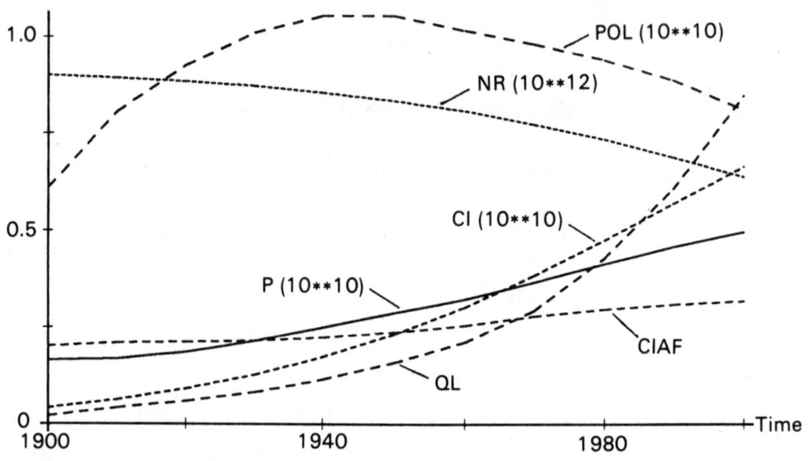

8.15 SENSITIVITY

In the simulation studies of this chapter, we have assumed that the systems are deterministic, that their parameters are known and fixed. But parameters are random variables, with means and variances, and can be known only to within some tolerance.

We should expect the dynamic behavior of a simulation model to be sensitive to the values of the parameters. As seen in the arms race model, a change in the values of coefficients (parameters) can change the behavior of an entire system from overdamped to oscillatory and then to explosive. Naturally, changes in some parameters affect system behavior more than changes in others. We should study the sensitivity of the behavior of simulation models to every parameter in a systematic way.

8.16 SUMMARY

To obtain the solution of continuous simulation models, a simulation language should possess the following capabilities: integration, clipping, delay, tabular interpolation, printing and plotting of data, and the generation of the forcing functions as inputs. In this chapter, we have discussed all of these except the last. We have presented a dynamic model for the arms race, described by a two-level (two-state) system, referred to as a second-order system. The behavior of more complex systems is often compared to the second-order system response. Simulation allows us to ask and solve the What if . . . ? question. In the next chapter, we will look at some commercially available packages for continuous simulation languages.

8.17 REVIEW PROBLEMS

1. What is system dynamics? What are the fundamental building blocks of dynamic systems?
2. Describe integration.
3. The exponential process has been blamed for various troubles in the world. Plot the exponential function $f(x) = e^x$ for values of $x = 1, 2, \ldots, 100$. Can your computer handle numbers as large as e^{100}? What is the ratio e^2/e^1? What is e^{x+1}/e^x?
4. Using rectangular integration, calculate the area under the curve $y = x^2$ for $0 \le x \le 1$. Compare your results with the theoretical value $x^3/3$ for $0 \le x \le 1$. Use $\Delta x = 0.01$, 0.02, 0.03, 0.04, and 0.05. Is your error a linear function of Δx?
5. The following Lotka-Volterra equations describe the interaction of two species in an environment with limited resources:[12,19]

$$\frac{ds_1}{dt} = (b_1 - d_1)s_1 - \alpha s_1 s_2 - c_1 s_1^2$$

$$\frac{ds_2}{dt} = b_2 \alpha s_1 s_2 - d_2 s_2 - c_2 s_2^2$$

where b_1 and b_2 = birth coefficients
d_1 and d_2 = death coefficients
c_1 and c_2 = competition coefficients
α = the probability coefficient of a member of s_1 dying from contact with a member of s_2.

Let $b_1 = 0.2$, $d_1 = 0.1$, $\alpha = 0.005$, $c_1 = 0.01$, $b_2 = 0.2$, $d_2 = 0.1$, $c_2 = 0.05$, $s_1(0) = 15$, and $s_2(0) = 5$.

   a. Construct the system trajectories in the s_1-s_2 plane.
   b. What are the equilibrium populations of s_1 and s_2?
   c. Simulate the system for 30 years.
   d. Simulate the system, but this time limit the population of s_1 to 20.

6. Consider a population, N, composed of couples with advanced degrees. These couples seek employment in rural areas where employment opportunities are limited. The probability of one member of a couple getting a job in Δt units of time is $\alpha\Delta t$. Once one member is employed, the probability of the second member getting a job is increased by $\beta\Delta t$. Each couple can be in one of three states:

   1. Both members are unemployed.
   2. One member is employed and the other is unemployed.
   3. Both members are employed.

The process may be described by three differential equations for the probability of being in state 1, 2, or 3:

$$\frac{dp_1}{dt} = -2\alpha p_1$$

$$\frac{dp_2}{dt} = 2\alpha p_1 - (\alpha + \beta)p_2$$

$$\frac{dp_3}{dt} = (\alpha + \beta)p_2$$

Let $N = 100$, $\alpha = 0.01$, $\beta = 0.01$, $p_1(0) = 1.0$, $p_2(0) = 0$, and $p_3(0) = 0$. Simulate the process and calculate the expected number of employed individuals after five time units.

7. The diffusion equations for a population of two groups, A and B, are

$$\frac{dx_A}{dt} = ax_A(1 - x_A) + bx_B(1 - x_A)$$

$$\frac{dx_B}{dt} = cx_A(1 - x_B) + dx_B(1 - x_B)$$

where x_A and x_B are the proportions of hearers in groups A and B, respectively. Let $a = \frac{5}{6}$, $b = \frac{1}{6}$, $c = \frac{1}{6}$, $d = \frac{5}{6}$, and $x_A(0) = x_B(0) = 0.1$.

a. What do the diffusion equations assume?
b. Simulate the process for five time units.
c. What are the equilibrium values of x_A and x_B?
d. Construct the state trajectories in the x_A-x_B plane.

8. Consider the Richardson model:

$$\frac{dx_1}{dt} = -\alpha x_1 + kx_2 + g$$

$$\frac{dx_2}{dt} = lx_1 - \beta x_2 + h$$

a. Construct a state trajectory for x_1 and x_2 in the x_1-x_2 plane if $g = h = 0$, $\alpha = \beta = 0.2$, and $k = l = 0.1$.
b. Assume that α is function of x_1 and is as shown in Figure 8.23(a), and that β is a function of x_2 and is as shown in Figure 8.23(b). Simulate the system for 20 years if $k = l = 0.1$, $g = 2$, and $h = 1$. Let $x_1(0) = 20$ and $x_2(0) = 10$.
c. Assume that the growth of x_1 cannot exceed 20 units and that the same applies to x_2. Simulate the system assuming the conditions of part (b).
d. Simulate the system if x_1 and x_2 are replaced by a transportation delay of x_1 and x_2 of one time unit.
e. Repeat part (d) but with a continuous delay of one time unit.

9. Assume the following supply and demand functions:

d = demand:	25	20	15	10	5	4
s = supply:	1	4	7	13	22	31
p = price:	1	2	3	4	5	6

a. By simulation, determine the equilibrium price if

$$\frac{\Delta p}{\Delta t} = \frac{1}{2}(d - s)$$

where $p(0) = 2$.

b. Assume that excess supply (demand) causes producers to reduce (increase) the quantity supplied in the next period according to

$$\frac{\Delta s}{\Delta t} = \frac{1}{2}(d - s)$$

Simulate the process if $p(0) = 2$.

FIGURE 8.23 Problem 8

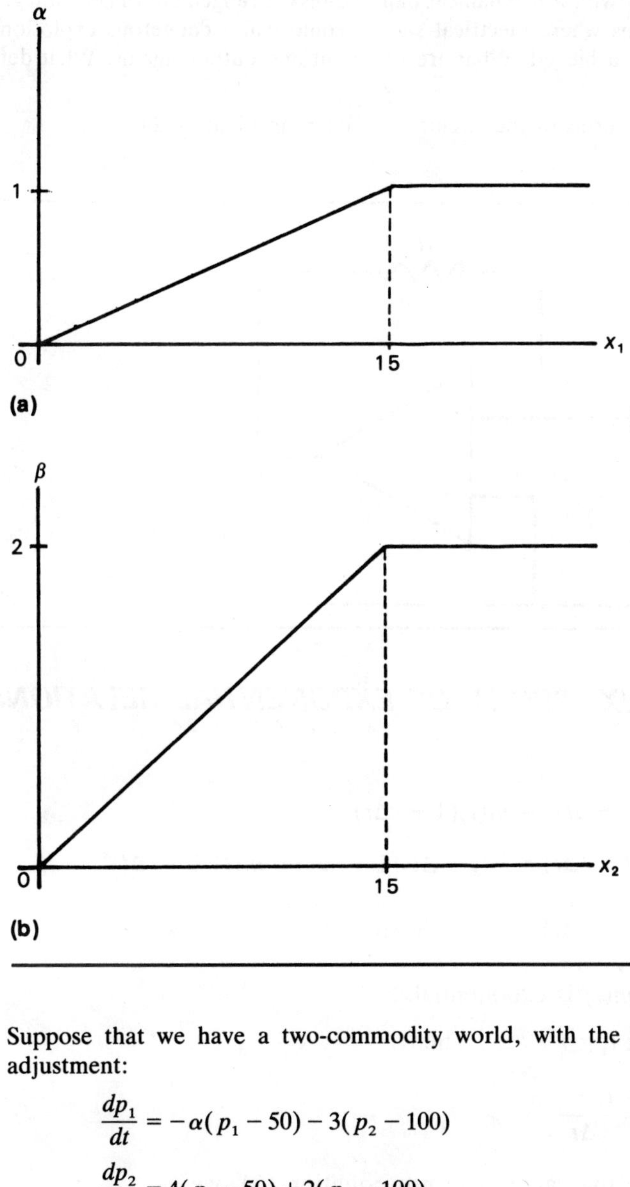

10. Suppose that we have a two-commodity world, with the following equations for price adjustment:

$$\frac{dp_1}{dt} = -\alpha(p_1 - 50) - 3(p_2 - 100)$$

$$\frac{dp_2}{dt} = 4(p_1 - 50) + 2(p_2 - 100)$$

where α is given by the following relationship:

α:	2	2.5	2.8	2.9	3.0	3.1	3.2
p_1:	0	10	20	30	40	50	60

a. Simulate the system with $p_1(0) = 5$ and $p_2(0) = 70$.
b. Repeat with $p_1(0) = 55$ and $p_2(0) = 115$.

11. In analog computers built during the early years (before you were born), integration was accomplished with a mechanical ball-and-disk arrangement. These integrators still appear in applications where electrical signals could cause dangerous explosions. Describe how integration is achieved. What are the input and output signals. What determines the gain constant?

12. Find $v_0(t)$ in terms of the circuit parameters in Figure 8.24.

FIGURE 8.24 Problem 12

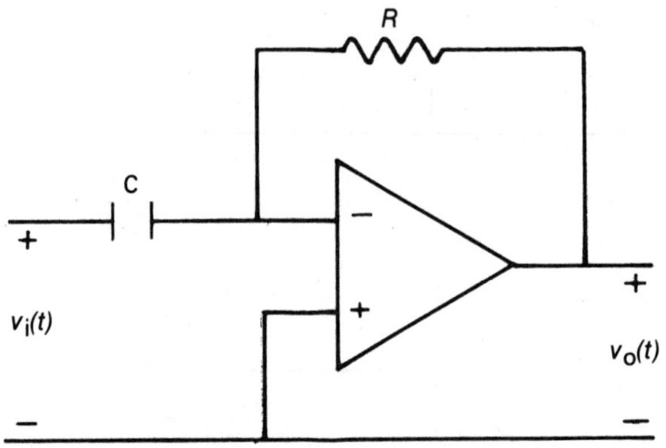

8.18 APPENDIX: PROOF OF EXPONENTIAL RELATIONSHIP

Given:

$$m(t+\Delta t) = m(t)(1+r\Delta t) \tag{1}$$

$$m(t+2\Delta t) = m(t+\Delta t)(1+r\Delta t) = m(t)(1+r\Delta t)^2 \tag{2}$$

$$\vdots$$

$$m(t+k\Delta t) = m(t)(1+r\Delta t)^k \tag{3}$$

Show that $m(t)$ is exponential.

Proof: Let $t_1 = t + k\Delta t$. Then

$$k = \frac{t_1 - t}{\Delta t} \tag{4}$$

Substituting the value of k into equation (3), we get

$$m(t_1) = m(t)(1+r\Delta t)^k \tag{5}$$

Equation (5) can be written as

$$m(t_1) = m(t)a^{r(t_1-t)} \tag{6}$$

where

$$a = (1 + r\Delta t)^{1/r\Delta t} \tag{7}$$

Equation (6) shows that $m(t_1)$ is a smooth exponential even when Δt is finite.

Next, we will show that as $r\Delta t$ approaches zero, a approaches e, and equation (6) becomes

$$m(t_1) = m(t)\exp[r(t_1 - t)] \tag{8}$$

This is easily shown if we let x equal $r\Delta t$. Then equation (7) becomes

$$a = (1 + x)^{1/x} \tag{9}$$

As $r\Delta t$ approaches zero, x approaches zero. Therefore,

$$\lim_{x\to 0} a = a = \lim_{x\to 0} (1 + x)^{1/x} = e \tag{10}$$

Substituting the value of a from equation (10) into equation (6) gives equation (8).

The proof for equation (10) is as follows. We will show that

$$\lim_{x\to 0} (1 + x)^{1/x} = e \tag{11}$$

Let

$$y = \ln(1 + x)^{1/x} = \frac{\ln(1 + x)}{x} \tag{12}$$

and

$$\lim_{x\to 0} y = \lim_{x\to 0} \frac{\ln(1 + x)}{x} \tag{13}$$

$$= \lim_{x\to 0} \frac{\frac{1}{1 + x}}{x} = 1$$

From equation (12), we obtain

$$e^y = (1 + x)^{1/x} = e^1 = e$$

Therefore,

$$\lim_{x\to 0} (1 + x)^{1/x} = e \tag{14}$$

8.19 APPENDIX: SECOND-ORDER SYSTEM RESPONSE AND STATE VARIABLES

Consider the following second-order differential equation:

$$\frac{d^2x(t)}{dt^2} + 2\zeta w_n \frac{dx(t)}{dt} + w_n^2 x(t) = u(t) \tag{1}$$

where w_n = the natural frequency
ζ = the damping factor

We can convert this equation to state variable form if we choose the states to be

$$x_1(t) = x(t)$$
$$x_2(t) = \frac{dx(t)}{dt} \tag{2}$$

Then

$$\dot{x}_1(t) = \dot{x}(t)$$
$$\dot{x}_2(t) = \frac{d}{dt}\frac{dx(t)}{dt} = \frac{d^2x(t)}{dt^2} \tag{3}$$

Solving equation (1) for $d^2x(t)/dt^2$, we obtain

$$\frac{d^2x(t)}{dt^2} = -2\zeta w_n \frac{dx(t)}{dt} - w_n^2 x(t) + u(t) \tag{4}$$

The state equations then become

$$\dot{x}_1(t) = x_2(t)$$
$$\dot{x}_2(t) = -w_n^2 x_1(t) - 2\zeta w_n x_2(t) + u(t)$$

which can be written as

$$\begin{bmatrix} \dot{x}_1 \\ \dot{x}_2 \end{bmatrix} = \begin{bmatrix} 0 & 1 \\ -w_n^2 & -2\zeta w_n \end{bmatrix} \begin{bmatrix} x_1 \\ x_2 \end{bmatrix} + \begin{bmatrix} 0 \\ 1 \end{bmatrix} u(t) \tag{5}$$

The behavior of $x(t)$ versus time is well documented in engineering literature. This is essentially a special case of a method for converting an nth-order differential equation to a system of n first-order equations.

The values of w_n and ζ determine the behavior: w_n determines the speed of the response, and ζ determines the form of the response. Figure 8.25 shows the behavior of this system.

FIGURE 8.25 Response of a Second-Order System

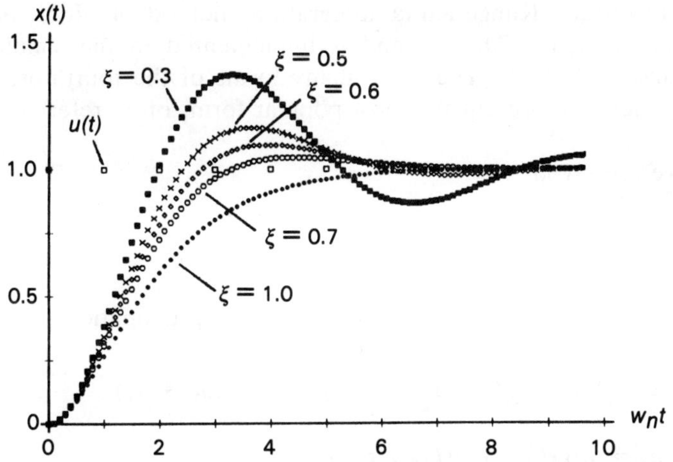

Let us convert the arms race model equations,

$$\dot{x}_1(t) = -\alpha x_1(t) + kx_2(t) \tag{6}$$

$$\dot{x}_2(t) = lx_1(t) - \beta x_2(t) \tag{7}$$

into the form of equation (1). Taking the derivative of both sides of equation (6) results in

$$\ddot{x}_1(t) = -\alpha \dot{x}_1(t) + k\dot{x}_2(t) \tag{8}$$

By substituting equations (6) and (7) into equation (8), we get

$$\ddot{x}_1(t) = -\alpha(-\alpha x_1(t) + kx_2(t)) + k(lx_1(t) - \beta x_2(t))$$

Solving for $x_2(t)$ from equation (6) and substituting this value into equation (8) yields

$$\ddot{x}_1(t) = -(\alpha + \beta)\dot{x}_1(t) - (\alpha\beta - kl)x_1(t)$$

and

$$\ddot{x}_1 + (\alpha + \beta)\dot{x}_1 + (\alpha\beta - kl)x_1(t) = 0$$

This is in the same form as equation (1), with w_n^2 replaced by $(\alpha\beta - kl)$ and $2\zeta w_n = \alpha + \beta$. The conditions on α, β, k, and l in the six categories can now be checked.

8.20 APPENDIX: RUNGE-KUTTA INTEGRATION METHOD

The fourth-order Runge-Kutta integration method is often used to simulate continuous systems. The method is implemented in the simulation languages ACSL and NDTRAN. There are many forms of the fourth-order Runge-Kutta method. Here we present the most popular form, often referred to as the classic form.

If we are given

$$\frac{dx(t)}{dt} = \dot{x}(t) = f(x(t),\ u(t),\ t)$$

then we can compute $x((k+1)\Delta t)$ using the approximation

$$x((k+1)\Delta t) = x(k\Delta t) + \frac{1}{6}(a_1 + 2a_2 + 2a_3 + a_4)$$

where $a_1 = \Delta t f(x(k\Delta t), u(k\Delta t), t = k\Delta t)$

$$a_2 = \Delta t f\left(x\left(k\Delta t + \frac{a_1}{2}\right), u\left(k\Delta t + \frac{1}{2}\Delta t\right), t = \left(k\Delta t + \frac{\Delta t}{2}\right)\right)$$

$$a_3 = \Delta t f\left(x\left(k\Delta t + \frac{a_2}{2}\right), u\left(k\Delta t + \frac{1}{2}\Delta t\right), t = \left(k\Delta t + \frac{\Delta t}{2}\right)\right)$$

$$a_4 = \Delta t f(x(k\Delta t + \Delta t), u(k\Delta t + a_3), t = (k+1)\Delta t)$$

This method is much more accurate than the Euler method for the same integration increment but is a little more difficult to program and uses more computation time. The per-step error can be shown to be of order Δt^5 as compared to the per-step error of order Δt for the Euler method.[11]

8.21 REFERENCES

1. Adams, F.G., and Burmeister, B., "Economic Models," *IEEE Transactions on Systems, Man, and Cybernetics*, vol. SM C-3, no. 1 (January 1973): 19–27.
2. Banks, S.P., *Control Systems Engineering* (Englewood Cliffs, NJ: Prentice-Hall, 1986).
3. Blake, K., and Gordon, G., "System Simulation with Digital Computers," *IBM Systems Journal* 3, no. 1 (September 1964): 14–20.
4. Clark, J., and Cole, S., *Global Simulation Models: A Comparative Study* (New York: Wiley, 1975).
5. Dorf, R.C., *Modern Control Systems*, 4th ed. (Reading, MA: Addison-Wesley, 1986).
6. Elgerd, Olle I., *Control Systems Theory* (New York: McGraw-Hill, 1967).
7. Forrester, J.W., *Industrial Dynamics* (Cambridge, MA: MIT Press, 1961).
8. Forrester, J.W., *Urban Dynamics* (Cambridge, MA: MIT Press, 1967).
9. Forrester, J.W., *Principles of Systems* (Cambridge, MA: Wright-Allen, 1968).

10. Forrester, J.W., *World Dynamics* (Cambridge, MA: Wright-Allen, 1971).

11. James, M.L.; Smith, G.M.; and Wolford, J.C., *Applied Numerical Methods for Digital Computation with FORTRAN and CSMP* (New York: Dun-Donnelly, IEP, 1977).

12. Lotka, A.J., *Elements of Physical Biology* (New York: Dover, 1956).

13. Martin, F.F., *Computer Modeling and Simulation* (New York: Wiley, 1968).

14. Naylor, Thomas H.; Balintfy, Joseph L.; Burdick, Donald S.; and Chu, Kong, *Computer Simulation Techniques* (New York: Wiley, 1966).

15. Richardson, Lewis F., *Arms and Insecurity*, edited by Nicolas Rashevsky and Ernesto Trucco (Pittsburgh: Boxwood Press, 1960).

16. Ross, Shepley L., *Differential Equations* (New York: Blaisdell, 1964).

17. Sage, Andrew P., *Methodology for Large-Scale Systems* (New York: McGraw-Hill, 1977).

18. Schmidt, J.W., and Taylor, R.E., *Simulation and Analysis of Industrial Systems* (Homewood, IL: Richard D. Irwin, 1970).

19. Volterra, V., *Memoria della Royale Accademia Nazionale dei Lincei*, vol. 2, no. 31 (1926). Translated by M.E. Wells in *Animal Ecology*, edited by R.N. Chapman (New York: McGraw-Hill, 1931).

9 Continuous-System Simulation Languages

9.1 INTRODUCTION

We have defined a continuous-time system as one in which the behavior develops continuously over time and can occur at every instant of time. We also used differential equations and state variable formulations to develop a lumped model for continuous systems. In the state variable formulations, the derivatives of the states were the important relationships. In Chapter 8, we saw that the integration operation relates the rates to the states and is the fundamental operation needed for continuous simulation. Therefore, it should not be surprising that integration is the basic component of continuous-system simulation languages (CSSLs), such as ACSL[11], CSMP[9,14], DYNAMO[12], and NDTRAN.[6] In this Chapter, we will discuss the principles of operation of CSSLs and simulate the examples of Chapter 8 using these languages.

9.2 PRINCIPLES OF OPERATION

We will illustrate the principles of CSSLs using the arms race model presented in Chapter 8. We described our model by the following equations:

$$\begin{aligned} \dot{x}_1 &= -\alpha x_1 + kx_2 + g \\ \dot{x}_2 &= lx_1 - \beta x_2 + h \\ y_1 &= x_1 \\ y_2 &= x_2 \end{aligned} \tag{9.1}$$

where β, k, and l = the constants
α = the parameter
g and h = the inputs
$\dot{x}_1$ and $\dot{x}_2$ = the derivatives
x_1 and x_2 = the states
y_1 and y_2 = the outputs
$x_1(0) = x_{10}$ and $x_2(0) = x_{20}$ = the initial conditions

CSSLs are basically implementations of Figure 9.1. They differ, however, in the implementation of steps 9, 10, 11, and 13. In step 9, some languages store all the values of the states, derivatives, outputs, and time-dependent relationships. Others store only the values for the plot or print intervals. Some store this information in a file while others store it in memory.

FIGURE 9.1 Steps in Continuous-System Simulation

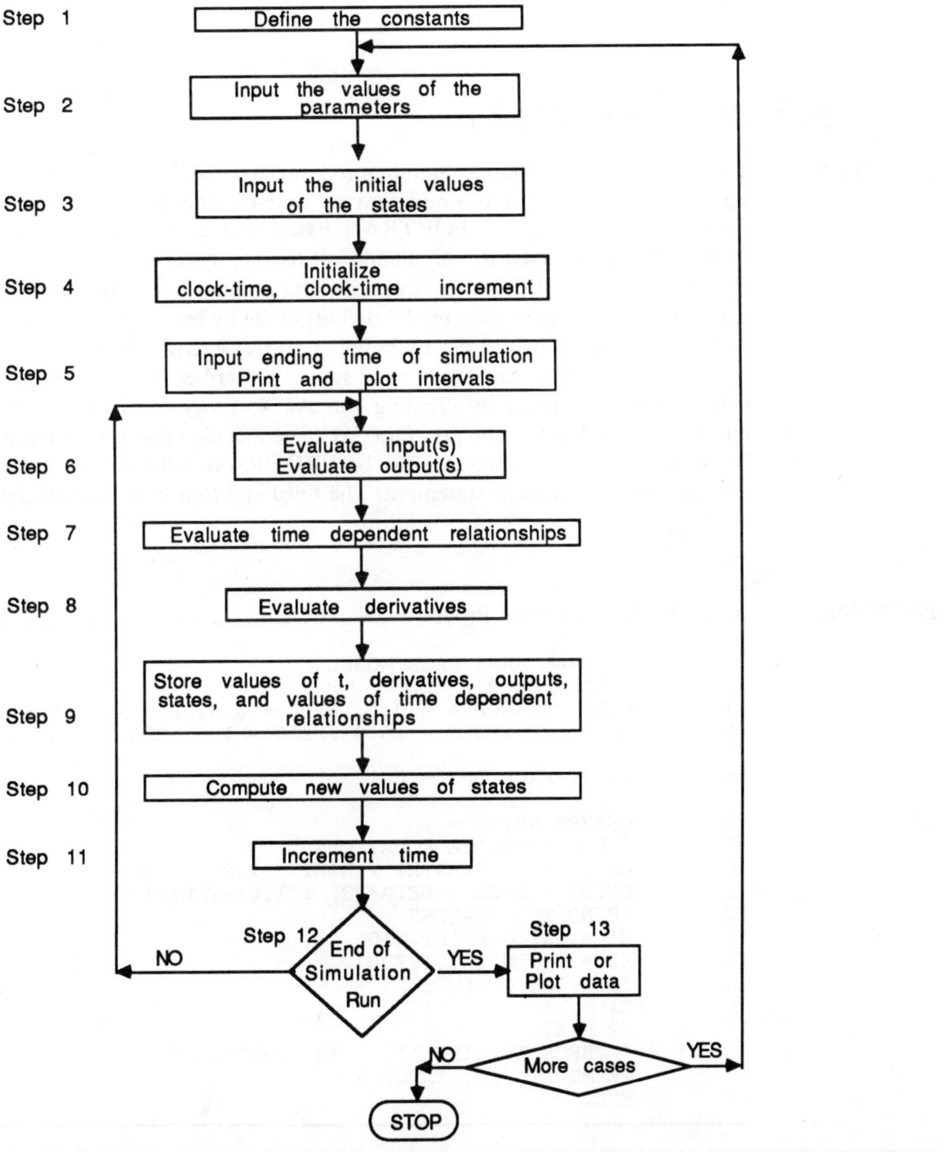

In step 10, some languages have only one (the Euler) integration routine, while others have many from which we can choose. In step 11, depending on the integration scheme chosen, some languages increment time in fixed increments, while others have variable time increments.

The printouts and plots of step 13 are obtained from the data stored in step 9. If all values are stored for every time increment and the print or plot intervals are multiples of the integration increment, we must search the data file. Some languages provide alphanumeric plots, while others provide both alphanumeric and graphic plots.

9.3 CSSL EXAMPLES

EXAMPLE 9.1 The Advanced Continuous Simulation Language (ACSL, pronounced "axle") was designed for modelling and simulation of time-dependent linear and nonlinear dynamic systems. The language is FORTRAN based and uses most of FORTRAN's features. ACSL can also be used for simulation of discrete systems.

Highlights of ACSL are free-form input and integration capability. To use the language, we must provide a model definition file by building a CSL type file and create an ACSL run-time command file by building a DATA type file.

Let us simulate the arms race system described by equations (9.1). The model definition is completed by creating the ACSL program shown in Program 9.1. (Line numbers are not part of the program. We have included them for reference purposes only.) In ACSL, the first statement must be PROGRAM followed by a character string. The second line is a comment statement. The third and fourth statements define the parameter

PROGRAM 9.1 ACSL Model Definition Program

```
1        PROGRAM ARMS RACE MODEL
2        " EXAMPLE 9.1 "
3        CONSTANT ALPHA = 0.5, BETA = 0.2, L = 1.0, K = 0.2
4        CONSTANT TSTOP = 10.0 , X10 = 1.0, X20 = 1.0
5        "                  "
6        CINTERVAL CINT = 0.1
7        "                  "
8        NSTEPS NSTEP = 1
9        " DEFINE STATE EQUATIONS "
10       DX1DT = - ALPHA*X1 + K*X2 + 2.0*STEP(0)
11       DX2DT = L*X1 - BETA*X2  + 1.0*STEP(0)
12       "FIND THE STATES"
13       X1 = INTEG(DX1DT,X10)
14       X2 = INTEG(DX2DT,X20)
15       " DEFINE OUTPUTS "
16       Y1 = X1
17       Y2 = X2
18       " CHECK IF SIMULATION IS COMPLETE "
19       TERMT (T .GE. TSTOP )
20       END
```

α and constants β, k, l, x_{10}, x_{20}, and TSTOP. TSTOP (a name inherently defined by ACSL) specifies the time at which the simulation should stop, and x_{10} and x_{20} are the initial values of x_1 and x_2 (initial conditions), respectively. The sixth statement defines the communication interval and calls it CINT.

The communication interval is the size of the time increment for which results will be stored or displayed. In ACSL, results are stored in files called RRR.DAT and FOR020.DAT. The communication interval is not the integration increment; however, they can be made equal, as shown in the eighth statement. NSTEPS is the ACSL statement that defines the number of integration (calculation) intervals per communication interval. The calculation interval is called NSTEP.

The 10th and 11th statements describe the derivative equations in (9.1). Input g is defined as a step of magnitude 2.0 beginning at time zero, and h is defined as a step of magnitude 1.0 beginning at time zero.

The 13th and 14th lines define x_1 as the integral of $\dot{x}_1$ and x_2 as the integral of $\dot{x}_2$ using the initial condition of x_{10} for x_1. The 16th and 17th lines define the outputs as $y_1 = x_1$ and $y_2 = x_2$. The 19th line,

TERMT(T.GE.TSTOP)

indicates that the simulation should be terminated when $t \geq$ TSTOP. The last statement of an ACSL program must be an END statement. A summary of ACSL model definition statements is included in Appendix D of reference 11.

The run-time commands for this example are shown in Program 9.2 (again, line numbers are for reference only). These statements were defined in a data type file.

The SET TITLE statement assigns the title of the headings for all printer and plotter output data. The PREPAR statement directs the ACSL simulator to save the values of the program variables T (time), DX1DT, DX2DT, X1, X2, Y1, and Y2 for every communication interval in files RRR.DAT and FOR020.DAT. The statement

OUTPUT "NCIOUT" = 20 ,T, DX1DT, DX2DT, Y1, Y2

PROGRAM 9.2 ASCL Run-Time Command File for the Arms Race Model

```
 SET TITLE = " ARMS RACE MODEL "
 SET PRNPLT =.TRUE.
 PREPAR T, DX1DT, DX2DT, X1, X2, Y1, Y2
 OUTPUT "NCIOUT" = 20 ,T,  DX1DT, DX2DT, Y1, Y2
START
RANGE "ALL"
PLOT "XAXIS" = T, DX1DT, DX2DT, Y1, Y2 , ...
     "XHI" = 10.0, "XLO" = 0.0, "XTAG" = "TIME", ...
     "HI" =20.0, "LO" = 0.0, "TAG" = " Y1,Y2,DX1DT,DX2DT "
" CHANGE VALUE OF ALPHA "
SET ALPHA = 0.2
START
RANGE "ALL"
PLOT "XAXIS" = T, Y1, Y2, ...
     "XHI" = 10.0, "XLO" = 0.0, "XTAG" = "TIME "
     "HI" = 20.0, "LO" = 0.0, "TAG" ="Y1, Y2 "
STOP
```

displays the values of T, DX1DT, DX2DT, Y1 and Y2 for every 20th communication interval when the simulation is started. The statement START executes the simulation.

The statement RANGE "ALL" requests that the minimum and maximum values of all variables listed in the PREPAR statement be printed out. The PLOT statement indicates that the x-axis should be time and that the y-axis should be DX1DT, DX2DT, Y_1, and Y_2. The three periods indicate a continuation statement. LO and HI indicate the minimum and maximum value of the ordinates, and TAG labels the ordinate axis.

Comments can be included with double quotation marks as in line 10. The next statement changes the value of parameter α to 0.2. The simulation is run again by line 12. The last statement is STOP. See Figure 9.2 for the output results.

FIGURE 9.2 Output Results of Program 9.1 (a) Results with $\alpha = 0.5$ (b) Plots with $\alpha = 0.5$ (c) Results with $\alpha = 0.2$ (d) Plots with $\alpha = 0.2$

```
SET TITLE = " ARMS RACE MODEL "
SET PRNPLT =.TRUE.
PREPAR.T, DX1DT, DX2DT, X1, X2, Y1, Y2
OUTPUT "NCIOUT" = 20 ,T,  DX1DT, DX2DT, Y1, Y2
START
        T 0.                 DX1DT 1.70000000          DX2DT 1.80000000
       Y1 1.00000000            Y2 1.00000000

        T 2.00000000         DX1DT 1.36902000          DX2DT 3.58197000
       Y1 3.90197000            Y2 6.60002000

        T 4.00000000         DX1DT 1.59173000          DX2DT 4.81767000
       Y1 6.81881000            Y2 15.0057000

        T 6.00000000         DX1DT 2.00050000          DX2DT 6.19337000
       Y1 10.3877000            Y2 25.9718000

        T 8.00000000         DX1DT 2.54601000          DX2DT 7.90940000
       Y1 14.9108000            Y2 40.0072000

        T 10.0000000         DX1DT 3.24652000          DX2DT 10.0908000
       Y1 20.6747000            Y2 57.9194000

RANGE "ALL"
        T 0.          10.0000000
    DX1DT 1.36498000  3.24652000
    DX2DT 1.80000000  10.0908000
       X1 1.00000000  20.6747000
       X2 1.00000000  57.9194000
       Y1 1.00000000  20.6747000
       Y2 1.00000000  57.9194000
```

(a)

FIGURE 9.2 Output Results of Program 9.1 (Continued)

```
PLOT "XAXIS" = T, DX1DT, DX2DT, Y1, Y2 , ...
     "XHI" = 10.0, "XLO" = 0.0, "XTAG" = "TIME", ...
     "HI" =20.0, "LO" = 0.0, "TAG" = " Y1,Y2,DX1DT,DX2DT "
DRAWING PLOT NUMBER      1
```

```
DX1DT A  0.     0.400000  0.800000  1.200000  1.600000  2.000000  2.400000  2.800000  3.200000  3.600000  4.000000
DX2DT B  0.     2.000000  4.000000  6.000000  8.000000  10.00000  12.00000  14.00000  16.00000  18.00000  20.00000
   Y1 C  0.     4.000000  8.000000  12.00000  16.00000  20.00000  24.00000  28.00000  32.00000  36.00000  40.00000
   Y2 D  0.     2.000000  4.000000  6.000000  8.000000  10.00000  12.00000  14.00000  16.00000  18.00000  20.00000
    T XAXIS .
```

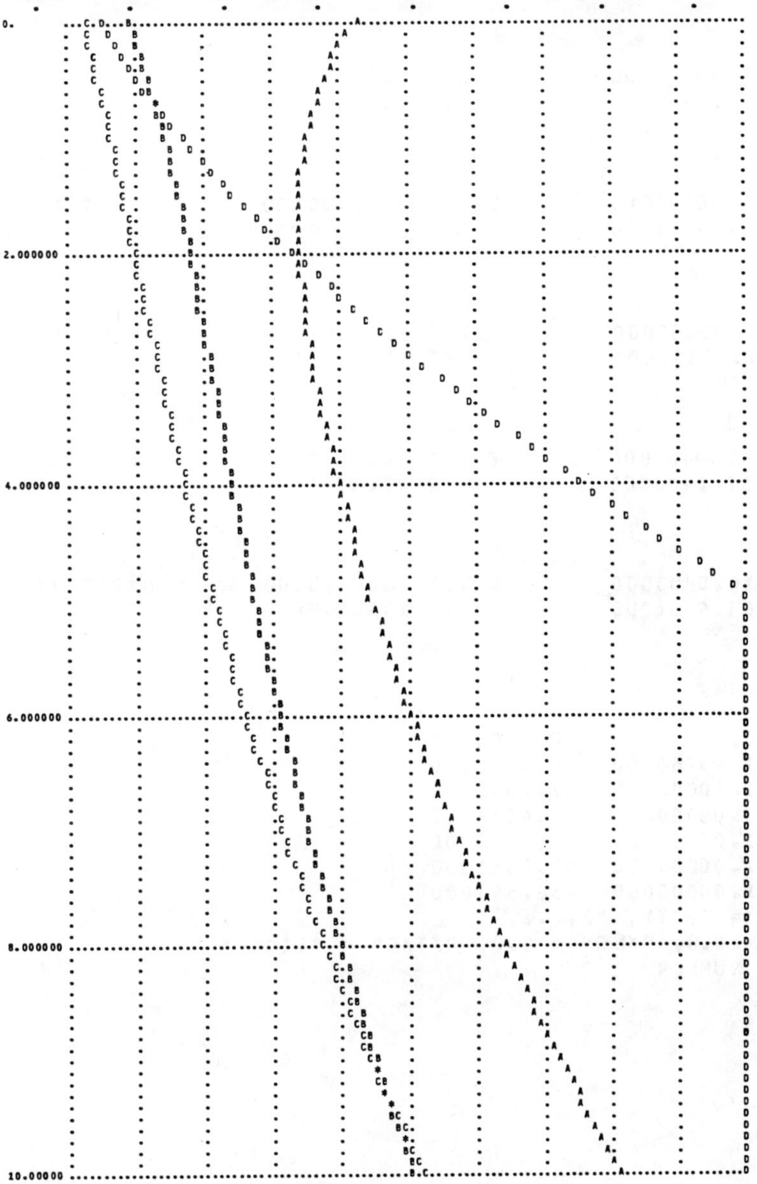

(b)

continues

FIGURE 9.2 Output Results of Program 9.1 (Continued)

```
" CHANGE VALUE OF ALPHA "
SET ALPHA = 0.2
START
         T 0.                  DX1DT 2.00000000          DX2DT 1.80000000
        Y1 1.00000000             Y2 1.00000000


         T 2.00000000          DX1DT 2.46322000          DX2DT 4.77561000
        Y1 5.29854000             Y2 7.61463000


         T 4.00000000          DX1DT 3.81499000          DX2DT 8.32988000
        Y1 11.4311000             Y2 20.5060000


         T 6.00000000          DX1DT 6.19362000          DX2DT 13.7944000
        Y1 21.2350000             Y2 42.2031000


         T 8.00000000          DX1DT 10.1380000          DX2DT 22.6542000
        Y1 37.2403000             Y2 77.9304000


         T 10.0000000          DX1DT 16.6173000          DX2DT 37.1532000
        Y1 63.4631000             Y2 136.549000


RANGE "ALL"
         T 0.          10.0000000
     DX1DT 1.99766000  16.6173000
     DX2DT 1.80000000  37.1532000
        X1 1.00000000  63.4631000
        X2 1.00000000  136.549000
        Y1 1.00000000  63.4631000
        Y2 1.00000000  136.549000
PLOT "XAXIS" = T, Y1, Y2, ...
     "XHI" = 10.0, "XLO" = 0.0, "XTAG" = "TIME "
DRAWING PLOT NUMBER    2
```

(c)

FIGURE 9.2 Output Results of Program 9.1 (Continued)

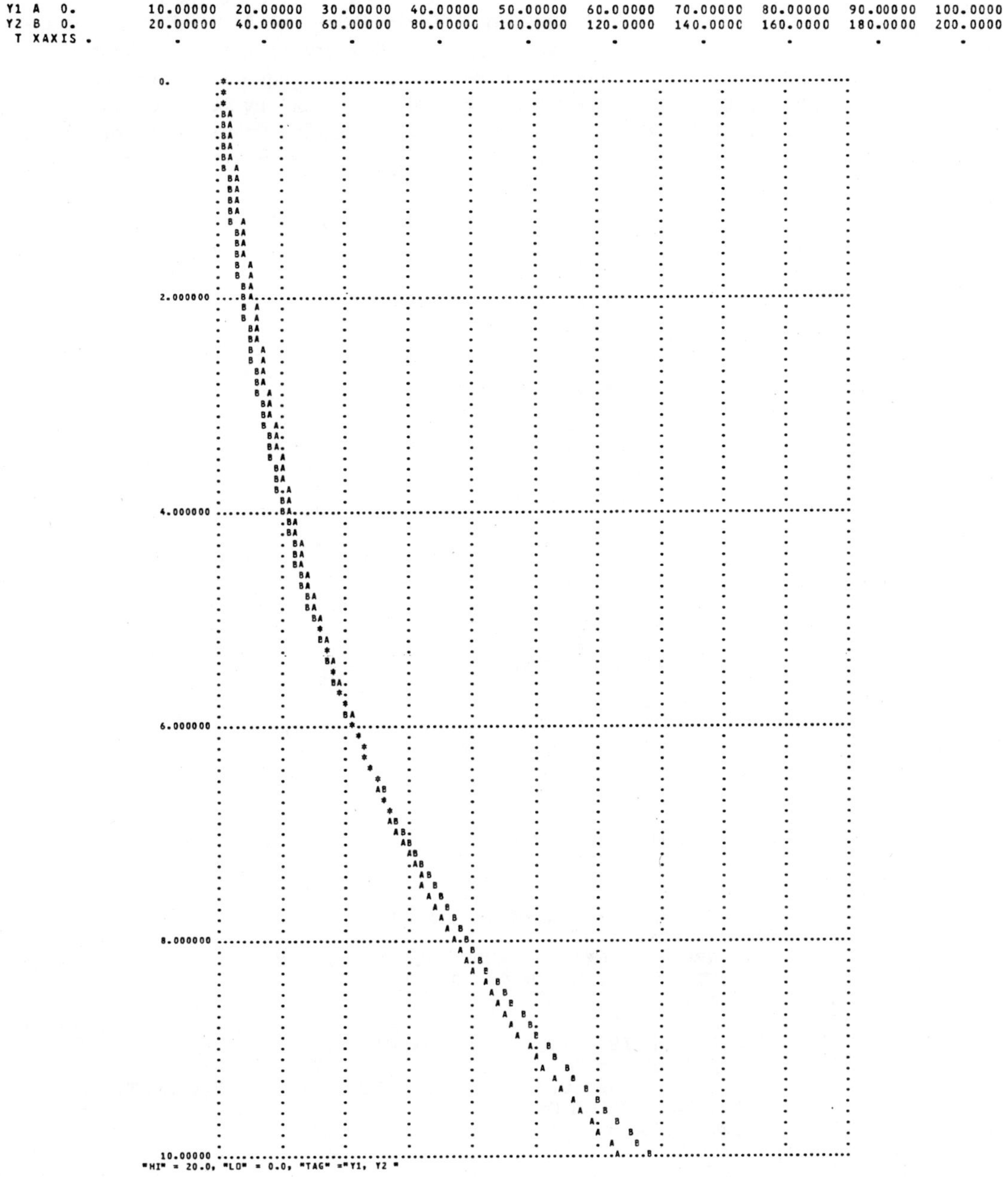

(d)

EXAMPLE 9.2 In this example, we present Program 9.3 in a form that highlights the structure for CSSLs shown in Figure 9.1. Starting with the initial section, we note that the variables described in the statements do not change during the simulation run. In the dynamic section, we include all statements whose values change with every communication interval. In the derivative section, we include all variables needed to compute the derivatives of the states of the model. Finally, in the terminal section, we include any and all computations that must be done after the simulation reaches its termination criterion. The run-time commands are the same as in Example 9.1 and, as you would expect, our output results are also the same.

PROGRAM 9.3 (a) ACSL Simulation Program for the Arms Race Model (b) ACSL Command File

```
PROGRAM ARMS RACE MODEL
    INITIAL
      "EXAMPLE 9.2, ARMS RACE MODEL "
       CONSTANT ALPHA = 0.5, BETA = 0.2, L = 1.0, K = 0.2
       CONSTANT TSTOP = 10.0 , X10 = 1.0, X20 = 1.0
       "               "
       CINTERVAL CINT = 0.1
       "               "
       NSTEPS NSTEP = 1
    END$"OF INITIAL"
    DYNAMIC
          DERIVATIVE
            " DEFINE STATE EQUATIONS "
            DX1DT =- ALPHA * X1 + K * X2 + STEP(0)
            DX2DT = L * X1 - BETA * X2  + 1.5*STEP(0)
            "FIND THE STATES"
            X1 = INTEG(DX1DT,X10)
            X2 = INTEG(DX2DT,X20)
          END$"OF DERIVATIVE"
     " DEFINE OUTPUTS "
     Y1 = X1
     Y2 = X2
     " CHECK IF SIMULATION IS COMPLETE "
     TERMT (T .GE. TSTOP )
   END$"OF DYNAMIC"
   TERMINAL
    "          "
    END$"OF TERMINAL"
END$"OF PROGRAM"
```

(a)

```
 SET TITLE = " ARMS RACE MODEL "
 SET PRNPLT =.TRUE.
 PREPAR T, DX1DT, DX2DT, X1, X2, Y1, Y2
 OUTPUT "NCIOUT" = 20 ,T,  DX1DT, DX2DT, Y1, Y2
START
RANGE "ALL"
PLOT "XAXIS" = T, DX1DT, DX2DT, Y1, Y2 , ...
     "XHI" = 10.0, "XLO" = 0.0, "XTAG" = "TIME", ...
     "HI" =20.0, "LO" = 0.0, "TAG" = " Y1,Y2,DX1DT,DX2DT "
" CHANGE VALUE OF ALPHA "
SET ALPHA = 0.2
START
RANGE "ALL"
PLOT "XAXIS" = T, Y1, Y2, ...
     "XHI" = 10.0, "XLO" = 0.0, "XTAG" = "TIME "
     "HI" = 20.0, "LO" = 0.0, "TAG" ="Y1, Y2 "
STOP
```

(b)

PROGRAM 9.4 CSMP Simulation Program for the Arms Race Model

```
TITLE EXAMPLE 9.3
* THIS IS A COMMENT
PARAM K=(.2,.3,.4)
CONST L=1.,A=.5,B=.2
DX1=K*X2-A*X1+2*STEP(0.)
DX2=L*X1-B*X2+STEP(0.)
 X1=INTGRL(1.,DX1)
 X2=INTGRL(1.,DX2)
*
TIMER DELT=.01,FINTIM=10.,PRDEL=1.,OUTDEL=.1
PRINT X1,DX1,X2,DX2
PRTPLOT X1,X2
FINISH X2=20.
METHOD RKSFX
* RUNGE KUTTA FOURTH ORDER
LABEL WORLD WAR I DATA
END
STOP
```

EXAMPLE 9.3 Here, we present Program 9.4, a CSMP program for the simulation of our arms race model. Using a PARAM statement, we can run the simulation for three different values of k (0.2, 0.3, and 0.4). With the INTGRL statement, we define x_1 and x_2 as the integrals of $\dot{x}_1$ and $\dot{x}_2$, respectively, each with initial condition 1.

The TIMER statement specifies the integration increment to be 0.01 unit of time, the final time (FINTIM) for the simulation to be 10.0 units, the printout interval for the PRINT statement (PRDEL) to be 1.0 unit of time, and the plot interval PRTPLOT (OUTDEL) (print plot) to be 0.1 unit. The FINISH statement stops the simulation when x_2 reaches a value of 20. Otherwise, by default, the simulation will stop when time is greater than or equal to 10. This program uses the fourth-order Runge-Kutta integration method presented in Chapter 8. The LABEL statement labels the output plots with World War I DATA.

EXAMPLE 9.4 Here, we present a DYNAMO program for the arms race model using data starting with the year 1909 (see Program 9.5). In DYNAMO, we use combinations of letters and numbers to label system variables, with no more than five characters per name. The label TIME is used strictly to refer to time in our model. TIME.K refers to the specific moment in time that a calculation is made. TIME.J and TIME.L, respectively, are moments in time that occur before and after TIME.K. In other words, JK represents a time interval in the past, and KL represents a time interval in the future, as shown in Figure 9.3. All intervals are of uniform length, DT, which we, as users, determine.

States of the system are referred to as levels, L, while the derivatives or rates are shown as R. Initial values are denoted by N, and all constants by C. Specifications of DT, PRINT, and PLOT intervals, and the length of the simulation run, are described by SPEC statements.

The DYNAMO simulation language has been used almost exclusively to simulate system dynamics problems. This, however, should not lead us to the impression that DYNAMO is a necessary tool for the solution of system dynamics problems. System dynamics problems can be simulated by using most scientifically oriented computer languages. DYNAMO is a useful language in formulating the model of the system and should not be dismissed if the package is available.

PROGRAM 9.5 DYNAMO Simulation Program for the Arms Race Model

```
*       ARMS RACE MODEL
N       TIME=1909
N       D1=115.3
N       D2=83.9
L       D1.K=D1.J+DT*(R1.JK-R2.JK)+167.97*DT
L       D2.K=D2.J+DT*(R3.JK-R4.JK)-334.37*DT
R       R1.KL=C1*D2.K
R       R2.KL=C2*D1.K
R       R3.KL=C3*D1.K
R       R4.KL=C4*D2.K
C       C1=-4.94
C       C2=2.19
C       C3=-0.63
C       C4=4.84
PLOT    D1=*,D2=$
PRINT D1,D2,R1,R2,R3,R4
SPEC    DT=.01,LENGTH=5/PRTPER=.1/PLTPER=.1
RUN 1
```

FIGURE 9.3 Time in DYNAMO

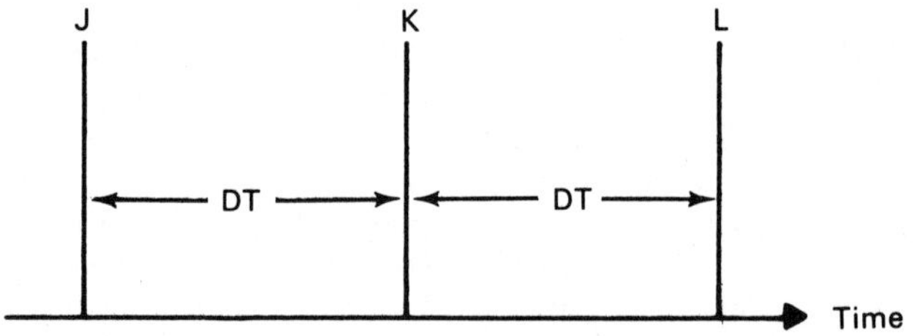

EXAMPLE 9.5 In this example, we present Program 9.6, a Notre Dame Translator (NDTRAN2) simulation program for the arms race model. The NDTRAN2 simulation processor is similar to the DYNAMO processor. Looking at Program 9.6, we see that the first statement is a TITLE statement. This title appears on every page of the printout. Next, we see several control statements:

NOSTATS—This deletes the statistical summary.

NOWARN—This deletes the warning messages only for those cases when the processor is bound to execute the program.

NARROW—This tells the processor to assume a narrow page width. Narrow width is 72 characters. Default is 121.

RKINT—Select the fourth-order Runge–Kutta (RKINT) integration algorithm. You can also use EULER (Euler integration) or ABINY (Adams-Bashforth predictor corrector).

PROGRAM 9.6 NDTRAN2 Simulation Program for the Arms Race Model

```
TITLE NDTRAN EXAMPLE -ARMS RACE RESPONSE
* NOSTATS
* NOWARN
* NARROW
* RKINT
L X1.K=INTGRL(DX1DT.JK)
L X2.K=INTGRL(DX2DT.JK)
R DX1DT.KL=-ALPHA*X1.K+K*X2.K+STEP(1.0,0.0)
R DX2DT.KL=L*X1.K-BETA*X2.K+STEP(1.5,0.0)
N X1=0.5
N X2=0.5
C K=0.25
C BETA=0.2
C L=0.15
C ALPHA=0.1
PARM DT=0.1
PARM START=0.0
PARM STOP=20.0
PARM PLTPER=1.0
PARM PRTPER=1.0
NOTE TIME,  X1,      X2
PRINT X1,X2
PLOT X1,X2
```

The next six statements set up the state equations. We tell the processor what the output of the integrator is (level) and what is integrated (rate). We also describe the rates and initial conditions. There are standard codes for each of these: L for levels, R for rates, and N for initial conditions.

Notice that there are no excess spaces anywhere. Beware! The subscripts (after the periods) refer to time intervals. K refers to the present time. In integrating, we integrate over the previous interval, from J to K, to get to the present values at K. We declare constants by starting a statement with a C.

In our example, we must specify the integration interval (DT), the starting time (START), the ending time (STOP), the plotting period (PLTPER), and the printing period (PRTPER). Here, we are plotting and printing every 10 computation intervals.

```
PARM DT=0.1
PARM START=0.0
PARM STOP=20.0
PARM PLTPER=1.0
PARM PRTPER=1.0
```

Next, we tell the processor what to print and plot:

```
NOTE TIME,  X1,  X2
PRINT X1,X2
PLOT X1,X2
```

All we have to do to get a short title on the printout and plots is to put a comment (NOTE) just before PRINT and PLOT.

There are several different variable types in NDTRAN, and knowledge of these types is crucial to getting NDTRAN programs to work correctly. Let us examine this issue by looking at the context in which each variable type occurs in a program.

The first variable type is a level variable. A level variable is the output of an integration process. Many engineers would prefer to think of level variables as states. In NDTRAN, we generate level variables in an integration with

```
L VARL.K=INTGRL(VARLD.JK)
```

Here, VARL is a level variable that is the output of an integrator that is integrating VARLD, which is the derivative of VARL. Note that the subscripts must be as shown in the INTGRL call.

The other kind of variable appearing in the INTGRL call is a rate variable. Within the arguments for INTGRL, only rate variables are legal.

Three other kinds of variables can be used: auxiliary variables, supplementary variables, and constants. We will present all five variable types, giving restrictions on their use, and their peculiarities.

Level variables. Level variables appear on the left side of calls on INTGRL and may not appear on the left side of any other type of equation. "L" statements are always calls on INTGRL and always define a single level variable. Level variables can appear on the right side of rate, auxiliary, and supplementary equations.

Rate variables. Rate variables are the only type permitted in calls on INTGRL. Further, only additions and subtractions of rate variables can occur there. Rate variables appear on the left side of "R" statements but cannot appear on the right side of rate equations or auxiliary equations. They are permitted on the right side of supplementary equations.

Constants. Constants are defined in "C" statements and can be used on the right side of rate, auxiliary, and supplementary equations, but not in L statements.

Auxiliary variables. Auxiliary variables are defined in "A" statements, and such variables can be used throughout a program wherever level variables are permitted, except on the left side of an L statement.

Supplementary variables. Supplementary variables are defined in "S" statements and are used only for output generation. S statements are unique since they may contain rate variables on the right side.

Figure 9.4 gives the output results of Program 9.6.

Plotting in NDTRAN is easy. It takes only a PLOT command to get a complete plot of the data generated by the program. In the simplest form, the plot command is

```
PLOT X
```

If this command is included at the end of an NDTRAN source program, it will generate a plot of x versus time. The plot will be generated with automatic scaling.

Two or more variables can be plotted on the same plot. This capability raises the question of whether we want the same scale for all variables. If we want all variables on the same scale, then we use

```
PLOT X,Y,Z
```

FIGURE 9.4 Output Results of Program 9.6

TIME	X1	X2
0.000	0.500	0.500
1.000	1.697	1.918
2.000	3.106	3.258
3.000	4.695	4.559
4.000	6.441	5.851
5.000	8.329	7.157
6.000	10.351	8.492
7.000	12.504	9.869
8.000	14.785	11.298
9.000	17.196	12.788
10.000	19.741	14.344
11.000	22.422	15.974
12.000	25.245	17.683
13.000	28.216	19.476
14.000	31.342	21.359
15.000	34.630	23.337
16.000	38.087	25.415
17.000	41.723	27.599
18.000	45.545	29.894
19.000	49.564	32.307
20.000	53.789	34.843

X=X1 A=X2

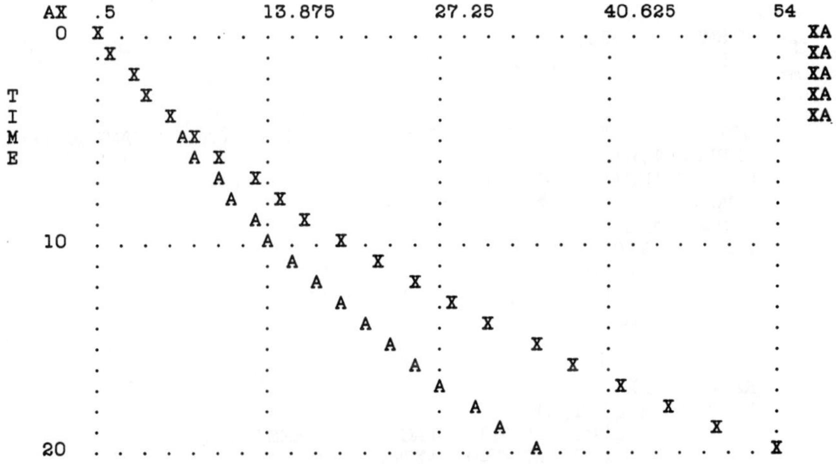

However, if we want all variables to be plotted with different scales (chosen automatically), we use

PLOT X/Y/Z

If we want to specify the scales used for the different variables, we use, for example,

PLOT X(0,10)/Y(−5,5)/Z(0,20)

It is also possible to set the plotting symbol for any particular variable. For example, if we want to use "*" for X, and other symbols for Y and Z, then we might use,

PLOT X = * / Y = + / Z = 0

This command gives automatic scaling for each variable, using the symbol indicated when the plot is generated.

EXAMPLE 9.6 In this example, we present Program 9.7, an ACSL simulation program for the information diffusion system of Chapter 2, using three different values of the conversion coefficients k (0.9, 0.5, and 0.1).

PROGRAM 9.7 ACSL Simulation Program for the Information Diffusion System

```
PROGRAM SPREAD OF RUMOR SIMULATION
CONSTANT K=0.9 , TSTOP = 1.0 , HO = 1.0 , N = 100
CINTERVAL CINT = 0.01
NSTEPS NSTEP = 10
   DHDT = K * HT * (N-HT)
        HT = INTEG(DHDT,HO)
        IT = N-HT
TERMT(T.GE.TSTOP)
END


SET TITLE ="DIFFUSION PROCESS IN A POPULATION SAMPLE (K=.9,.5,.1)"
PREPAR T,HT
OUTPUT 'NCIOUT'=1,T,HT
START
RANGE 'ALL'
PLOT 'XAXIS'=T,HT,...
     'HI'=100.,'LO'=0.,'TAG'='HEARERS',...
     'XHI'=0.3,'XLO'=0.,'XTAG'='TIME'
'CHANGE VALUE OF K'
SET K=.5
START
RANGE 'ALL'
PLOT 'XAXIS'=T,HT,...
     'HI'=100.,'LO'=0.,'TAG'='HEARERS',...
     'XHI'=0.3,'XLO'=0.,'XTAG'='TIME'
'CHANGE VALUE OF K'
SET K=.1
START
RANGE 'ALL'
PLOT 'XAXIS'=T,HT,...
     'HI'=100.,'LO'=0.,'TAG'='HEARERS',...
     'XHI'=0.3,'XLO'=0.,'XTAG'='TIME'
STOP
```

EXAMPLE 9.7 Here, we present Program 9.8, for a simulation of the RC circuit shown in Figure 9.5(a). PULSE is an ACSL function describing $v_i(t)$ as shown in Figure 9.5(b).

PROGRAM 9.8 ACSL Simulation Program for the RC Circuit in Figure 9.5(a)

```
PROGRAM PULSE RESPONSE OF RC CIRCUIT
  INITIAL
     CONSTANT R=10000,C=10.E-6,TSTOP=5.,IC=0.
     CINTERVAL CINT=.1
     COEF=1./(R*C)
   NSTEPS NSTEP=1
 END$"END OF INITIAL"
   DYNAMIC
     DERIVATIVE
        DVODT=COEF*(PULSE(0.,2.,.6)-VO)
        VO=INTEG(DVODT,IC)
    END$"OF DERIVATIVE"
     TERMT(T.GE.TSTOP)
  END$"OF DYNAMIC"
TERMINAL
END$"OF TERMINAL"
END $"OF PROGRAM"

SET TITLE ="TIME RESPONSE OF RC CIRCUIT"
PREPAR T,DVODT,VO
OUTPUT 'NCIOUT'=10 ,T,DVODT,VO
START
RANGE 'ALL'
PLOT 'XAXIS'=T,VO,DVODT,...
     'HI'=10.,'LO'=-10.,'TAG'='VO,DVODT',...
     'XHI'=5.,'XLO'=0.,'XTAG'='TIME'
'CHANGE VALUE OF CAPACITOR'
SET C=20.0E-6
START
STOP
```

FIGURE 9.5 (a) RC Circuit (b) Pulse Train

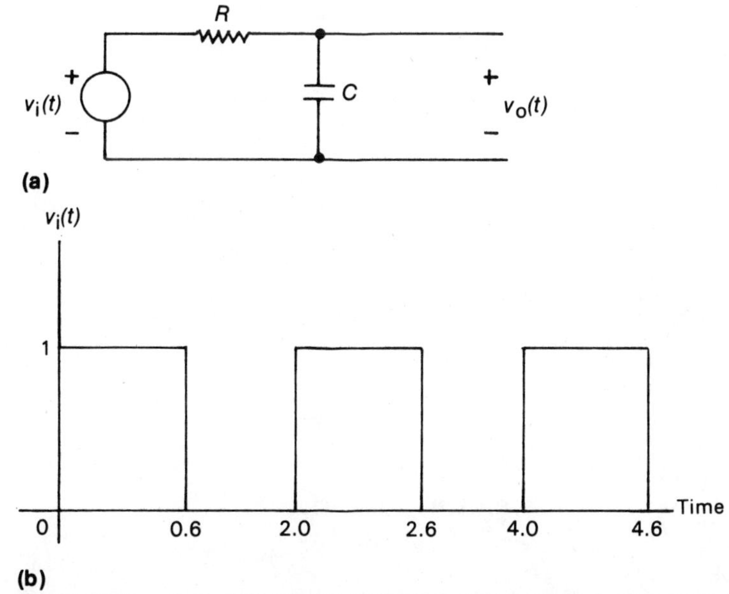

FIGURE 9.6 Bang-Bang Control System

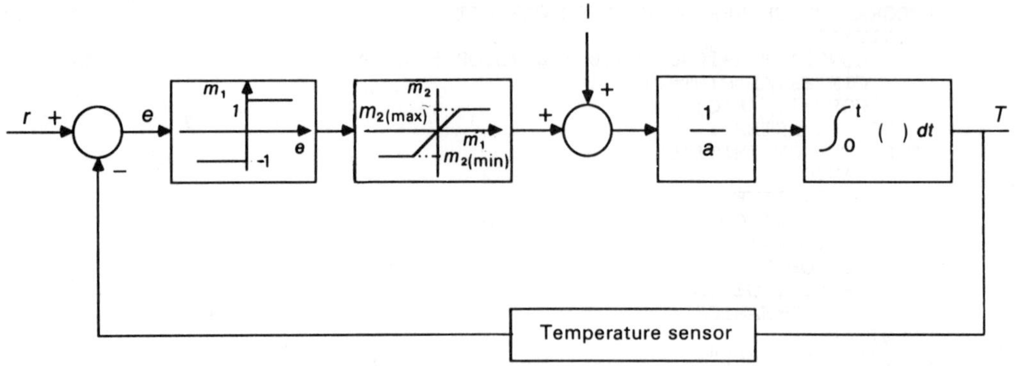

EXAMPLE 9.8 The block diagram of Figure 9.6 represents a control system for an environmental chamber. Quantity r is the set point or the desired temperature, expressed as a voltage. The temperature of the chamber, T, is measured and expressed as a voltage. The difference, e, is then used to adjust the energy input to the system by either heating or cooling. That is, if e is greater than zero, heat is added to the system; if e is less than zero, heat is removed.

Expressed mathematically, the output of this bang-bang controller is represented as follows.

If $e \geq 0$, then $m_1 = 1$ and the heat is turned on to its maximum value, $m_{2(\max)}$.
If $e < 0$, then $m_1 = -1$ and the heat is removed at its highest rate, $m_{2(\min)}$.

PROGRAM 9.9 ACSL Simulation Program for the Bang-Bang Control System

```
PROGRAM CONTROL OF 1ST-ORDER SYSTEM WITH BANG-BANG CONTROLLER
CONSTANT A=1.,M2MAX=1.,M2MIN=-1.,L=0,R0=1.,IC=0.
CINTERVAL CINT=.5
NSTEPS NSTEP=10
DERIVATIVE
CDOT=M3/A
R=R0*STEP(0.)-R0*STEP(10.)
E=R-C
M1=FCNSW(E,-1.,1.,1.)
M2=BOUND(M2MIN,M2MAX,M1)
M3=L+M2
C=INTEG(CDOT,IC)
TERMT (T .GE. 20.)
END

OUTPUT "NCIOUT"=1,T,R,C
PREPAR T,M1,M2,R,C,E
RANGE "ALL"
PLOT "XAXIS"=T,R,C,E,M1,M2,"XHI"=20.
RANGE "ALL"
START
STOP
```

The load or noise, l, represents the uncontrollable external factors which affect the temperature of the chamber. The dynamic behavior of the chamber itself is described by the equation $dT/dt = (1/a)m_3$, where a is a function of the chamber's size, shape, insulation, and so on, and $m_3 = l + m_2$.

Program 9.9 is an ACSL program for simulating this system. The FCNSW (function switch) statement implements the bang-bang controller. The BOUND statement represents the control element implementation.

EXAMPLE 9.9 Let us look at simulating a system described by the following equations:

$$\dot{x}_1 = \frac{1}{\tau_1} x_1 + u$$

$$\dot{x}_2 = \frac{-1}{\tau_2} x_2 + y_1$$

$$\dot{x}_3 = \frac{-1}{\tau_3} x_3 + y_2$$

$$\vdots$$

$$\dot{x}_n = \frac{-1}{\tau_m} x_n + y_{n-1}$$

$$y_1 = c_1 x_1$$

$$y_2 = c_2 x_2$$

$$\vdots$$

$$y_n = c_n x_n$$

As we can see, the state equations have the same form, and hence the code for obtaining the outputs y_i is repetitive. Let us build a macro (in NDTRAN2) that can be used to simulate this system.

To build the macro, we note that every macro has three main parts:

The macro title and argument list

The macro body

A "macro end" statement, MEND

In the macro statement list, the variables that appear are dummy variables as they appear in FORTRAN SUBROUTINE programs. Let us call our macro "TC." Writing the macro, we would have

```
MACRO TC(IN,OUT,TAU,C)
L STATE.K = INTGRL(IN.JK-STATED.JK)
N STATE = 0
R STATED.KL = STATE. K/TAU
A OUT.K = STATE*C
MEND
```

Looking at the macro, we find that we have an internal state, STATE. The derivative of STATE is STATED and appears on the left side of a rate equation. STATED is integrated in a level equation with a zero initial condition. The output of the TC macro is the value of the state, multiplied by C.

If we have a macro (either one we have defined, or one of the built-in macros) that we wish to use, then we need to know how to make use of it. The command that gives access to a macro is EXPND. Use of the command

```
EXPND TC(IN,OUT1,TAU,C)
```

will expand all of the code for the macro at the point the EXPND statement appears.

Use of the EXPND statement is best illustrated by an example. In this example, two TC blocks are used.

```
  ⋮
U.K=STEP(1.0,0.0)
EXPND TC(U,Y1,TAU1,C1)
EXPND TC(Y1,Y2,TAU2,C2)
```

The input to the first TC block is a constant (step input) of one. The output of the first TC block forms the input to the second TC block.

EXAMPLE 9.10 If we need to simulate a system with a time delay, we can use a special function in NDTRAN that automatically generates an approximation to a time delay. To use this facility, we need to write

```
EXPND DELAY1(OUT,IN,DEL)
```

This calls a built-in macro in NDTRAN, which generates the code for a one-state approximation to a time delay. The variables OUT, IN, and DEL that appear in the macro call are defined as follows.

OUT is the output variable from the time delay and is a rate variable. It can only be used later as an argument in an INTGRL call.

IN is the input variable into the time delay and is also a rate variable. It must be defined in an R statement to be used as an argument in the call on the DELAY macro.

DEL is the time delay in time units. DEL needs to be defined in a C statement.

There is also a macro DELAY3(OUT,IN,DEL) that approximates the same delay with the same arguments but that generates a three-state approximation to the delay. This routine is a better approximation to a delay than DELAY1 but will require use of more CPU time.

To use these macros, we need only the EXPND command in our source. Invoking the macro will result in generation of additional code in our program when we get a run-listing. The internally generated code will be marked with a "+" in the listing, indicating that the NDTRAN processor added it for us.

9.4 COMPUTER-AIDED DESIGN AND SIMULATION

Computer-aided control system design (CACSD) has emerged as an indispensable tool for design engineers, freeing them from the complex and monotonous details of design and allowing time for formulating, manipulating, and analyzing system models. CACSD systems are also being used in the educational environment as tools to help students learn and understand system responses and the process of design.[8] In either context, the available CACSD systems can be analyzed and compared on the basis of their ability to aid the designer or student. One way to compare them is to ask questions regarding each CACSD system:[7]

What is its objective?
What functions does it perform?
How is it implemented?
What type of command format does it use?
What special features does it include?
What type of documentation exists?

Here we will focus on TOTAL, BUCOP, ACSL, and NDTRAN. We will present a comparison and discussion of these systems on the basis of some of the questions posed above.

TOTAL

TOTAL is an interactive control systems package, developed by the Air Force Institute of Technology.[4] Its objective is to aid in the analysis and simulation of control systems. The command format was designed to enable TOTAL to be used with the speed of a familiar hand calculator. TOTAL has the capability of manipulating polynomials to reduce block diagrams and matrices to handle state variable solutions. It handles single-input, single-output and multi-input, multi-output systems with storage space to accommodate up to 50th-order polynomials and up to 10×10 matrices.

The functions this package can perform include polynomial operations (add, multiply, divide, factor, and expand), matrix operations (add, multiply, invert, transpose, obtain eigenvalues, and copy), root locus plots and design, frequency or time response analysis, block diagram reduction, and transformation between continuous and digital domains.

TOTAL was implemented by the Air Force on the VAX 11/780. The source code is written in FORTRAN and requires 212K of user memory.

Once the system has been defined, TOTAL can be used to perform the time response, frequency response, or root locus analysis. For each type of analysis, both plots and tabular listings of the response due to one of five possible inputs are available. For time response analysis, the user can list the desired time response data including rise time, peak time, settling time, and final value. A root locus plot can be obtained with a damping factor or gain of interest marked. Some of the frequency response options include listings and plots of magnitude or phase, plots using a linear or log frequency axis, listing of points of interest (including all response peaks, zero-dB crossings, 180° crossings, break frequencies, and asymptotes), Nyquist plots, and Nichols plots. The ease of changing variables makes it simple for the engineer to observe the change in any or all of these responses when any one of the system parameters is altered.

Commands in TOTAL are executed through a menu system. Each command is called by number with related commands grouped together by tens. A total or partial list of the 100 commands can be obtained at any time for easy reference. This method gives experienced users the ability to quickly enter a sequence of commands with which they are familiar. The disadvantage comes to the inexperienced users who are unable to associate the option numbers with their related commands.

One special feature of TOTAL is its calculator mode. This mode, which is modelled after a HP-45 calculator, may be entered at any time simply by typing a "C." In this mode, the user may make the same calculations possible on a hand calculator without disrupting or changing any variables already stored outside the calculator mode. A HP-45 user's manual would be a helpful addition to the TOTAL documentation to assist those persons unfamiliar with the necessary reverse Polish notation.

Another special feature of TOTAL is the ability to print multiple plots on the same set of axes. This capability is particularly useful to the engineer when system response sensitivity due to system parameter variation is being studied. When the session is ended, all work done in the current TOTAL session is automatically stored in a file called MEMORY.

TOTAL's documentation consists of an on-line help system, a TOTAL user's manual, and a set of example problems.

BUCOP

BUCOP (Bucknell University Control Operations Package) is an interactive software package for computer-aided design of linear control systems.[1–3] This package has been developed primarily by undergraduate students at Bucknell University. Its primary objective is to acquaint students with the relationship between systems and their mathematical models.

BUCOP, like TOTAL, models systems in transfer function or state variable form. The package performs time response analysis, frequency response analysis, polynomial root finding, and root locus analysis. Tabular prints and plots are available for all of the above.

BUCOP uses a hierarchical, menu-driven structure. Upon entering BUCOP, the main menu is displayed. Each of the major commands within this menu leads to the secondary menu consisting of the specific commands for that section. The unfamiliar user finds BUCOP easy to apply because of the clear progression of the menus and the connection between the command names and their functions.

One convenient feature of BUCOP is that all computations are stored in one of four files: a state variable file, a transfer function file, a frequency response file, and a root locus file. Because the designer controls the names of these files, a permanent record of different design experiments can be kept. Two of BUCOP's additional capabilities are the Routh-Hurwitz stability test and the generation of a Nyquist plot. The menu format used in this package easily lends itself to continued development of new sections.

BUCOP has a user's manual and a useful on-line help system which gives a short description of all commands. An example using BUCOP is shown on pp. 281–291.

Both TOTAL and BUCOP are used to model linear control systems. The next two packages, ACSL and NDTRAN, are used for both linear and nonlinear simulations.

ACSL

ACSL differs from TOTAL and BUCOP in that it is normally operated in batch mode.[11] To use the language, the user must provide a model definition (by building a .CSL type file) and create an ACSL run-time command file (by building a .DAT type file). The model definition contains the statements derived from a block diagram or mathematical expressions which define the relationships between the system parameters. Beyond these model statements, the .CSL file must also contain a few syntactical lines such as the program header, integration interval, termination condition, and program end statement. The run-time file contains the commands that control how the model is to be exercised. These commands include commands for printing or plotting the output and for changing parameter values.

ACSL contains many convenient built-in operators and run-time commands. In addition to all the functions of the FORTRAN library such as SQRT (square root), AMOD (modulus), and so on, the ACSL operators include INTEG (integration), QNTZR (quantization), UNIF (uniform random number generation), BOUND (function bounding), and DELAY (time delay). When using the integration operation, it is necessary to supply the initial condition of the variable and the calculation interval over which the variable is updated. Vector integration can also be performed by using the operator INTVC.

One of the most useful ACSL operators is the macro statement that allows the user to expand the existing language capability by defining new operators as the need arises. Macro sections are similar to subroutines or functions in FORTRAN. Once defined, they can be called from any place within the ACSL program. Within a macro section, variables and statement labels may be locally generated. Macro definitions may invoke other macros to an unlimited level of nesting.

Once the model has been translated and executed, exercising of the model is regulated by the ACSL run-time commands contained in the .DAT file. These commands include the START, STOP, and CONTINUE commands, all output commands, and any variable changes for an additional run. Data may be output in lists or plots. ACSL is capable of multiple variable plots with the x- and y-scales automatically chosen or specified by the user. Logarithmic scales are also available.

No on-line help is available for ACSL because of its batch mode operation. However, an extensive user's guide and reference manual details the syntax and definition of all operators and run-time commands. The manual also includes extensive examples, a general discussion of how to build the necessary files, and explanations of the error messages that may be generated during the translation phase of the program.

NDTRAN

NDTRAN (Notre Dame Translator) is a continuous and stochastic system-modelling interpreter designed for use on small computer systems. Like BUCOP, NDTRAN was developed primarily by undergraduates, at the University of Notre Dame. The most recent version, NDTRAN2, can be used to simulate linear and nonlinear systems, perform time response analysis, and output listings or plots of the time response data.[6]

NDTRAN is an exclusively batch program. To perform a system simulation, we must write a source file. This file contains all the equations derived from the system flow or block diagram, which describe the relationship between the system parameters. As in ACSL, the equations must be in integral or differential form. The file also contains the commands that control information output, printing, or plotting. NDTRAN can perform and output the time response of the system to any specified input.

NDTRAN defines several types of variables used in modelling a system. Level or state variables result from an accumulation or integration and depend on both present and past values. Rate variables are derivatives of the level variables. These rate variables represent a policy decision and are always defined with reference to a time period. Supplementary variables are defined to be used only in output. Other types of variables include auxiliary variables used in defining other variables and constants. The restrictions on the ways in which and places where certain variables types can be used and the strict subscripting notation can be confusing to the infrequent user.

Other statements contained in the source file are control statements that dictate the execution of many of NDTRAN's useful options. One such option is the choice of integration method to be used. Three methods are available: Euler, Runge-Kutta, and Adams-Bashforth, which is the default method. Another useful option is the cross-reference list. When chosen, this option prints a list of where all variables are defined and where they are used in the program. This list can be beneficial to the user in debugging a complex program. Other options include a diagnostic warning option, an output size option, a statistics option, and a documentation option.

Another type of statement contained in the source file is the parameter statement. This statement is used to define the integration interval, the starting and stopping times of execution, and the printing or plotting periods.

Like ACSL, NDTRAN contains the Macro statement. NDTRAN users can also take advantage of the RERUN statement to make a number of different runs of the same program by changing constants, integration step size, or integration method.

NDTRAN contains a number of useful predefined functions such as PULSE, STEP, CLIP, and NOISE. PULSE and STEP generate the functions that the names imply. That is, CLIP can be used to limit a response to a specified value, and NOISE can be used to generate a standard uniformly distributed random number.

Although NDTRAN includes no graphics capabilities, it can be used to produce dot matrix plots of one or more variables versus time. The scales for each variable may be defined or automatically determined. We can also plot one variable against another.

Since NDTRAN is not interactive, it contains no on-line help facility, but the documentation that exists is quite thorough. The project directors have written a primer for NDTRAN which includes an explanation of NDTRAN's syntax and capabilities, and some detailed examples of possible uses. This manual also contains a list of error messages that NDTRAN may provide during execution.

Like all CACSD packages, TOTAL, ACSL, BUCOP, and NDTRAN2 serve the general purpose of simplifying the design task. CACSD systems are being used in industry to relieve the engineer of the details of design, and in education to help teach students control theory. As computer-aided design of control systems becomes more popular with engineers and educators, design packages will continue to be created and refined.

9.5 SUMMARY

We have presented the basic principles of operation for continuous-system simulation languages (CSSLs) and some example simulations using these languages. It is apparent that CSSLs require minimal coding effort. In addition, we have compared four computer-aided control system design (CACSD) packages that can be used for simulation of continuous systems.

9.6 REVIEW PROBLEMS

1. Describe the basic steps in continuous-system simulation.
2. Continuous simulation languages can be used to compute integrals. Write a program to compute the probability distribution function for the standard normal distribution,

$$f(x) = \frac{1}{\sqrt{2\pi}} \exp\left(\frac{-x^2}{2}\right)$$

3. Continuous simulation languages can be used to obtain numerical solutions of differential equations. Obtain the numerical solution to the equation

$$\frac{d^3x}{dt^3} + 5\frac{d^2x}{dt^2} + 15\frac{dx}{dt} + 50x = 5$$

Assume that x and all its derivatives are zero at $t = 0$.

4. Obtain the solution of the equation

$$\ddot{x} + (1 - \varepsilon)x\dot{x} + x = 0$$

Let $\varepsilon = 0.5$, $x(0) = 1$, and $\dot{x}(0) = 0$.

5. See Figure 9.7. Develop a simulation program to predict the position of mass M_1. To set this system in motion, we hold mass M_2 fixed, move mass M_1 a distance a to the left, and then release the two masses. Let $M_1 = M_2 = 2$ kilograms, $k_1 = k_2 = k_3 = 2$ newtons per meter, and $a = 1$.

FIGURE 9.7 Problem 5

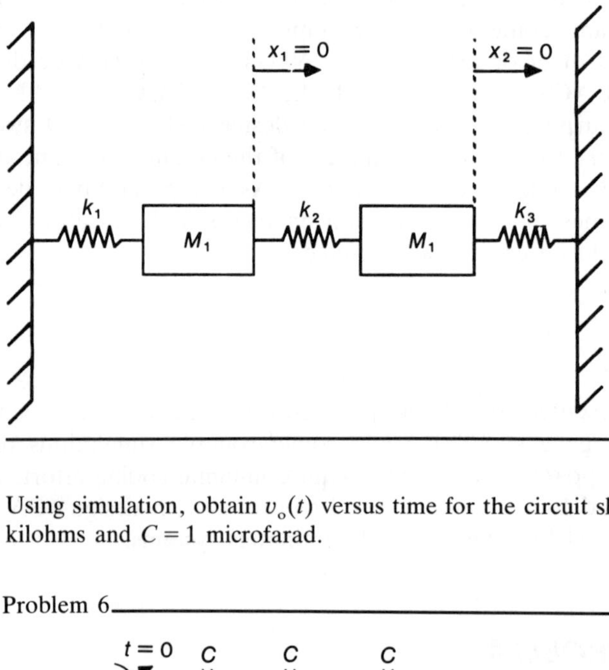

6. Using simulation, obtain $v_o(t)$ versus time for the circuit shown in Figure 9.8. Let $R = 100$ kilohms and $C = 1$ microfarad.

FIGURE 9.8 Problem 6

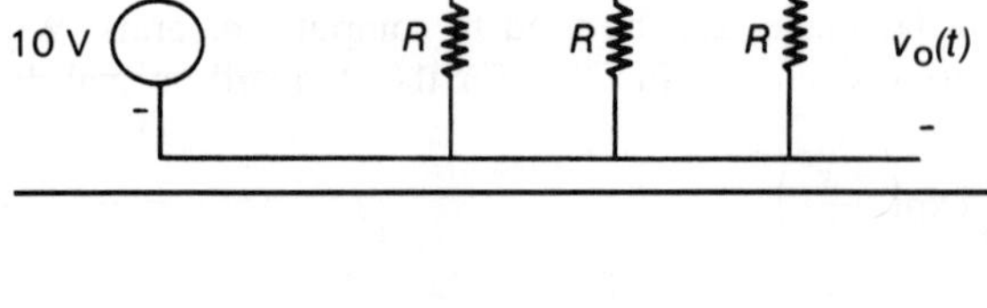

9.7 APPENDIX: MORE SIMULATION PROGRAMS

In this appendix, we present simulation programs for the examples given in Chapter 8. The ACSL plots are provided by Edward L. Mitchell of Mitchell and Gauthier Associates.

ACSL Simulation Program for the Arms Race Model

```
PROGRAM ARMS RACE MODEL
    INITIAL
      "APPENDIX FOR CHAPTER 9 EXAMPLE , ARMS RACE MODEL "
       CONSTANT ALPHA = 0.5, BETA = 0.2, L = 1.0, K = 0.2
       CONSTANT TSTOP = 50.0 , X10 = 1.0 , X20 = 1.0
       "              "
       CINTERVAL CINT = 0.1
       "              "
       NSTEPS NSTEP = 1
    END$"OF INITIAL"
    DYNAMIC
          DERIVATIVE
            " DEFINE STATE EQUATIONS "
            DX1DT =- ALPHA * X1 + K * X2 + 1.0 * STEP(0)
            DX2DT = L * X1 - BETA * X2  + 1.0 * STEP(0)
            "FIND THE STATES"
            X1 = INTEG(DX1DT,X10)
            X2 = INTEG(DX2DT,X20)
          END$"OF DERIVATIVE"
     " DEFINE OUTPUTS "
     Y1 = X1
     Y2 = X2
     " CHECK IF SIMULATION IS COMPLETE "
     TERMT (T .GE. TSTOP )
   END$"OF DYNAMIC"
   TERMINAL
    "            "
    END$"OF TERMINAL"
END$"OF PROGRAM"
```

ACSL Command File

```
 SET TITLE = " ARMS RACE MODEL "
 SET PRNPLT =.TRUE.
 PREPAR T, Y1, Y2
 OUTPUT "NCIOUT" = 20 ,T, Y1, Y2
"                                 "
 PROCED RUN $ START
RANGE "ALL"
PLOT "XAXIS" = T, Y1, Y2 , ...
     "XHI" = 40.0, "XLO" = 0.0, "XTAG" = "TIME", ...
     "TAG" = " Y1,Y2 "
END
```

continues

ACSL Command File (Continued)

```
"                                 "
" UNDAMPED OR OSCILLATORY SYSTEM "
SET ALPHA = 0.5, BETA = -0.5 , K = -0.8 , L = 0.8 $ RUN
"                            "
" DAMPED OSCILLATORY SYSTEM "
"                            "
SET ALPHA = 0.2 , BETA = 0.3, K = -1.2 , L = 1.0 $ RUN
"                                           "
" CRITICALLY DAMPED SYSTEM "
SET ALPHA = 0.5 , BETA = 0.5 , K = 0.0, L= 1.0 $ RUN
"                                         "
" OVER CRITICALLY DAMPED SYSTEM "
"                                 "
SET ALPHA = 0.5 , BETA = 0.5 , K = 0.2, L = 1.0 $ RUN
"                                              "
" EXPLOSIVE OSCILLATORY SYSTEM "
"                                "
SET ALPHA = +0.2 , BETA = -0.3 , K = -1.0 , L = 1.0 $ RUN
"                                                    "
" EXPLOSIVE SYSTEM "
"                   "
SET ALPHA = - 0.5, BETA = 0.5 , K = - 0.24 , L = 1.0 $ RUN
"                                                    "
STOP
```

Oscillatory Case

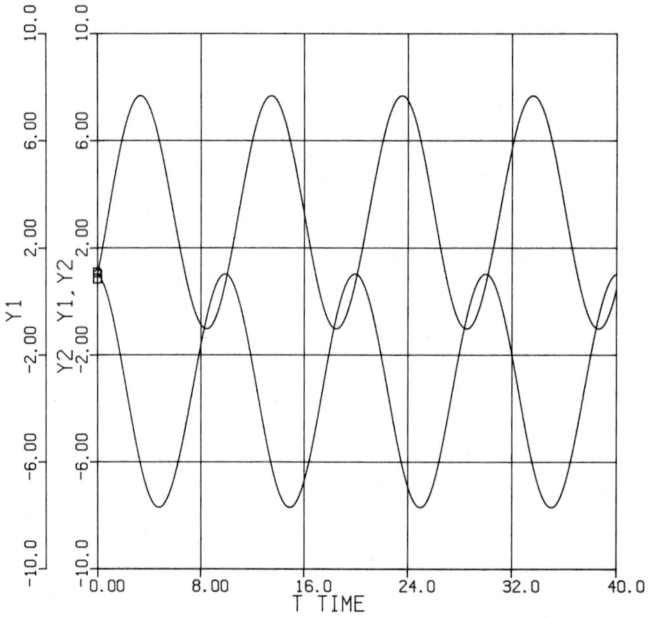

ACSL Command File (Continued)

Underdamped Case

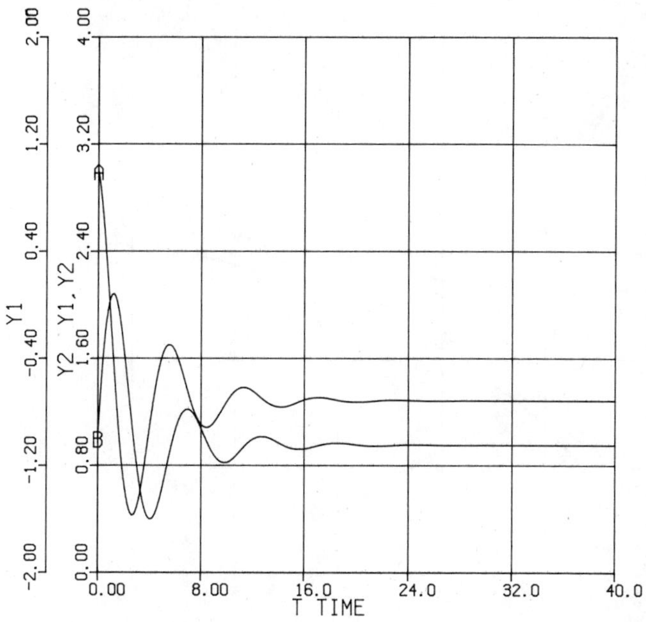

Critically Damped Case

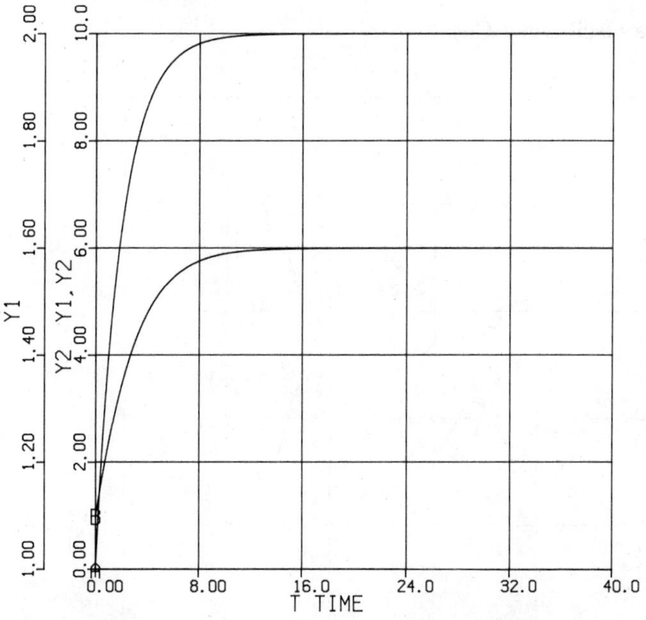

continues

ACSL Command File (Continued)

Overdamped Case

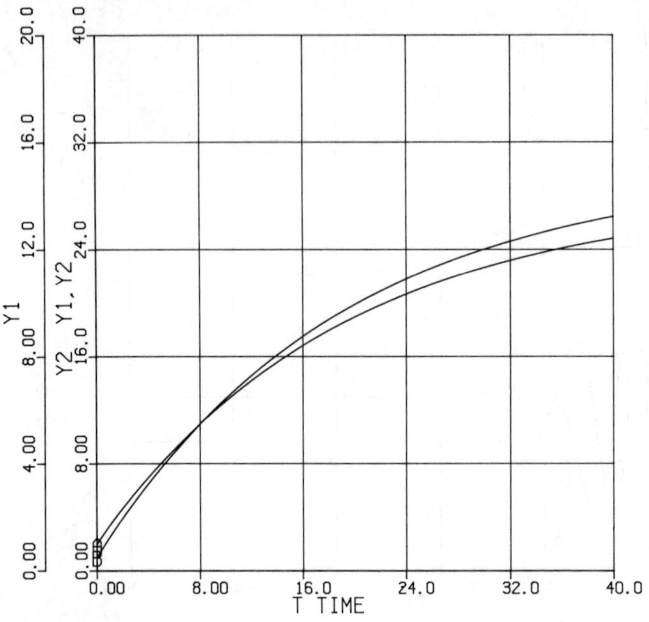

Explosive Oscillatory Case

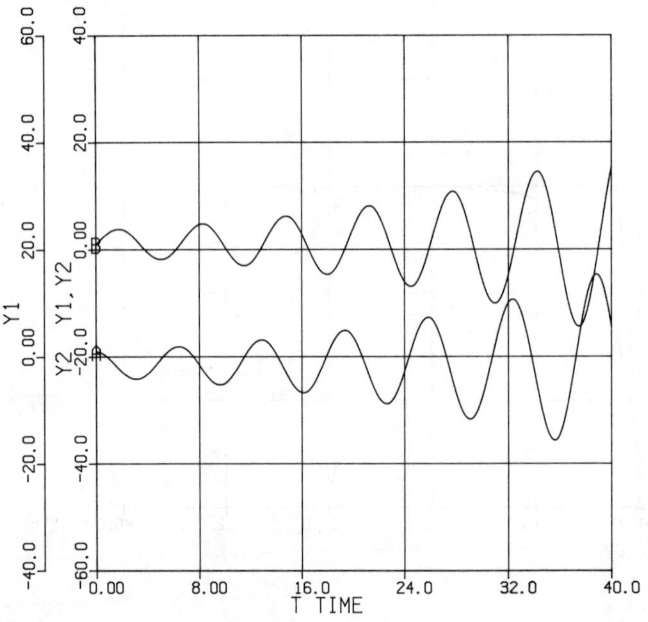

ACSL Command File (Continued)

Explosive Case

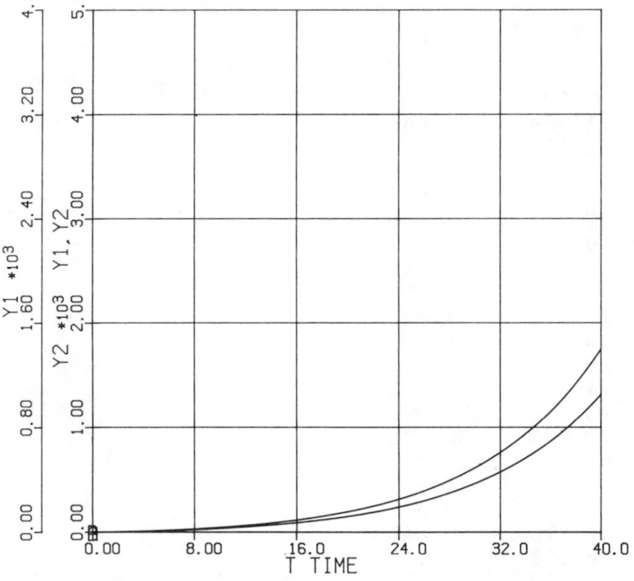

NDTRAN2 Program for the Arms Race Model: Comparative Plots

```
TITLE NDTRAN EXAMPLE -ARMS RACE RESPONSE
* NOSTATS
* NOWARN
* NARROW
* RKINT
L X1.K=INTGRL(DX1DT.JK)
L X2.K=INTGRL(DX2DT.JK)
R DX1DT.KL=-ALPHA*X1.K+K*X2.K+STEP(1.0,0.0)
R DX2DT.KL=L*X1.K-BETA*X2.K+STEP(1.5,0.0)
N X1=0.2
N X2=0.1
C K=0.25
C BETA=0.2
C L=0.15
C ALPHA=0.1
PARM DT=0.1
PARM START=0.0
PARM STOP=20.0
PARM PLTPER=1.0
PARM PRTPER=1.0
NOTE TIME, X1,      X2
PRINT X1,X2
PRINT X1.*
PRINT X2.*
PLOT X1,X2
PLOT X1.*
PLOT X2.*
RERUN CHANGE VALUE OF ALPHA
C ALPHA=0.2
RERUN CHANGE VALUE OF ALPHA
C ALPHA=0.25
RERUN CHANGE VALUE OF ALPHA
C ALPHA=0.4
```

continues

Output Results: Comparative Plots

Printout of x_1 and x_2 versus Time for $\alpha = 0.10$

TIME	X1	X2
0.000	0.200	0.100
1.000	1.334	1.545
2.000	2.691	2.899
3.000	4.235	4.205
4.000	5.940	5.496
5.000	7.791	6.796
6.000	9.777	8.121
7.000	11.895	9.485
8.000	14.141	10.898
9.000	16.516	12.370
10.000	19.023	13.907
11.000	21.666	15.516
12.000	24.450	17.203
13.000	27.380	18.972
14.000	30.462	20.829
15.000	33.705	22.781
16.000	37.114	24.830
17.000	40.700	26.985
18.000	44.470	29.248
19.000	48.433	31.628
20.000	52.600	34.129

Printout of x_1 versus Time for $\alpha = 0.10$, 0.20, 0.25, 0.40

TIME	X1.1	X1.2	X1.3	X1.4
0.000	0.200	0.200	0.200	0.200
1.000	1.334	1.264	1.231	1.139
2.000	2.691	2.448	2.339	2.049
3.000	4.235	3.711	3.484	2.914
4.000	5.940	5.024	4.642	3.727
5.000	7.791	6.367	5.796	4.485
6.000	9.777	7.728	6.936	5.189
7.000	11.895	9.098	8.056	5.841
8.000	14.141	10.472	9.151	6.444
9.000	16.516	11.845	10.220	7.000
10.000	19.023	13.215	11.262	7.514
11.000	21.666	14.580	12.275	7.987
12.000	24.450	15.938	13.260	8.423
13.000	27.380	17.290	14.218	8.825
14.000	30.462	18.634	15.148	9.196
15.000	33.705	19.971	16.051	9.537
16.000	37.114	21.299	16.928	9.851
17.000	40.700	22.620	17.780	10.141
18.000	44.470	23.932	18.607	10.408
19.000	48.433	25.236	19.409	10.654
20.000	52.600	26.532	20.188	10.881

Output Results: Comparative Plots (Continued)

Printout of x_2 versus Time for $\alpha = 0.10, 0.20, 0.25, 0.40$

TIME	X2.1	X2.2	X2.3	X2.4
0.000	0.100	0.100	0.100	0.100
1.000	1.545	1.542	1.540	1.535
2.000	2.899	2.876	2.865	2.835
3.000	4.205	4.135	4.103	4.021
4.000	5.496	5.341	5.274	5.105
5.000	6.796	6.509	6.390	6.100
6.000	8.121	7.650	7.459	7.013
7.000	9.485	8.770	8.488	7.854
8.000	10.898	9.873	9.482	8.626
9.000	12.370	10.963	10.442	9.338
10.000	13.907	12.042	11.372	9.993
11.000	15.516	13.111	12.272	10.596
12.000	17.203	14.172	13.146	11.152
13.000	18.972	15.224	13.993	11.664
14.000	20.829	16.269	14.814	12.135
15.000	22.781	17.307	15.612	12.569
16.000	24.830	18.338	16.385	12.969
17.000	26.985	19.362	17.136	13.338
18.000	29.248	20.379	17.865	13.677
19.000	31.628	21.390	18.572	13.990
20.000	34.129	22.394	19.259	14.278

Plot of x_1 and x_2 versus Time for $\alpha = 0.10$

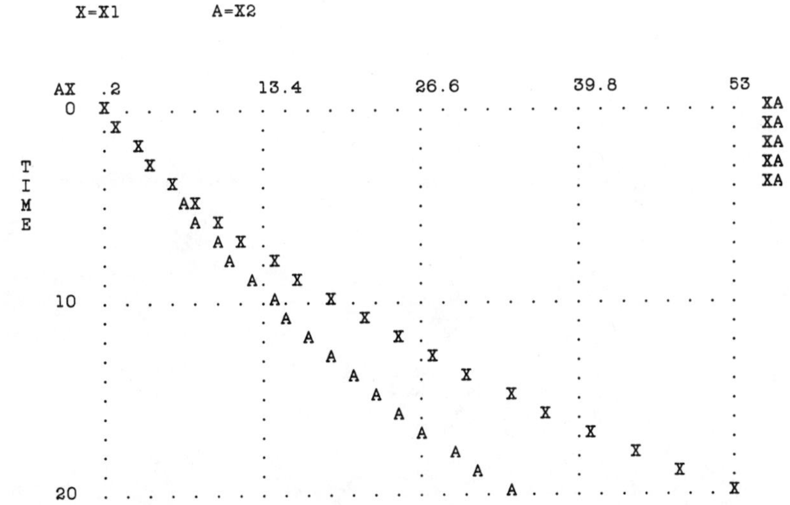

continues

Output Results: Comparative Plots (Continued)

Plot of x_1 versus Time for $\alpha=0.10, 0.20, 0.25, 0.40$

```
      1=X1.1       2=X1.2       3=X1.3       4=X1.4


4321  .2           13.4              26.6            39.8             53
   0  1 . . . . . . . . . . . . . . . . . . . . . . . . . . . . . . . . . 1234
      .1            .                .               .                .  1234
      . 21          .                .               .                .  234
T     .  31         .                .               .                .  12,34
I     .   421       .                .               .                .  23
M     .    4321     .                .               .                .
E     .    4 32 1   .                .               .                .
      .     4 32  1 .                .               .                .
      .      4 3 2  .1               .               .                .
      .      4   32 .  1             .               .                .
  10  . . . . 4 . 3 2 . . 1 . . . . . . . . . . . . . . . . . . . . . .
      .       4    3.2        1      .               .                .
      .        4    3  2         1   .               .                .
      .        4    .3  2            .1              .                .
      .         4   . 3  2           .   1           .                .
      .         4   .  3  2          .       1       .                .
      .         4   .   3  2         .          1    .                .
      .          4  .    3   2       .               .1               .
      .          4  .     3    2     .               .    1           .
      .          4  .     3       2. .               .        1       .
  20  . . . . . .4. . . . .3. . . 2 . . . . . . . . . . . . . . . . . 1
```

Plot of x_2 versus Time for $\alpha=0.10, 0.20, 0.25, 0.40$

```
      1=X2.1       2=X2.2       3=X2.3       4=X2.4


4321  .1           8.825             17.55           26.275           35
   0  1 . . . . . . . . . . . . . . . . . . . . . . . . . . . . . . . . .  1234
      . 1           .                .               .                .  1234
      .   1         .                .               .                .  1234
T     .     21      .                .               .                .  234
I     .       21    .                .               .                .  234
M     .         21  .                .               .                .  234
E     .          421.                .               .                .  23
      .           4321               .               .                .
      .             4321             .               .                .
      .             .4 2  1          .               .                .  23
  10  . . . . . . . . 4 32. 1 . . . . . . . . . . . . . . . . . . . . .
      .             .  4  32   1     .               .                .
      .             .   4  3 2    1. .               .                .
      .             .    4  3 2    . 1               .                .
      .             .    4    3 2  .    1            .                .
      .             .     4    3  2 .       1        .                .
      .             .      4    3 .2           1     .                .
      .             .      4    3.   2               .1               .
      .             .       4   .3    2          .        1           .
      .             .       4   . 3     2        .            1       .
  20  . . . . . . . . . . . . . .4. . . .3. . 2 . . . . . . . . . . .1.
*STOP*
```

NDTRAN2 Program for the Arms Race Model: Phase Plane Plots

```
TITLE NDTRAN EXAMPLE -ARMS RACE RESPONSE
* NOSTATS
* NOWARN
* NARROW
* RKINT
L X1.K=INTGRL(DX1DT.JK)
L X2.K=INTGRL(DX2DT.JK)
R DX1DT.KL=-ALPHA*X1.K+K*X2.K+STEP(1.0,0.0)
R DX2DT.KL=L*X1.K-BETA*X2.K+STEP(1.5,0.0)
N X1=0.2
N X2=0.1
C K=0.25
C BETA=0.2
C L=0.15
C ALPHA=0.1
PARM DT=0.1
PARM START=0.0
PARM STOP=20.0
PARM PLTPER=1.0
PARM PRTPER=1.0
NOTE TIME,  X1,     X2
PRINT X1,X2
PLOT X1,X2
PLOT X1//X2
```

NDTRAN2 Output Results: Phase Plane Plots

Printout of x_1 and x_2 versus Time

TIME	X1	X2
0.000	0.200	0.100
1.000	1.334	1.545
2.000	2.691	2.899
3.000	4.235	4.205
4.000	5.940	5.496
5.000	7.791	6.796
6.000	9.777	8.121
7.000	11.895	9.485
8.000	14.141	10.898
9.000	16.516	12.370
10.000	19.023	13.907
11.000	21.666	15.516
12.000	24.450	17.203
13.000	27.380	18.972
14.000	30.462	20.829
15.000	33.705	22.781
16.000	37.114	24.830
17.000	40.700	26.985
18.000	44.470	29.248
19.000	48.433	31.628
20.000	52.600	34.129

continues

NDTRAN2 Output Results: Phase Plane Plots (Continued)

Plot of x_1 and x_2 versus Time

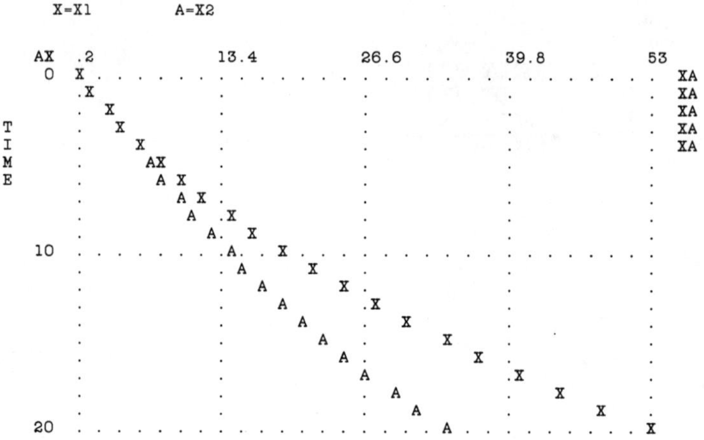

Plot of x_1 versus x_2

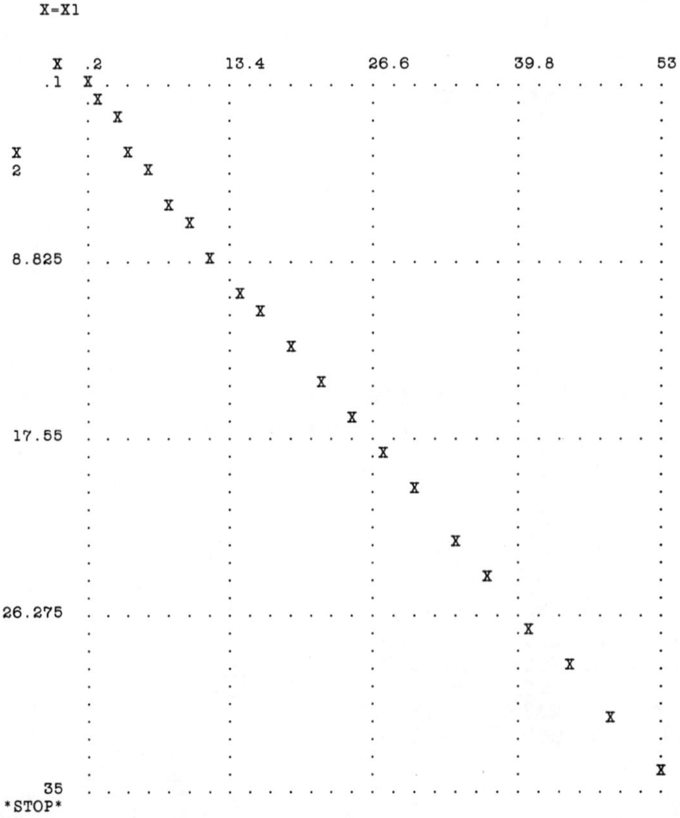

Graphical Representation of the Arms Race Model in STELLA

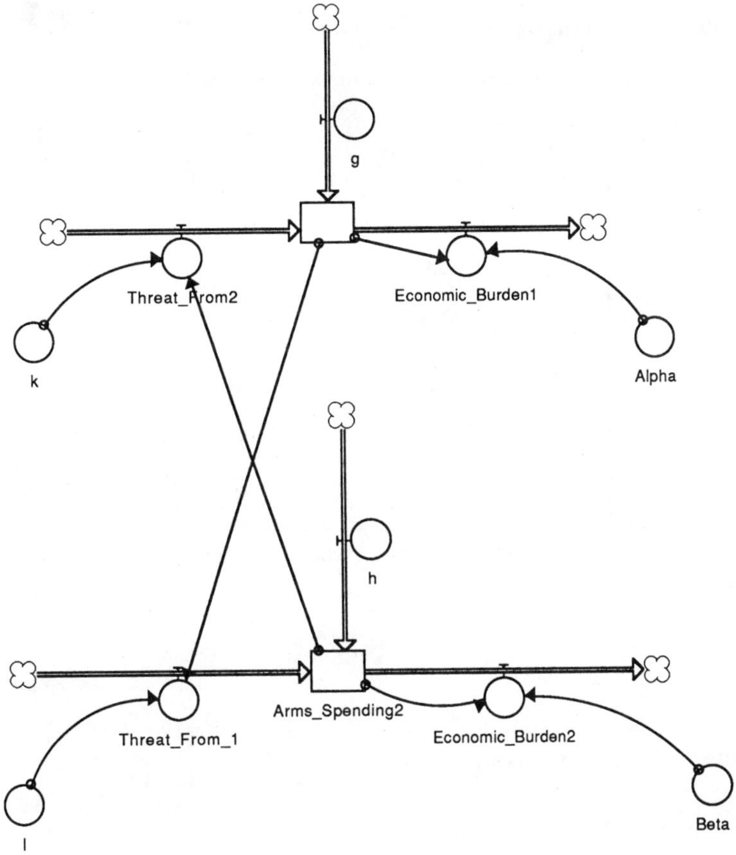

STELLA Program

```
□ Arms_Spending1 = Arms_Spending1 + Threat_From2 - Economic_Burden1 +
  g
  INIT(Arms_Spending1) = 2
□ Arms_Spending2 = Arms_Spending2 + Threat_From_1 - Economic_Burden2
  h
  INIT(Arms_Spending2) = 1
○ Alpha = .2
○ Beta = .2
○ Economic_Burden1 = Alpha*Arms_Spending1
○ Economic_Burden2 = Beta*Arms_Spending2
○ g = 2
○ h = 1
○ k = .1
○ l = .1
○ Threat_From2=k*Arms_Spending2
○ Threat_From_1 = l*Arms_Spending1
```

STELLA Output Results

Plot of x_2 versus x_1

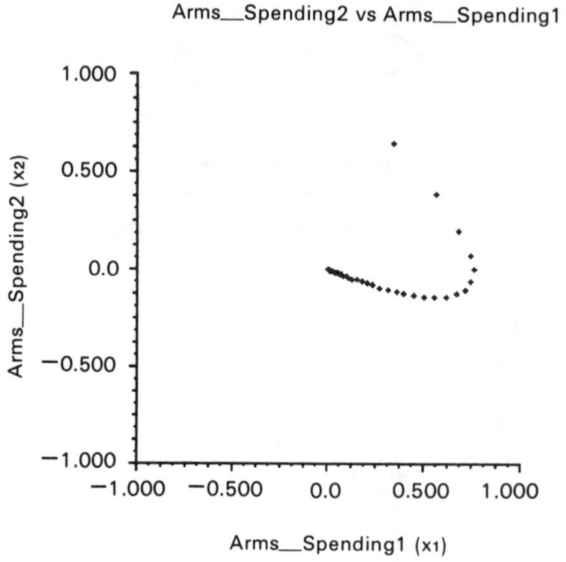

Plot of x_1 and x_2 versus Time

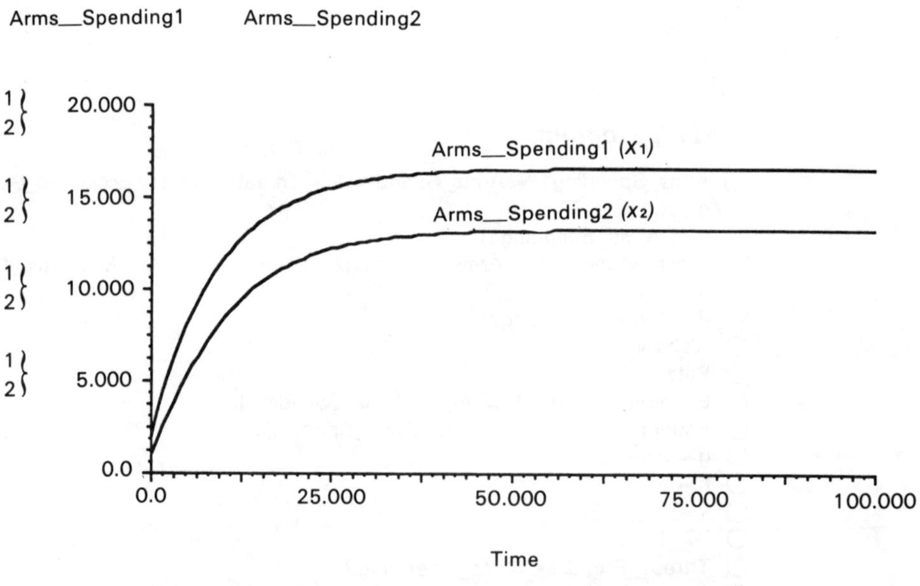

BUCOP Simulation of the Arms Race Model—Select State Variable Formulation

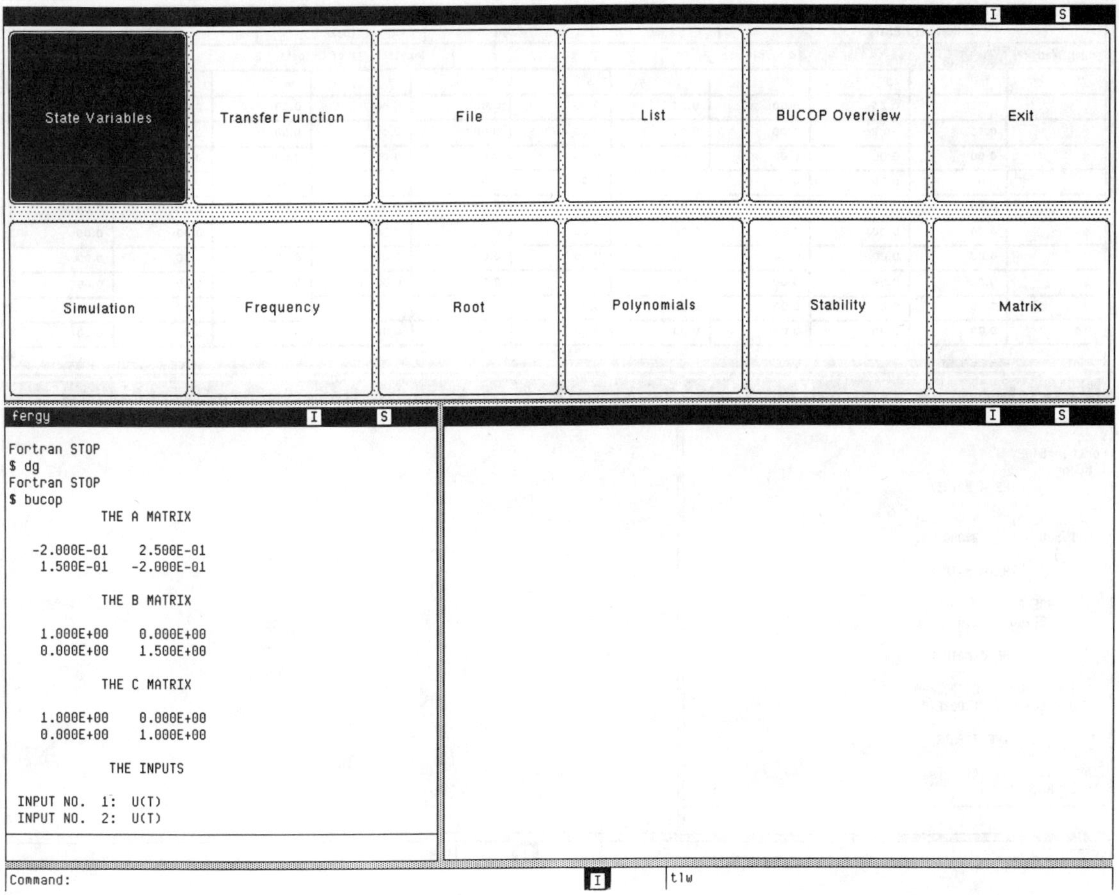

continues

BUCOP Simulation of the Arms Race Model—Select A Matrix

I S

Input State Variables

Keyboard Entry	File	Copy	Clear	Identity

Which Matrix? A B C | # of States: 2 | Inputs | # of Outputs: 2

	1	2	3	4	5	6	7	8	9	10
1	-0.20	0.25	0.00	0.00	0.00	0.00	0.00	0.00	0.00	0.00
2	0.15	-0.20	0.00	0.00	0.00	0.00	0.00	0.00	0.00	0.00
3	0.00	0.00	0.00	0.00	0.00	0.00	0.00	0.00	0.00	0.00
4	0.00	0.00	0.00	0.00	0.00	0.00	0.00	0.00	0.00	0.00
5	0.00	0.00	0.00	0.00	0.00	0.00	0.00	0.00	0.00	0.00
6	0.00	0.00	0.00	0.00	0.00	0.00	0.00	0.00	0.00	0.00
7	0.00	0.00	0.00	0.00	0.00	0.00	0.00	0.00	0.00	0.00
8	0.00	0.00	0.00	0.00	0.00	0.00	0.00	0.00	0.00	0.00
9	0.00	0.00	0.00	0.00	0.00	0.00	0.00	0.00	0.00	0.00
10	0.00	0.00	0.00	0.00	0.00	0.00	0.00	0.00	0.00	0.00

Return

fergy I S

```
Fortran STOP
$ dg
Fortran STOP
$ bucop
            THE A MATRIX

   -2.000E-01    2.500E-01
    1.500E-01   -2.000E-01

            THE B MATRIX

    1.000E+00    0.000E+00
    0.000E+00    1.500E+00

            THE C MATRIX

    1.000E+00    0.000E+00
    0.000E+00    1.000E+00

             THE INPUTS

 INPUT NO.  1:  U(T)
 INPUT NO.  2:  U(T)
```

I S

Command: I tlw

BUCOP Simulation of the Arms Race Model—Select B Matrix

Input State Variables

Keyboard Entry	File	Copy	Clear	Identity

Which Matrix? A B C | # of States: 2 | Inputs | # of Outputs: 2

	1	2	3	4	5	6	7	8	9	10
1	1.00	0.00	0.00	0.00	0.00	0.00	0.00	0.00	0.00	0.00
2	0.00	1.50	0.00	0.00	0.00	0.00	0.00	0.00	0.00	0.00
3	0.00	0.00	0.00	0.00	0.00	0.00	0.00	0.00	0.00	0.00
4	0.00	0.00	0.00	0.00	0.00	0.00	0.00	0.00	0.00	0.00
5	0.00	0.00	0.00	0.00	0.00	0.00	0.00	0.00	0.00	0.00
6	0.00	0.00	0.00	0.00	0.00	0.00	0.00	0.00	0.00	0.00
7	0.00	0.00	0.00	0.00	0.00	0.00	0.00	0.00	0.00	0.00
8	0.00	0.00	0.00	0.00	0.00	0.00	0.00	0.00	0.00	0.00
9	0.00	0.00	0.00	0.00	0.00	0.00	0.00	0.00	0.00	0.00
10	0.00	0.00	0.00	0.00	0.00	0.00	0.00	0.00	0.00	0.00

Return

fergy

```
Fortran STOP
$ dg
Fortran STOP
$ bucop
           THE A MATRIX

   -2.000E-01     2.500E-01
    1.500E-01    -2.000E-01

           THE B MATRIX

    1.000E+00     0.000E+00
    0.000E+00     1.500E+00

           THE C MATRIX

    1.000E+00     0.000E+00
    0.000E+00     1.000E+00

            THE INPUTS

 INPUT NO.  1:  U(T)
 INPUT NO.  2:  U(T)
```

Command: tlw

continues

BUCOP Simulation of the Arms Race Model—Select C Matrix

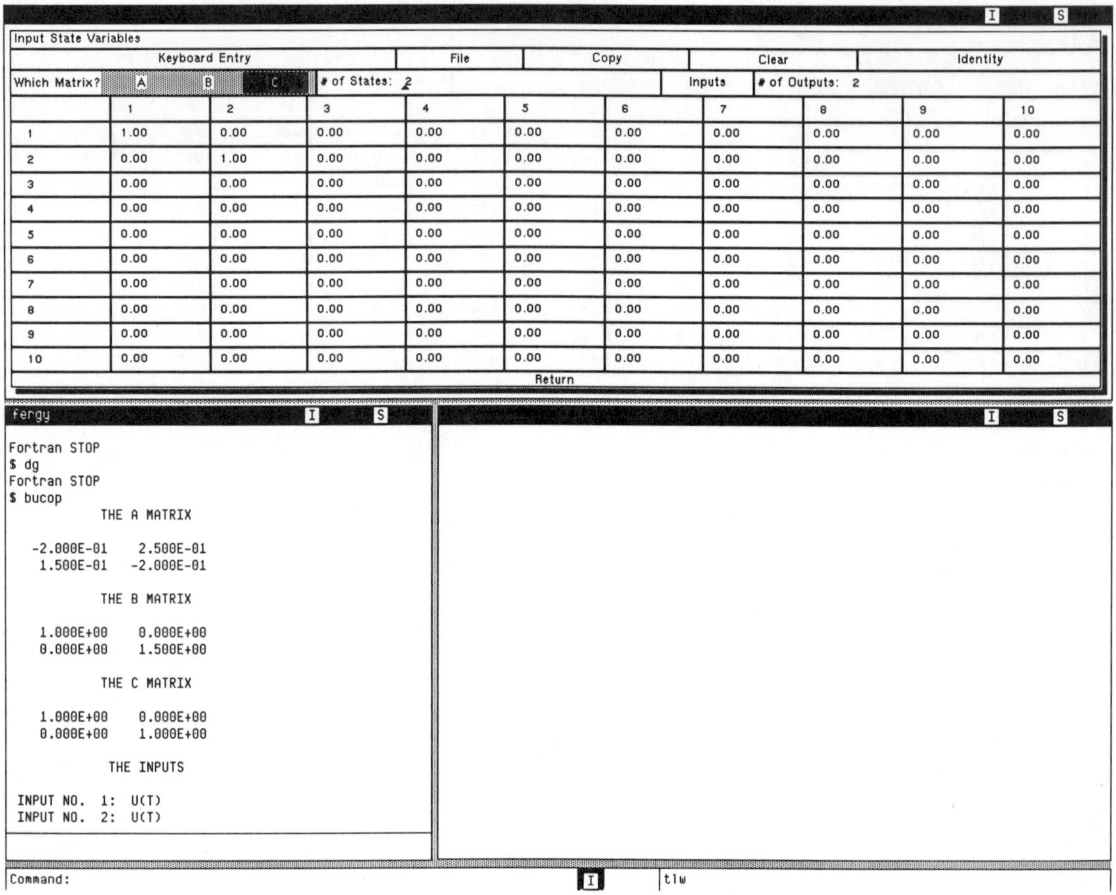

BUCOP Simulation of the Arms Race Model—Select Simulation Option

continues

BUCOP Simulation of the Arms Race Model—Choose Runge–Kutta Integration Method

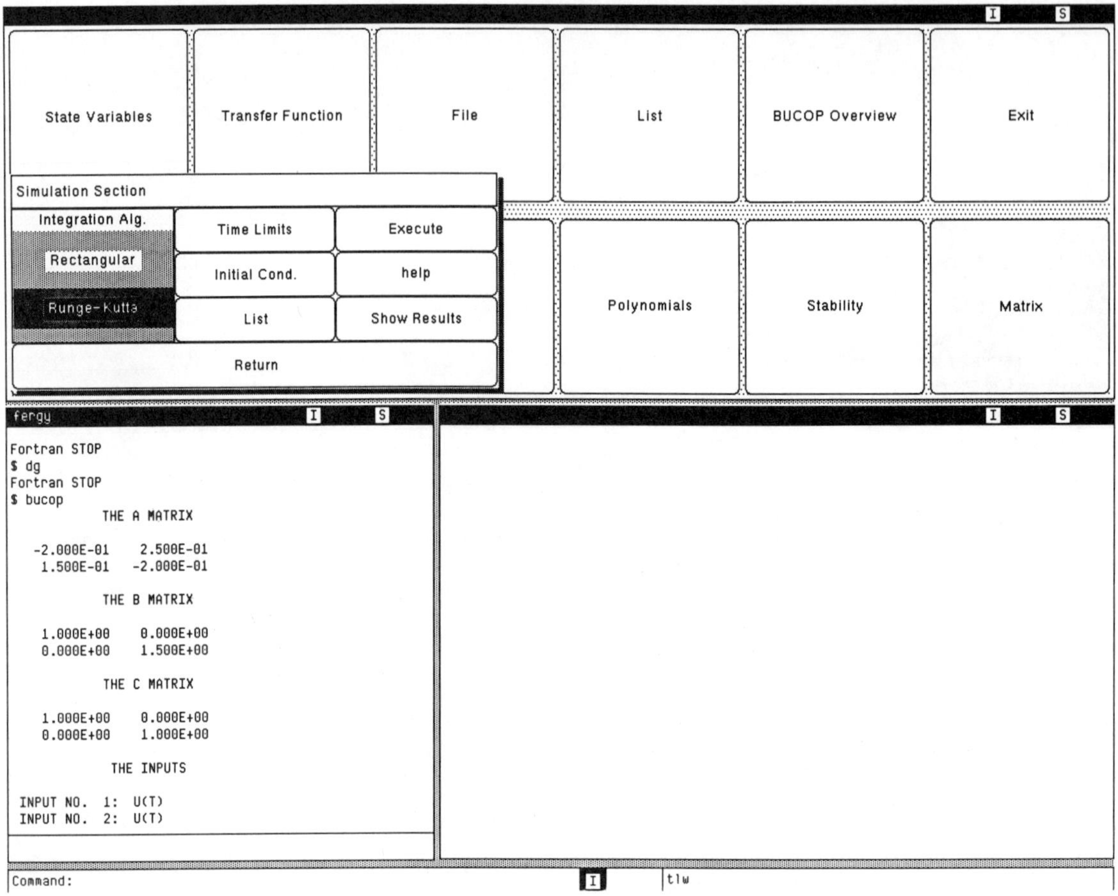

BUCOP Simulation of the Arms Race Model—Enter Start Time, End Time, and Integration Increment

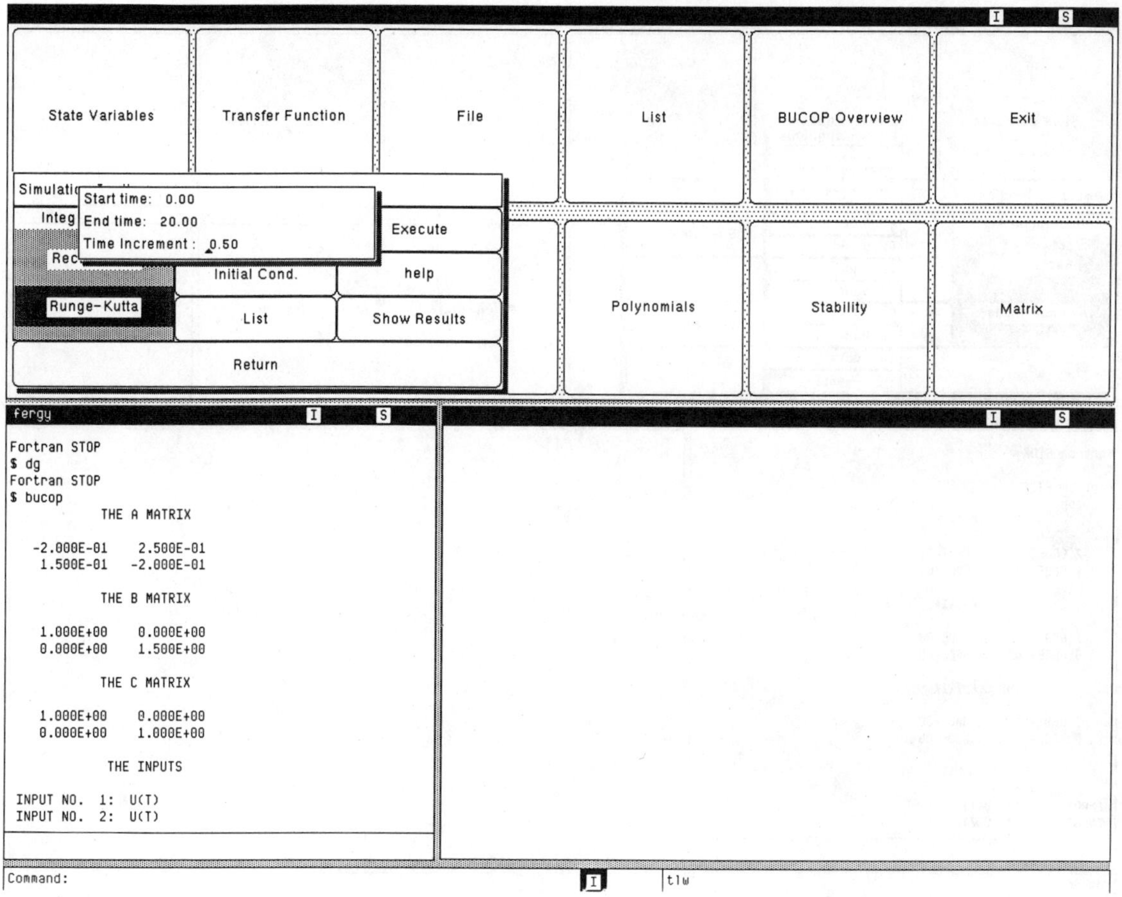

continues

BUCOP Simulation of the Arms Race Model—Enter Initial Conditions for the States

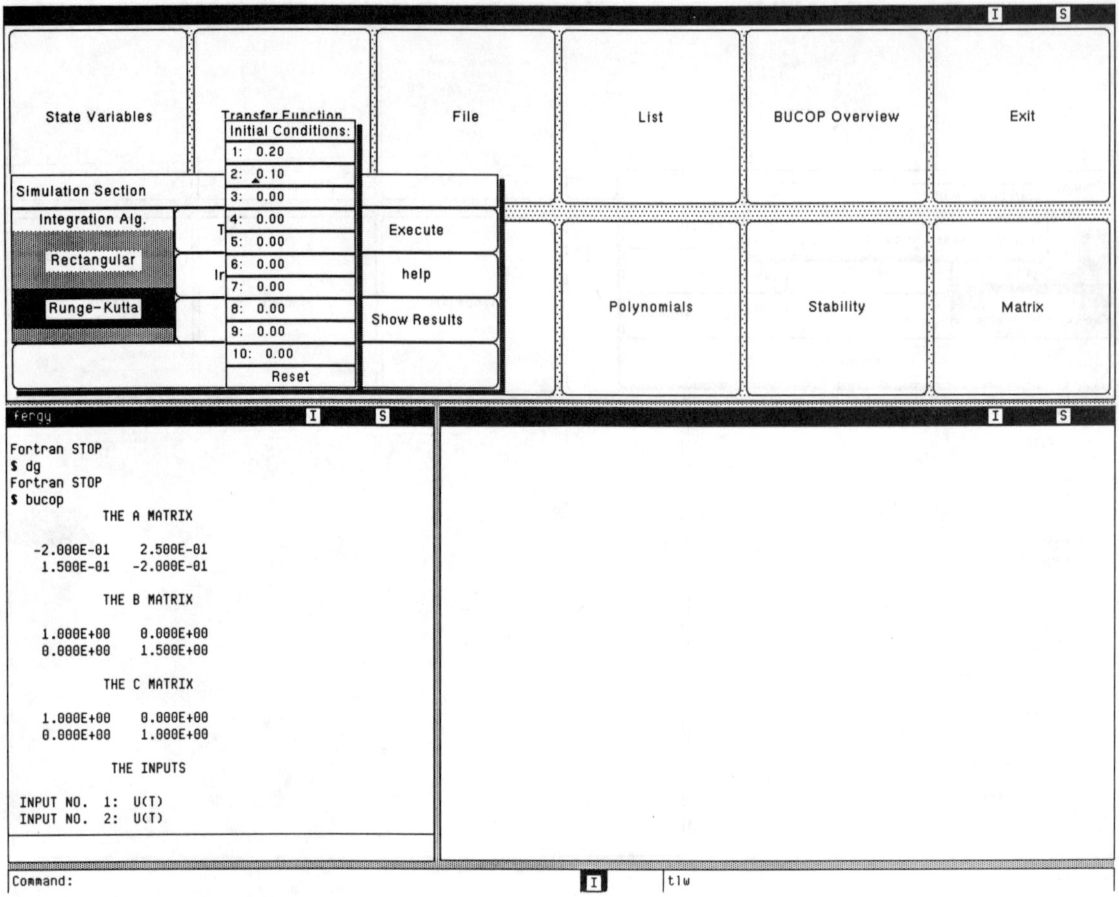

BUCOP Simulation of the Arms Race Model—Plot Output 1 (x_1) and Output 2(x_2)

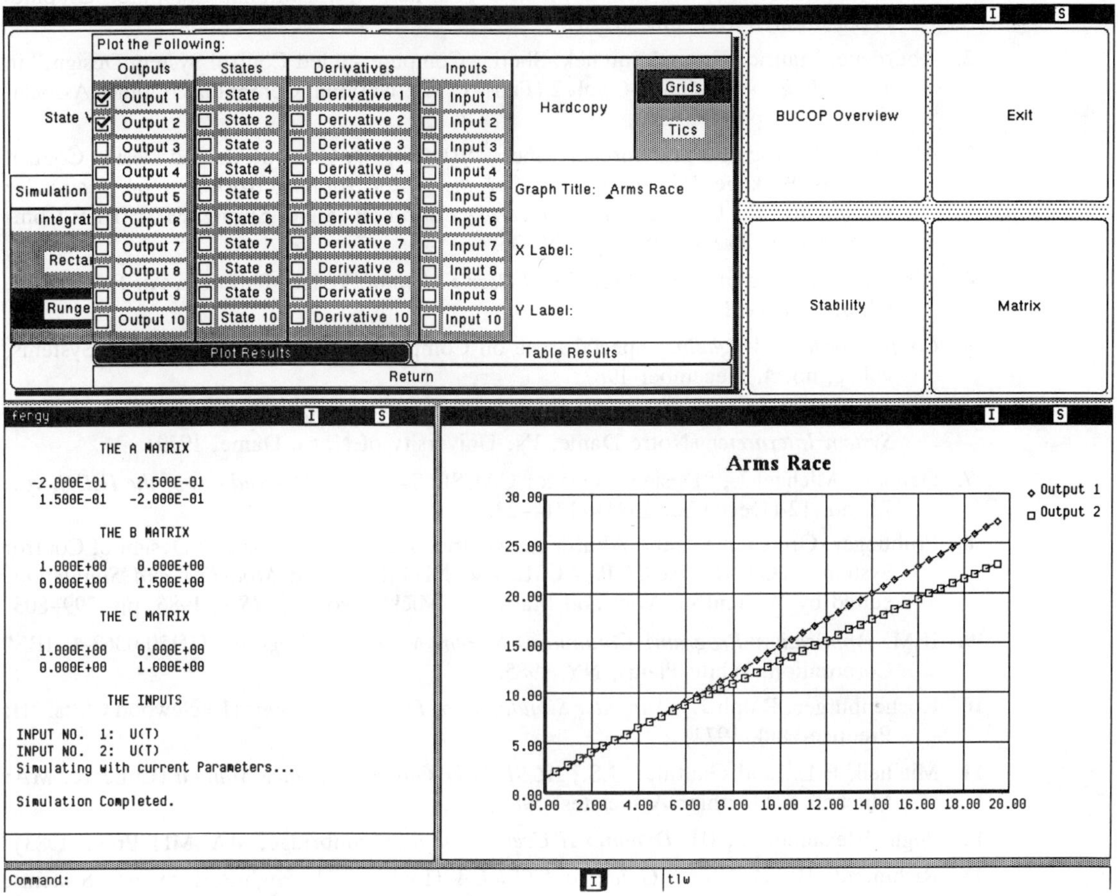

9.8 REFERENCES

1. Aburdene, Maurice F., and Consugar, Joseph, "Computer-aided Control System Design," in *Proceedings of NCGA '85*, vol. 3 (Fairfax, VA: National Computer Graphics Association, 1985), pp. 365–69.
2. Aburdene, Maurice F., and Surchek, Sheri, "Computer-aided Control System Design," in *Proceedings of NCGA '84*, vol. 2 (Fairfax, VA: National Computer Graphics Association, 1984), pp. 195–200.
3. Aburdene, Maurice F., and Surchek, Sheri, "BUCOP: A Computer-aided Design Control Systems Package," in *Proceedings of the Third Annual Workshop on Interactive Computing: CAE: Electrical Engineering Education* (Silver Spring, MD: IEEE Computer Society Press, 1984), pp. 26–36.
4. Air Force Institute of Technology, *Total User's Manual*, Wright-Patterson Air Force Base, Ohio, June 1981.
5. *Control Systems Magazine*, Special Issue on Computer-aided Design of Control Systems, vol. 2, no. 4, December 1982.
6. Davisson, William I., and Uhran, John J., Jr., *A Primer for NDTRAN: A Continuous System Interpreter* (Notre Dame, IN: University of Notre Dame, 1979).
7. Denham, Michael J., "Design Issues for CACSD Systems," *Proceedings of the IEEE*, vol. 72, no. 12 (December 1984): 1714–23.
8. Fruhlinger, Gretchen L., and Aburdene, Maurice F., "Computer-aided Design of Control Systems: TOTAL, BUCOP, ACSL, and NDTRAN," in *Modeling and Simulation*, edited by William G. Vogt and Marlin H. Mickle, vol. 16, *ISA*, 1986, pp. 799–803.
9. IBM, *Application Program: Continuous System Modeling Program*, GH20-0367-4, IBM Corporation, White Plains, NY, 1985.
10. Kochenburger, Ralph J., *Computer Simulation of Dynamic Systems* (Englewood Cliffs, NJ: Prentice-Hall, 1972).
11. Mitchell, E.L., and Gauthier, J.S., *ACSL User Guide/Reference Manual* (Concord, MA: Mitchell & Gauthier Associates, 1981).
12. Pugh, Alexander L., III, *Dynamo II User's Manual* (Cambridge, MA: MIT Press, 1983).
13. Richmond, B., *A User's Guide to STELLA* (Lyme, NH: High-Performance Systems, 1985).
14. Speckhart, H., and Green, W.H., *A Guide to Using CSMP* (Englewood Cliffs, NJ: Prentice-Hall, 1976).

10 Parameter Estimation

10.1 INTRODUCTION

The most difficult problem in the simulation of systems is identifying the system structure. Three types of information are used to develop the structural characteristics of the system: knowledge, opinion, and speculation. Various methods have been suggested to identify system structure. The "brainstorming" method calls for the formation of a committee, including a leader and a secretary. The committee postulates a structure to be verified by simulation while convened in a room with only a blackboard. A variation on brainstorming, the "brainwriting" method, has members of the committee write up their ideas and submit them to each other for discussion. This can be done easily with electronic mail. Another widely used technique is the Delphi method, in which each participant submits information and opinions; group members concurrently review one another's submissions.[2]

As you can see, structural identification depends greatly on the analyst's knowledge of the system. Thus far, in our examples dealing with physical science, engineering, and social systems, we have turned to the basic laws governing the behavior of each system in order to define system structure. In Chapter 4, we discussed the regression method for estimating system parameters, where the regression equation represents a relationship between x and y. In this chapter, we will address methods of parameter estimation for systems whose structure has already been defined or assumed.

10.2 PARAMETER ESTIMATION OF GROWTH AND DECAY PROCESSES

The structure of simple systems exhibiting exponential growth or decay is described by an equation of the form

$$\frac{dP(t)}{dt} = kP(t) \qquad \textbf{(10.1)}$$

where $kP(t)$ = the rate of change of population over the interval dt
k = the rate of growth or decay per unit population per unit time

TABLE 10.1

t	$P(t)$	k
0	1.0	—
1	2.7	0.99
2	7.3	0.99
3	20.0	0.99
4	54.6	1.00
5	148.4	0.99

The parameter of this system is k. As you recall, the value of k determines the behavior of the system. Our job is to estimate k.

Equation (10.1) has an analytic solution of the form

$$P(t) = P(0)e^{kt} \tag{10.2}$$

Solving for k yields

$$k = \frac{1}{t} \ln \frac{P(t)}{P(0)} \qquad (t \neq 0) \tag{10.3}$$

If the historical data for $P(t)$ are known, a table such as Table 10.1 can be formed.

With the data from Table 10.1, a least squares fit can be obtained for k as a function of t. If the constant term in the regression equation is large with respect to the other coefficients of the regression equation, then we consider the value of

FIGURE 10.1 Graphical Method for Parameter Identification

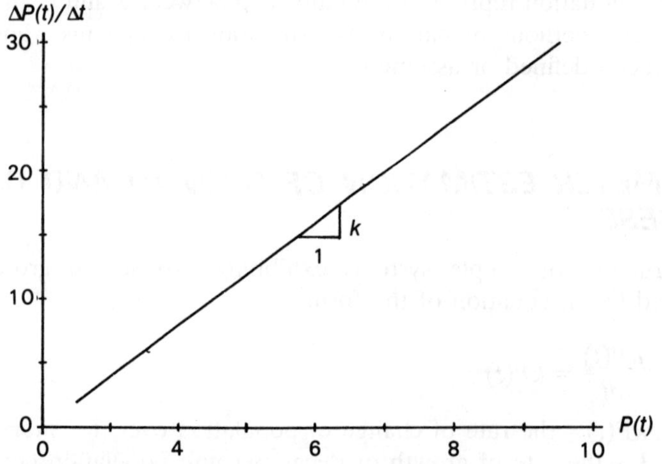

k to be constant. As an alternative to the least squares method, we can use the average value of k to estimate the parameter k. If the variance is small, we can assume that the average value is a good estimate of k. In our case, by inspection, $0.99 \approx 1.0$ is a good estimate.

A graphical technique to determine the value of k involves plotting $\Delta P(t)/\Delta t$ versus $P(t)$, as shown in Figure 10.1. If a straight line is a good approximation of the curve connecting the points on the graph, then the value of k is the slope of the line. This relationship can be verified by inspection of equation (10.1).

10.3 ESTIMATING THE LOGISTICS CURVE COEFFICIENT OF DIFFUSION

In Chapter 2, we developed the following equations for the information diffusion example:

$$\frac{dh}{dt} = khi \tag{10.4}$$

or

$$\frac{dh}{dt} = kh(N - h) \tag{10.5}$$

where h = the number of hearers
i = the number of ignorants
N = the total population
k = the conversion coefficient (parameter of interest)

Solving for k yields (see Section 10.8)

$$k = \frac{1}{Nt} \ln\left(\frac{f(t)}{f(0)} \, \frac{1 - f(0)}{1 - f(t)}\right) \tag{10.6}$$

where

$$f(t) = \frac{h(t)}{N} \qquad (h(0) \neq 0, \, t \neq 0) \tag{10.7}$$

From the observed data, we can use equations (10.6) and (10.7) to form a table similar to Table 10.1, and use the average value of k as an estimate of k.

Another approach would be to plot the values of k computed from equations (10.6) and (10.7) and then obtain a regression equation of k versus time. If the constant term of the regression equation is large compared to the other coefficients of the regression equation, we can use the constant term as our estimate of k.

EXAMPLE 10.1 Assume that the market share for compact disks obeys the growth model of the equation

$$\frac{ds}{dt} = ks(N - s) \tag{10.8}$$

where s is the percent market share.

Let us estimate the value of k in equation (10.8), with $N = 100$ and taking the year 1983 to be $t = 0$. We can then form the following from available data. Using the average value of k, we estimate k to be 5.91×10^{-3}.

t	$s(t)$ (%)	$k \times 10^3$
0	2	—
1	3	4.16
2	8	7.25
3	12	6.33

Using this estimate of k and the solution to equation (10.8), we can forecast the market share for the year 1990. The solution of equation (10.8) takes the form

$$s(t) = \frac{s(0)N}{s(0) + (N - s(0))e^{-kNt}} \tag{10.9}$$

and results in

$$s(t) = \frac{2(100)}{2 + (98)e^{-4.14}} \cong 56\%$$

This means that by 1990, compact disks will have 56% of the market.

10.4 PARAMETER ESTIMATION OF STATE VARIABLE SYSTEMS

In Chapter 3, we saw that the structure of a system can be represented by

$$\begin{aligned}
\dot{x}_1 &= a_{11}x_1 + a_{12}x_2 + \cdots + a_{1n}x_n + b_{11}u_1 + b_{21}u_2 + \cdots + b_{1m}u_m \\
\dot{x}_2 &= a_{21}x_1 + a_{22}x_2 + \cdots + a_{2n}x_n + b_{21}u_1 + b_{22}u_2 + \cdots + b_{2m}u_m \\
&\vdots \\
\dot{x}_n &= a_{n1}x_1 + a_{n2}x_2 + \cdots + a_{nn}x_n + b_{n1}u_1 + b_{n2}u_2 + \cdots + b_{nm}u_m \\
y_1 &= c_{11}x_1 + c_{12}x_2 + \cdots + c_{1n}x_n + d_{11}u_1 + d_{12}u_2 + \cdots + d_{1m}u_m \\
y_2 &= c_{21}x_1 + c_{22}x_2 + \cdots + c_{2n}x_n + d_{21}u_1 + d_{22}u_2 + \cdots + d_{2m}u_m \\
&\vdots \\
y_l &= c_{l1}x_1 + c_{l2}x_2 + \cdots + c_{ln}x_n + d_{l1}u_1 + d_{l2}u_2 + \cdots + d_{lm}u_m
\end{aligned} \tag{10.10}$$

These equations can be more concisely represented as

$$\begin{aligned}
\mathbf{x} &= [A]\mathbf{x} + [B]\mathbf{u} \\
\mathbf{y} &= [C]\mathbf{x} + [D]\mathbf{u}
\end{aligned} \tag{10.11}$$

The parameters of this system are the elements of matrices A, B, C, and D.

To control the system described by this structure, we need to estimate the value of the parameters in matrices A, B, C, and D. All estimation techniques assume that the values of the input vector $\mathbf{u}(t)$ are known. The literature abounds with different methods and techniques that assume, as known, combinations of the rates $\dot{\mathbf{x}}(t)$, the states $\mathbf{x}(t)$, and the outputs $\mathbf{y}(t)$. For example, if the model structure of equation (10.11) is assumed, then the least squares techniques of Chapter 4 and the optimization techniques of Chapter 11 can be used to estimate the unknown values in matrices A, B, C, and D from measurements of inputs $\mathbf{u}(t)$ and outputs $\mathbf{y}(t)$.

10.5 PARAMETER ESTIMATION OF STOCHASTIC SYSTEMS

The salesperson problem of Chapter 2 provides a good example for demonstrating the next method. Let us assume the following two conditions:

1. There exists a set of data partitioned into k mutually exclusive categories.
2. No information is available concerning the probability distribution of the data and whether one value of a particular parameter is preferable to another.

Sample data for the salesperson problem are given in Table 10.2. Our problem here is to identify the most plausible values for the probability of a sale, P_1, and the probability of no sale, $P_2 = (1 - P_1)$.

As background, let us first define the general likelihood function as

$$L(P_1, P_2, \ldots, P_k, n_1, n_2, \ldots, n_k) = C(P_1)^{n_1}(P_2)^{n_2} \cdots (P_k)^{n_k} \qquad \textbf{(10.12)}$$

where $n_1, n_2, \ldots, n_k$ = the number of occurrences of categories $1, 2, \ldots, k$, $\sum_{i=1}^{k} n_i = n$

$P_1, P_2, \ldots, P_k$ = the probabilities of categories $1, 2, \ldots, k$ occurring

C = a constant

TABLE 10.2 Sample Data for Salesperson System

Visit	*Sale*	*No Sale*
1	1	0
2	0	1
3	1	0
4	0	1
5	1	0
6	1	0
7	1	0
8	1	0
9	0	1
10	0	1
	6	4

Similarly, we define the maximum likelihood function as

$$L_M(\hat{P}_1, \hat{P}_2, \ldots, \hat{P}_k, n_1, n_2, \ldots, n_k) = C(\hat{P}_1)^{n_1}(\hat{P}_2)^{n_2} \cdots (\hat{P}_k)^{n_k} \quad \textbf{(10.13)}$$

where $\hat{P}_i$ is the value P_i that maximizes the right side of equation (10.12). Now let us apply the general likelihood function to the salesperson problem.

From Table 10.2, we obtain

$$L = C(P_1)^6(1 - P_1)^4 \quad \textbf{(10.14)}$$

The value $P_1 = 0.6$ maximizes the function L. Therefore, $\hat{P}_1 = 0.6$ is the most plausible or best supported value of P_1. It should not be surprising that this is also the observed relative frequency of a sale.

Let us now define the likelihood ratio:

$$R(P_1, P_2, \ldots, P_k) = \frac{L(P_1, P_2, \ldots, P_k, n_1, n_2, \ldots, n_k)}{L_M(\hat{P}_1, \hat{P}_2, \ldots, \hat{P}_k, n_1, n_2, \ldots, n_k)} \quad \textbf{(10.15)}$$

This ratio R, where $0 \leq R \leq 1$, is a measure of the relative plausibility of using values for the P_i other than those supported by the n sample observations. For the salesperson problem, the relative plausibility that $P_1 = 0.5$ is obtained from

$$R(P_1) = \frac{C(P_1)^6(1 - P_1)^4}{C(0.6)^6(0.4)^4}$$

Therefore,

$$R(P_1 = 0.5) = \frac{C(0.5)^6(0.5)^4}{C(0.6)^6(0.4)^4} = 0.817$$

Thus, if we estimated that the value of $P_1 = 0.5$, this estimate is 0.817 as probable as the value of $P_1 = 0.6$. Figure 10.2 is a plot of $R(P_1)$ versus P_1.

If the data show 60 sales and 40 no sales, Figure 10.2 also illustrates the relative likelihood function (dashed line). The estimate of P_1 is now clustered around 0.6; this precision (smaller standard deviation of the estimates) is a function of the larger sample size.

To determine the relative plausibility that P_1 is equal to 0.5, we calculate

$$R(0.5) = \frac{C(0.5)^{60}(0.5)^{40}}{C(0.6)^{60}(0.4)^{40}} = 0.133$$

This means that the probability of a sale being equal to 0.5 is not very plausible—approximately $\frac{1}{8}$ as probable as 0.6.

Thus far, we have assumed nothing about the distribution of the random variables, which are sale and no sale. Let us now turn to the case where a particular (e.g., Poisson) probability distribution is considered plausible from either the data or previous analysis of similar data.

FIGURE 10.2 Plot of Likelihood Ratio vs. P_1

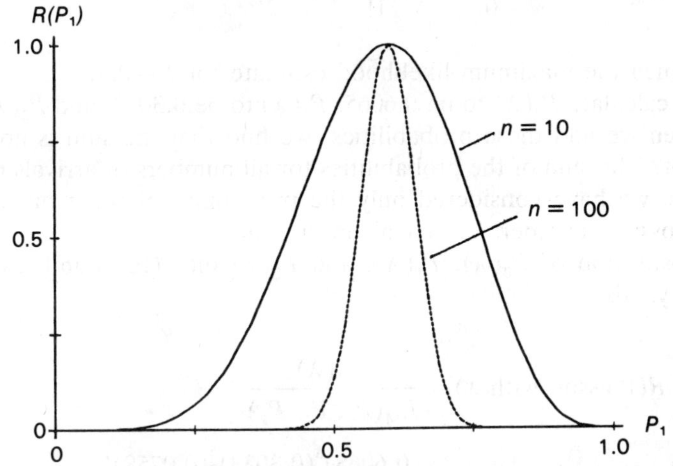

For example, the air traffic controller at a small rural airport kept records on the number of arrivals over a 10-hour period. The records are shown in Table 10.3. Based on this table, the likelihood function is

$$L = C(P_0)^6(P_1)^3(P_2)^1(P_3)^0 \tag{10.16}$$

the maximum likelihood estimates are $\hat{P}_0 = 0.6$, $\hat{P}_1 = 0.3$, $\hat{P}_2 = 0.1$, and $\hat{P}_3 = 0$; and the relative likelihood function is

$$R(P_0, P_1, P_2) = \frac{P_0^6 P_1^3 P_2^1 P_3^0}{(0.6)^6(0.3)^3(0.1)^1(1)} \tag{10.17}$$

If we consider the data to be Poisson distributed with parameter λ, then the likelihood function for the Poisson distribution is

$$\begin{aligned} L_P(\lambda) &= C\left(\frac{\lambda^0 e^{-\lambda}}{0!}\right)^6 \left(\frac{\lambda^1 e^{-\lambda}}{1!}\right)^3 \left(\frac{\lambda^2 e^{-\lambda}}{2!}\right)^1 \\ &= C\frac{\lambda^5 e^{-10\lambda}}{2} \end{aligned} \tag{10.18}$$

TABLE 10.3

Number of Arrivals in an Hour	*Number of Times Arrivals Recorded*
0	6
1	3
2	1
>2	0

and $$L_M(\hat{\lambda}) = C\left(\frac{\hat{\lambda}^0 e^{-\hat{\lambda}}}{0!}\right)^6 \left(\frac{\hat{\lambda}^1 e^{-\hat{\lambda}}}{1!}\right)^3 \left(\frac{\hat{\lambda}^2 e^{-\hat{\lambda}}}{2!}\right)^1 \tag{10.19}$$

from which the maximum likelihood estimate for $\hat{\lambda} = 0.5$.

We calculate $P_0(\hat{\lambda})$ to be 0.6065, $P_1(\hat{\lambda})$ to be 0.3033, and $P_2(\hat{\lambda})$ to be 0.0758. But when we add these probabilities, we find that the sum is not equal to one. Why not? The sum of the probabilities for all numbers of arrivals must equal one, but here we have considered only the probability of three out of the infinitely many possible numbers of arrivals in an hour.

Substitution of $P_0(\hat{\lambda})$, $P_1(\hat{\lambda})$, and $P_2(\hat{\lambda})$ into (the numerator of) equation (10.17) yields

$$\begin{aligned} R(\text{Poisson with } \hat{\lambda}) &= \frac{L_M(\hat{\lambda})}{L_M(\hat{P}_0, \hat{P}_1, \hat{P}_2)} \\ &= \frac{(0.6065)^6(0.3033)^3(0.0758)^1}{(0.6)^6(0.3)^3(0.1)^1} = 0.835 \end{aligned} \tag{10.20}$$

indicating that a Poisson probability distribution with $\hat{\lambda} = 0.5$ is very plausible for this set of data.

$R(\text{Poisson with } \hat{\lambda})$ must be interpreted with care. It represents the relative plausibility of using the Poisson distribution to fit the data as compared to using a general probability distribution to fit the data.

If a Poisson distribution is considered plausible for the number of arrivals, then the Poisson relative likelihood function is

$$\begin{aligned} R_P(\lambda) &= \frac{L_P(\lambda)(\text{Poisson with } \lambda)}{L_M(\hat{\lambda})(\text{Poisson with } \hat{\lambda})} \\ &= \prod_{i=1}^{k} \left(\frac{\lambda^i e^{-\lambda}}{\hat{\lambda}^i e^{-\hat{\lambda}}}\right)^{n_i} \end{aligned} \tag{10.21}$$

Plots of $R_P(\lambda)$ and $R(\lambda)$ versus λ are shown in Figure 10.3.

Now that we know how to develop the likelihood function, we will demonstrate how to obtain the maximum likelihood estimate for parameters. Let us show how to compute the $\hat{P}_i$, the maximum likelihood estimates of the P_i, in the equation

$$L_M(P_0, P_1, \ldots, P_k, n_0, n_1, \ldots, n_k) = C(\hat{P}_0)^{n_0}(\hat{P}_1)^{n_1} \cdots (\hat{P}_k)^{n_k} \tag{10.22}$$

To compute the values of P_i that maximize L_M, we take the partial derivatives of the log of L with respect to P_i, set these derivatives equal to zero, and solve for P_i. These values of P_i are the $\hat{P}_i$. The values that maximize $\ln(L)$ also maximize L

FIGURE 10.3 Plot of Likelihood Ratios

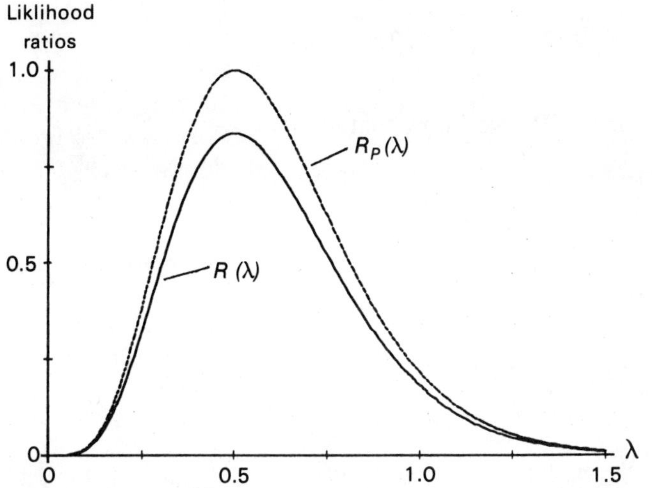

since $\ln(L)$ is montonic increasing. The likelihood function is given by

$$L = C(P_0)^{n_0}(P_1)^{n_1} \cdots (P_k)^{n_k} \tag{10.23}$$

and

$$\begin{aligned} \ln L(P_0, P_1, \ldots, P_k) &= \ln C + \ln(P_0)^{n_0} + \ln(P_1)^{n_1} + \cdots + \ln(P_k)^{n_k} \\ &= \ln C + n_0 \ln P_0 + n_1 \ln P_1 + \cdots + n_k \ln P_k \\ &= \ln C + n_0 \ln P_0 + n_1 \ln P_1 + \cdots \\ &\quad + n_k \ln[0 - (P_0 + P_1 + \cdots + P_{k-1})] \end{aligned} \tag{10.24}$$

Then

$$\frac{\partial \ln(L)}{\partial P_0} = \frac{n_0}{P_0} - \frac{n_k}{P_k} \tag{10.25}$$

Setting the right side of equation (10.25) equal to zero,

$$\frac{n_0}{P_0} - \frac{n_k}{P_k} = 0 \tag{10.26}$$

leads to

$$P_0 = \frac{n_0}{n_k} P_k \tag{10.27}$$

Similarly, we can show that

$$P_i = \frac{n_i}{n_k} P_k \qquad (i = 1, 2, \ldots, k) \tag{10.28}$$

To solve for the P_i, we know that

$$\sum_{i=0}^{k} P_i = 1 \tag{10.29}$$

Therefore,

$$\frac{n_0}{n_k} P_0 + \frac{n_1}{n_k} P_1 + \cdots + \frac{n_{k-1}}{n_k} P_{k-1} + P_k = 1 \tag{10.30}$$

and $$P_k = \frac{n_k}{n_0 + n_1 + \cdots + n_k} = \frac{n_k}{n} \tag{10.31}$$

and $$P_i = \frac{n_i}{n} \ (i = 0, 1, \ldots, k) \tag{10.32}$$

Next, let us compute the value of λ that maximizes $L_P(\lambda)$. As in the previous example,

$$L_P(\lambda) = C\left(\frac{e^{-\lambda}\lambda^0}{0!}\right)^{n_0}\left(\frac{e^{-\lambda}\lambda^1}{1!}\right)^{n_1}\cdots\left(\frac{e^{-\lambda}\lambda^k}{k!}\right)^{n_k} \tag{10.33}$$

which can be simplified to read

$$L_P(\lambda) = C\,\frac{e^{-\lambda(n_0+n_1+n_2+\cdots+n_k)}}{0!1!2!\cdots k!}\,\lambda^{n_1+2n_2+\cdots+kn_k} \tag{10.34}$$

which in turn can be written as

$$L_P(\lambda) = C\,\frac{e^{-\lambda n}\lambda^{\sum_{i=1}^{k} in_i}}{\prod_{i=1}^{k} i!} \tag{10.35}$$

If we take a log of L_P, we obtain

$$\ln(L_P(\lambda)) = \ln C - \lambda n + \sum_{i=0}^{k} in_i \ln\lambda - \ln\prod_{i=1}^{k} i! \tag{10.36}$$

Taking the derivative with respect to λ and equating the result to zero leads to

$$-n + \frac{1}{\lambda} \sum_{i=0}^{k} in_i = 0 \tag{10.37}$$

and

$$\lambda = \frac{\sum_{i=0}^{k} in_i}{n} \tag{10.38}$$

In the next example, we will apply the methods of the previous example to estimate the number of people with credit cards who pay their total monthly charges. A credit card company selects a random sample of 100 cardholders and counts the number of times the cardholders pay their total monthly bill over the course of a year. The sample data collected are

Number of Months a Cardholder Pays Total Monthly Bill	*Number of Cardholders, n_k, Who Pay Monthly Bills k Times*	kn_k
0	3	0
1	5	5
2	7	14
3	15	45
4	16	64
5	15	75
6	13	78
7	9	63
8	7	56
9	4	36
10	3	30
11	2	22
12	1	12
	100	500

If we assume a Poisson distribution for the number of months that a cardholder pays the total monthly bill, then $\hat{\lambda} = 500/100 = 5$ is a maximum likelihood estimate for this distribution.

We can compute the probability that the number of months in which a cardholder pays the total bill equals a specified value and compare it with our sampled data. For example, using the Poisson formula

$$P(k) = \frac{e^{-\lambda}\lambda^k}{k!}$$

with $\lambda = 5$, we obtain

k	*Probability of k*	100 × *Probability*
0	0.007	0.7
1	0.034	3.4
2	0.084	8.4
3	0.140	14.0
4	0.176	17.6
5	0.176	17.6
6	0.146	14.6
7	0.104	10.4
8	0.065	6.5
9	0.036	3.6
10	0.018	1.8
11	0.008	0.8
12	0.001	0.1

If we multiply this probability calculation by 100 (because we had a sample of 100) and compare entries with the observed data, we see that the number of cardholders who paid their total monthly bill five times is 18, compared to 15 cardholders in the sample. We encourage the reader to compare the rest of the values in the table.

10.6 SUMMARY

In this chapter, we presented methods for estimating the parameters of exponential growth and decay, logistics curves, and state variable models of systems. We also presented the maximum likelihood function methods of estimating parameters of a population using small sample sizes. The methods outlined should be used for analysis when the amount of sample data is small compared to the size of the total population under study.

10.7 REVIEW PROBLEMS

1. Distinguish among the "brainstorming," "brainwriting," and Delphi methods of identifying system structure.
2. The total U.S. urban population is expected to increase exponentially. Using the following data, calculate the growth coefficients. What is the doubling time?

Year	*Urban Population*
1970	118,446,566
1960	95,848,487
1950	69,249,148

Source: *Historical Statistics of the United States*, Part 1, Department of Commerce, Bureau of the Census, September 1975, p. 13.

3. According to Donald J. Bogue, the world population has been growing exponentially since 1650. From the data below, calculate the population growth rate.

Year	*Estimated World Total (Millions)*
1960	3008
1950	2509
1940	2249
1930	2015
1920	1811
1900	1590
1850	1131
1800	912
1750	711
1650	507

Source: Donald J. Bogue, *Principles of Demography* (New York: Wiley, 1969), p. 48.

4. The following data were obtained for the diffusion of radios in the United States. Calculate the diffusion coefficients. Hint: The population of the United States is not constant over the period of interest.

Year	*Households with Radio Sets*
1940	28,500
1941	29,300
1942	30,600
1943	30,800
1944	32,500
1945	33,100
1946	33,988
1947	35,900
1948	37,623
1949	39,300
1950	40,700
1951	41,900
1952	42,800
1953	44,800
1954	45,100
1955	45,900
1956	46,800
1957	47,600
1958	48,500
1959	49,450
1960	50,193

Source: *Historical Statistics of the United States*, Part 2, U.S. Department of Commerce, Bureau of the Census, September 1975.

5. The population of the United States exhibits characteristics of the logistics curve. From census data for 1810 to 1960, calculate the diffusion coefficient.

6. Show that the sample mean $\bar{Y}$ is a maximum likelihood estimator of the theoretical mean μ for a normally distributed random variable.
7. Show that the sample variance s^2 for a normally distributed random variable is also the maximum likelihood estimator of the theoretical variance σ^2.
8. The following data were obtained from a questionnaire with three mutually exclusive options:

   Five voted for option 1
   Four voted for option 2
   One voted for option 3

   Determine the maximum likelihood estimates for P_1, P_2, and P_3. What is the plausibility that the source had a Poisson distribution?
9. The following data were obtained from a questionnaire with three mutually exclusive options:

   Four voted for option 1
   Three voted for option 2
   Three voted for option 3

   What are the maximum likelihood estimates for P_1, P_2, and P_3? How plausible is the hypothesis that the source of the data was Poisson distributed?
10. Show that $\hat{\lambda}$ for the equation

$$L_P(\lambda) = C\frac{\lambda^5 e^{-10\lambda}}{2}$$

   is given by $\hat{\lambda} = 0.5$.
11. Show that the most plausible value for $\hat{P}_1$ for the equation

$$L(P_1) = C(P_1)^4(1 - P_1)^6$$

   is given by $\hat{P}_1 = 0.4$.

10.8 APPENDIX: SOLUTION FOR ESTIMATING THE LOGISTICS CURVE COEFFICIENT OF DIFFUSION

We begin with

$$\frac{dh(t)}{dt} = kh(t)(N - h(t)) \tag{1}$$

Dividing both sides by N yields

$$\frac{1}{N}\frac{dh(t)}{dt} = Nk\frac{h(t)}{N}\left(1 - \frac{h(t)}{N}\right) \tag{2}$$

Now, let $k' = Nk$ and $f(t) = h(t)/N$. Then

$$\frac{df(t)}{dt} = k'f(t)(1 - f(t)) \tag{3}$$

Rearranging, we get

$$\frac{df(t)}{f(t)(1 - f(t))} = k'\,dt \tag{4}$$

The left side of equation (4) can be written as

$$\frac{df(t)}{f(t)(1 - f(t))} = \frac{df(t)}{f(t)} + \frac{df(t)}{1 - f(t)} \tag{5}$$

Substituting equation (5) into equation (4), we get

$$\frac{df(t)}{f(t)} + \frac{df(t)}{1 - f(t)} = k'\,dt \tag{6}$$

Integrating both sides over the interval $(0, t)$ leads to

$$(\ln f(t) - \ln f(0)) - [\ln(1 - f(t)) - \ln(1 - f(0))] = k't \tag{7}$$

which yields

$$\ln \frac{f(t)}{f(0)} - \ln\left(\frac{1 - f(t)}{1 - f(0)}\right) = k't \tag{8}$$

which is equivalent to

$$\ln\left(\frac{f(t)}{f(0)}\,\frac{1 - f(0)}{1 - f(t)}\right) = k't \tag{9}$$

Solving for k', we obtain

$$k' = \frac{1}{t} \ln\left(\frac{f(t)}{f(0)}\,\frac{1 - f(0)}{1 - f(t)}\right) \qquad (t \neq 0) \tag{10}$$

and the diffusion coefficient is

$$k = \frac{k'}{N} = \frac{1}{Nt} \ln\left(\frac{f(t)}{f(0)}\,\frac{1 - f(0)}{1 - f(t)}\right) \tag{11}$$

10.9 REFERENCES

1. Borrie, John A., *Modern Control Systems* (Englewood Cliffs, NJ: Prentice-Hall, 1987).
2. Dalkey, N.C., "An Experimental Study of Group Opinion—The Delphi Method," *Futures* 1, no. 5 (September 1969): 408–26.
3. Dzhaparidze, K.O., *Parameter Estimation and Hypothesis Testing in Spectral Analysis of Stationary Time Series*, translated by Samuel Kotz (New York: Springer-Verlag, 1986).
4. Lindsey, J.K., *Inferences from Sociological Survey Data: A Unified Approach* (San Francisco: Jossey-Bass, 1973).
5. Mendel, J.M., *Lessons in Digital Estimation Theory* (Englewood Cliffs, NJ: Prentice-Hall, 1987).
6. Ratkowsky, David A., *Nonlinear Regression Modeling: A Unified Practical Approach* (New York: Marcel Dekker, 1983).
7. Soderstrom, Torsten, *Instrumental Variable Methods for System Identification* (Berlin: Springer-Verlag, 1983).
8. Sorenson, Harold Wayne, *Parameter Estimation: Principles and Problems* (New York: Marcel Dekker, 1980).

11 Parameter Optimization

11.1 INTRODUCTION

This chapter focuses on the use of simulation in making policy decisions, specifically for the selection of the most desirable value(s) of one or more controllable system parameters. The process of determining what is best is called optimization. Determination of the social security withholding rate is a good example of choosing the value of one parameter to "optimize" a system, in this case the social security system.

What is best depends on the *objectives* of the system designer. What is best for one person might be the worst for another person. For example, what do we mean by the optimum city size? According to A.B. Wolfe, the optimum population size is the size that gives the largest possible per capita product of consumer goods with the given available natural resources.[17] David Rasmussen defined the optimum size of a city as that which generates the most advantages from agglomeration.[14] Some of the factors that determine optimal city size are accessibility; health conditions; crime and safety; educational facilities; leisure and recreational activities; social institutions; community and family ties; political participation; psychological and mental health hazards; labor force; number of births, deaths, and population over 60; and government output.

To find an optimum, we must express our goal quantitatively in terms of a cost function (figure of merit, object function, or objective function). The optimization process either maximizes or minimizes this cost function. In this chapter, we will discuss optimization in terms of minimizing the cost function, since maximization can be accomplished by minimizing the negative, or inverse, of the cost function. We are limiting our discussion to finding a local minimum and not the global minimum.

11.2 COST FUNCTIONS

The following two examples help demonstrate the formulation of cost functions, which are essential for optimization of systems.

In our first example, we want to determine the "best" value of insulation thickness that will minimize the cost of heating our house. The heating cost includes the fuel cost as well as the cost of adding insulation amortized over a period of years.

From our basic physics courses, we know that if the heat supplied by the heat source equals the heat loss from the house, then the temperature inside the house remains constant. Usually a home heating system is "sized" for a particular temperature difference between the desired temperature inside the house and the temperature outside.

We also know that

$$Q_{\text{loss}} = \frac{k}{w}(T_{\text{h}} - T_{\text{a}})$$

where Q_{loss} = the heat loss
k = a constant that depends on the type of insulation
T_{a} = the outside temperature
T_{h} = the temperature inside the house
w = the insulation thickness

For simplicity and demonstration purposes, let us assume that $k = 1/2$ and that $T_{\text{h}} - T_{\text{a}} = 40°$. Now we can sketch Q_{loss} versus w, as shown in Figure 11.1.

We notice that each time we double the thickness, we reduce the heat loss by one half. This plot indicates that by increasing the insulation thickness from 1 unit to 2 units, we decrease Q_{loss} by 10 units; however, when we increase the thickness from 4 units to 8 units, we reduce the heat loss by only 2.5 units.

FIGURE 11.1 Plot of Heat Loss vs. Insulation Thickness

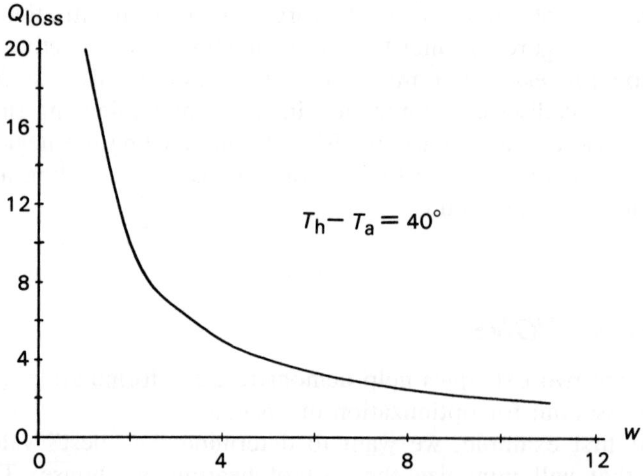

FIGURE 11.2 Plot of Fuel Cost and Insulation Cost vs. Insulation Thickness

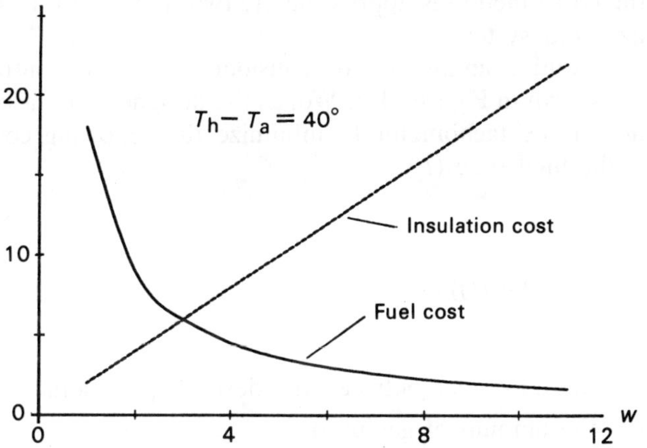

To keep the temperature difference ($T_h - T_a$) constant, the heat source must supply heat at a rate Q, equal to the rate of heat loss, Q_{loss}. If we assume that a unit of Q costs \$0.90, then the cost of fuel versus w, where $T_h - T_a$ is constant, is shown by the solid line in Figure 11.2. If we also assume that the amortized cost of insulation thickness is \$2.00 per unit insulation thickness (w), then the dashed line in Figure 11.2 shows the cost of insulation versus w.

Figure 11.3 shows the total cost of heating the house versus w, which we calculate by adding the fuel cost and the insulation cost from Figure 11.2. If we

FIGURE 11.3 Plot of Total Cost vs. Insulation Thickness

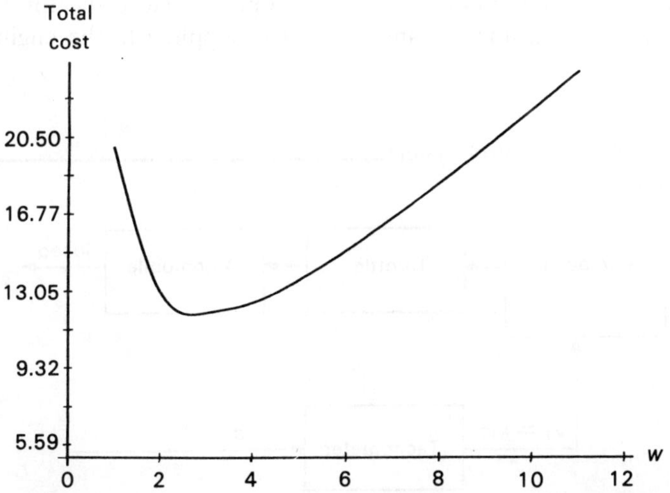

consider the total heating cost to be our cost function, then the value of w that minimizes the cost function is approximately two units. With this value for w, we have optimized our system.

For our second example, let us consider the cruise control system of an automobile as shown in Figure 11.4. We, as the designers, are asked to choose the characteristics of the tachometer to minimize the following cost function from time zero to the final time (t_f):

$$\int_0^{t_f} (Ae^2(t) + Bg^2(t))\, dt \tag{11.1}$$

where $e(t)$ = the difference between the desired speed and the actual speed
$g(t)$ = the amount of gas used
A and B = constants

As we can see, the cost function represents a compromise between minimizing the size of the error squared and minimizing the amount of gas used. The relative importance of these two goals is determined by the values of A and B.

The tachometer is a device that converts speed s to a voltage v_T, mathematically represented by

$$v_T = k_T s$$

where k_T is the conversion coefficient. The voltage of the tachometer is read by the computer through an analog-to-digital (A/D) converter. Based on this value, the computer uses a control strategy to open or close the throttle, respectively increasing or decreasing the amount of gas supplied to the engine.

FIGURE 11.4 Automobile Cruise Control System

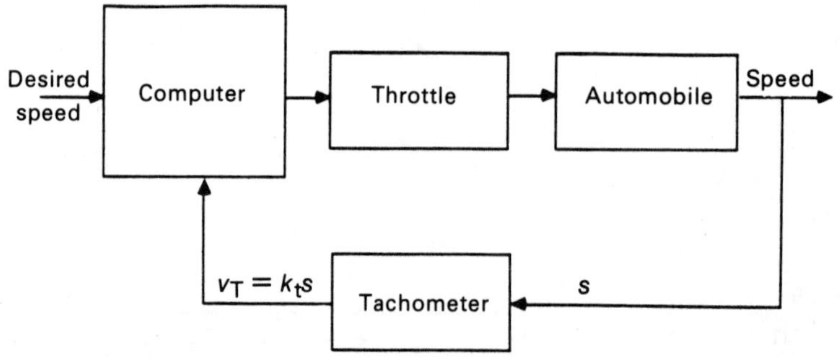

11.3 SINGLE-VARIABLE OPTIMIZATION METHODS

This section is devoted to minimizing the cost function with respect to a single parameter. We will cover Golden Sections and Powell's method, which are best demonstrated through examples. These methods are the basis of multiparameter optimization methods, which we will present in following sections. Our purpose is to provide an introduction to these useful and basic computer search techniques.

Let us assume that the cost functions $F(x)$ versus the controllable parameter x is as shown in Figure 11.5. The function $F(x)$ is convex and has a unique and global minimum $F(x^*)$. There are many algorithms that determine the value of x^* that will minimize $F(x)$. The algorithms fall into two categories: function comparison and function approximation.

Function Comparison Methods

The basic feature of function comparison methods is that the cost function values at estimates of x^* are used in comparison tests. Representative search algorithms are the Golden Sections, bisection, and Fibonacci. The difference between these algorithms is the way in which the new estimate of x^* (for which the cost function is evaluated) is obtained from the prior estimate of x^*.

The Golden Sections method is based on a search of the intervals within which the minimum lies. If L is the length of the interval within which the minimum is assumed to lie, then this interval is divided into two intervals of lengths $0.618L$ and $0.382L$. We divide the interval L to guarantee that the minimum lies within the interval of length $0.618L$. The ratio of the interval in which the minimum lies to the original interval is given by the constant 0.618. The

FIGURE 11.5 Plot of Cost Function vs. Controllable Parameter

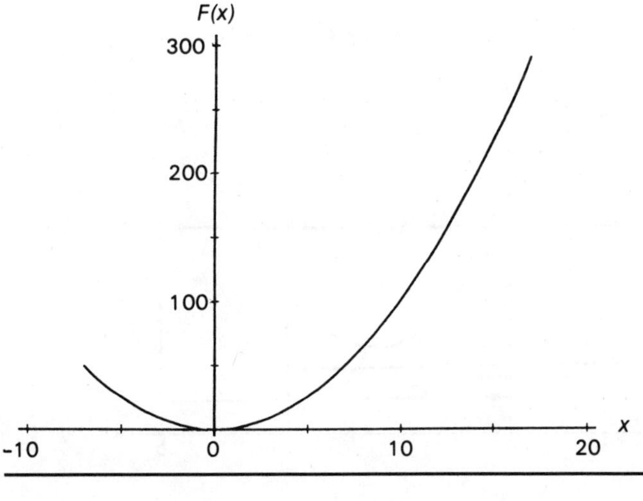

procedure continues until the value of x^* is found to within a tolerance. In short, the method divides an interval into two parts such that the ratio of the longer part to the original is equal to the ratio of the shorter part to the longer part. We represent this mathematically as

$$\frac{L_k}{L_{k-1}} = \frac{L_{k+1}}{L_k} = 0.618 \qquad (k = 2, 3, \ldots) \tag{11.2}$$

where L_{k-1} = the original
L_k = the larger interval
L_{k+1} = the shorter interval

There are two ways of choosing L_k and L_{k+1}, as shown in Figure 11.6.

The ratio 0.618 was known to Greek geometricians as the Golden Section. To implement the algorithm, we should do the following:

1. Determine the initial interval, L_1, by choosing two initial points (a, b) within which the minimum is assumed to lie.
2. Select the two succeeding intervals, $L_2/L_1 = 0.618$ and $L_2'/L_1 = 0.618$, which leads to two other points, c and d, such that $d - a = b - c = 0.618(b - a)$. (See Figure 11.6.)
3. Evaluate the cost function at a, b, c, and d and select the interval in which the minimum lies by comparing the cost function values. Choose region L_2 or L_2'. (See Figure 11.6.) For example, if $\min(F(a), F(c), F(d), F(b)) = F(c)$, choose L_2 because c is an interior point of L_2 but an end point of L_2'.

FIGURE 11.6 Golden Sections Method for Choosing Intervals

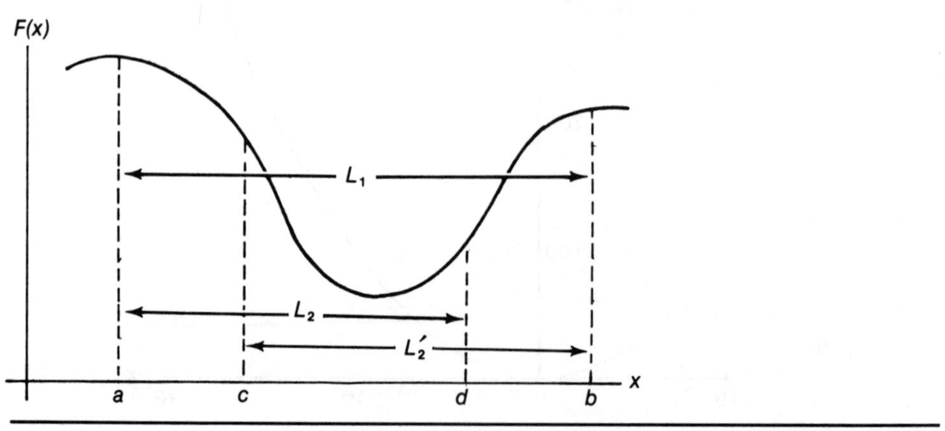

4. Repeat the procedure from step 2 until the interval size is such that $L_k \leq \varepsilon$. This is the tolerance. We should point out that the interval in which the minimum is known to lie is reduced by a factor of 0.618 after each iteration, and only two function evaluations are required after the first iteration. The length of the nth interval, L_n, is

$$L_n = (0.618)^{n-1} L_1$$

and the value of x^* is in the final interval.

To represent this method mathematically, we will use the following notation:

$$x_i^k = x_i \text{ at the } k\text{th iteration, with } k = 1, 2, \ldots, n.$$

Specifically,

$$x_1^2 = x_1 \text{ at the 2nd iteration}$$
$$x_2^2 = x_2 \text{ at the 2nd iteration}$$

where the initial boundaries of the search interval L_0 are (x_1^1, x_2^1).

The following is a step-by-step description of the algorithm for the Golden Sections search:

Step 1 Set the iteration count to $k = 1$. Choose the initial interval in which the minimum is assumed to lie such that the lower boundary $= x_1^1$ and the upper boundary $= x_2^1$. Define $L_1 = x_2^1 - x_1^1$. Evaluate $F(x_1^1)$ and $F(x_2^1)$. Then set the iteration count to $k = 2$.

Step 2 Compute

$$x_1^k = x_1^{k-1} + 0.618 L_{k-1}$$
$$x_2^k = x_2^{k-1} - 0.618 L_{k-1}$$

where x_1^k defines the upper boundary of the interval L_k with lower boundary x_1^{k-1}

x_2^k defines the lower boundary of the interval L_k' with upper boundary x_2^{k-1}

Evaluate $F(x_1^k)$ and $F(x_2^k)$. Sort $F(x_1^k)$, $F(x_2^k)$, $F(x_1^{k-1})$, and $F(x_2^{k-1})$ in descending order. Choose the interval containing the minimum value and compute L_k.

Set

$x_1^k =$ lower boundary of L_k

$x_2^k =$ upper boundary of L_k

Store $F(x_1^k)$ and $F(x_2^k)$ for comparison in the next iteration.

Step 3 If $L_k < \varepsilon$, then $x^* = x_1^k \cong x_2^k$ and stop. Otherwise, let $k = k + 1$ and go to step 2.

EXAMPLE 11.1 Let us show a numerical example for finding x^* that minimizes the function $F(x)$ shown in Figure 11.7.

Step 1 Let our initial choice of boundaries for the interval in which the minimum is assumed to lie be $a = x_1^1 = -1$ and $b = x_2^1 = 9$. Therefore, $L_1 = b - a = x_2^1 - x_1^1 = 10$.

Step 2 Let us choose boundaries for the new intervals by computing

$$x_2^2 = x_2^1 - 0.618L_1 = 9 - 0.618(10) = 3.82$$

$$x_1^2 = x_1^1 + 0.618L_1 = -1 + 0.618(10) = 5.18$$

Calculating $F(x_1^1) = 1$, $F(x_2^1) = 81$, $F(x_2^2) = 14.59$, and $F(x_1^2) = 26.83$, we find that $F(x_2^1) > F(x_1^2) > F(x_2^2) > F(x_1^1)$. Therefore, the minimum lies in the interval (x_1^1, x_2^2), and $L_2 = x_1^2 - x_1^1 = 5.18 - (-1) = 6.18$.

Now we set

$$x_1^2 = x_1^1 = -1 = \text{lower boundary of } L_2$$

$$x_2^2 = x_1^2 = 5.18 = \text{upper boundary of } L_2$$

Then we store $F(x_1^2) = 1$ and $F(x_2^2) = 26.83$.

FIGURE 11.7 Example 11.1

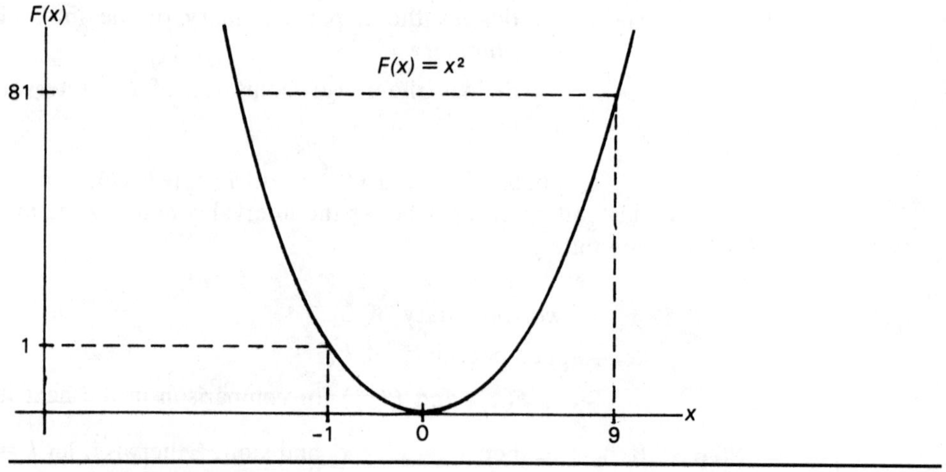

Step 3 Next, we compare the length of L_2 with ε, the predetermined value chosen to stop the algorithm, In our case, we will continue and repeat steps 2 and 3, with $k = 3$. The values then become

$$x_2^3 = x_2^2 - 0.618L_2 = 1.36$$
$$x_1^3 = x_1^2 + 0.618L_2 = 2.82$$
$$F(x_2^3) = 1.85$$
$$F(x_1^3) = 7.95$$
$$F(x_1^2) = 1.00$$
$$F(x_2^2) = 26.83$$
$$F(x_2^2) > F(x_1^3) > F(x_2^3) > F(x_1^2)$$
$$L_3 \text{ boundaries} = (-1, 2.82)$$

For practice, the reader is encouraged to repeat this procedure with $\varepsilon = 0.5$. This will require three additional iterations.

The bisection search method is similar to the Golden Sections technique except that each interval is divided into two equal intervals.

The Fibonacci series

$$(1, 1, 2, 3, 5, 8, 13, 21, 34, \ldots)$$

is described by the equation

$$F_n = F_{n-1} + F_{n-2} \qquad (n \geq 2) \tag{11.3}$$

where $F_0 = 1$ and $F_1 = 1$.

Let us define

$$r_j = \frac{F_{n-j-1}}{F_{n-j+1}} \qquad (1 \leq j \leq n-1) \tag{11.4}$$

For example, if $n = 9$, then

$$r_5 = \frac{F_3}{F_5} = \frac{2}{5}$$

The Fibonacci search method divides the initial interval into two intervals, as do the bisection and Golden Sections search methods, with the boundaries of each interval computed as follows:

$$x_1^k = (1 - r_{k-1})x_1^{k-1} + r_{k-1}x_2^{k-1} \tag{11.5}$$

$$x_2^k = r_{k-1}x_1^{k-1} + (1 - r_{k-1})x_2^{k-1} \tag{11.6}$$

EXAMPLE 11.2 Using the function $F(x)$ from Example 11.1 and the same choice of initial interval boundaries, $x_1^1 = -1$ and $x_2^1 = 9$, we find

$$r_1 = \frac{F_7}{F_9} = \frac{13}{34}$$

$$x_1^2 = 2.823$$

$$x_2^2 = 5.176$$

$$L_2 \text{ boundaries} = (-1, 5.176)$$

Repeating the procedure, we obtain

$$r_2 = \frac{F_6}{F_8} = \frac{8}{21}$$

$$x_1^3 = 1.597$$

$$x_2^3 = 2.823$$

$$L_3 \text{ boundaries} = (-1, 2.823)$$

Notice that $x_1^2 = x_2^3$. This characteristic of the Fibonacci search method, where one of the new boundary points will be equal to one of the previous boundary points, reduces the number of function evaluations required.

The bisection method is a special form of equation (11.5), with $r_{k+1} = r_k = 0.5$. The Golden Sections case of equation (11.6) uses $r_{k+1} = r_k = 0.618$.

Function Approximation Method: Powell's Method

A disadvantage of using function comparison methods is that we do not use all the information available to us, specifically the magnitude by which the cost functions differ at the interval boundaries. However, function approximation methods use this information to approximate the cost function by a simpler cost function whose minimum can be computed easily. For example, if we use a second-order polynomial $y(x)$ to approximate the cost function, we calculate the minimum point x^* of the second-order polynomial to be

$$x^* = \frac{1}{2}\left[\frac{(x_2^2 - x_3^2)F_1 + (x_3^2 - x_1^2)F_2 + (x_1^2 - x_2^2)F_3}{(x_2 - x_3)F_1 + (x_3 - x_1)F_2 + (x_1 - x_2)F_3}\right] \tag{11.7}$$

where x_1, x_2, and x_3 = points along the search direction for the minimum $F_1 = F(x_1)$, $F_2 = F(x_2)$, and $F_3 = F(x_3)$

If the old values (x_1, x_2, x_3) bracket the minimum, then the new x^* will also be in the interval where the minimum is known to lie. We will now describe the Powell algorithm to minimize a function $F(x)$.

When using the Powell method, we take the following steps:

1. Choose an initial starting x to be our first estimate of x^*, a direction for the search, and increment size h.
2. Evaluate $F(x)$ and $F(x+h)$.
3. If $F(x+h) > F(x)$, then evaluate $F(x-h)$. Otherwise, evaluate $F(x+2h)$. We now know three values of the function $F(x)$, values at $x_1 = x$, $x_2 = x + h$, and $x_3 = x + 2h$ or $x - h$.
4. Use equation (11.7) to find the new estimate of x^*. This value of x^* corresponds to a minimum of $y(x)$ when

$$\frac{(x_2 - x_1)F(x_1) + (x_3 - x_2)F(x_2) + (x_1 - x_2)F(x_3)}{(x_1 - x_2)(x_2 - x_3)(x_3 - x_1)} \geq 0$$

   Otherwise, it corresponds to a maximum of $y(x)$.
5. If x^* corresponds to a minimum of $y(x)$ and is greater than h from the three current points x_1, x_2, and x_3, discard the point furthest from x^*. Obtain a new current point by taking a step h from the point nearest to x^* in the direction that decreases $F(x)$. Go to step 4.
6. If x^* corresponds to a maximum of $y(x)$, discard the point that is furthest from x^*. Obtain a new current point by taking a step h from the discarded point that decreases $F(x)$. Go to step 4.
7. If point x^* is within a predetermined tolerance h of the three current points x_1, x_2, and x_3, assume that the smallest of the three values $F(x_1)$, $F(x_2)$, and $F(x_3)$ is the minimum, and x^* is the value of the corresponding x.
8. If point x^* corresponds to a minimum of $y(x)$ but steps 5 through 7 do not apply, discard the point with the highest function value $F(x)$, and use x^* as the third current point. Go to step 4. Exception: Point 3 has the highest function value but point 2 is discarded, because the minimum is between points 1 and 3, as shown in Figure 11.8.

FIGURE 11.8 Problem 8

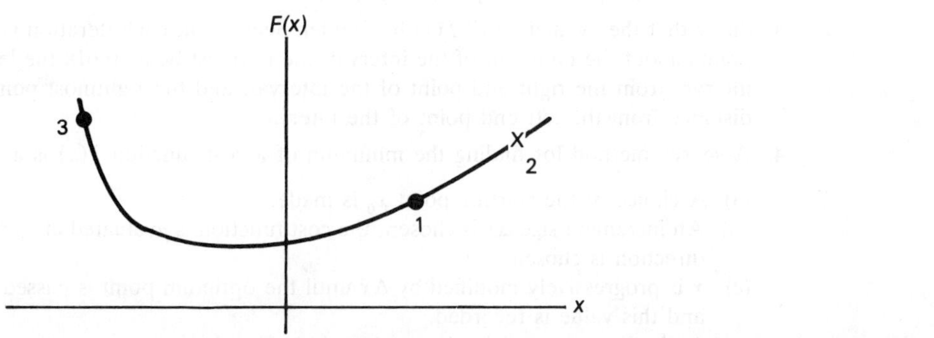

11.4 MULTIPARAMETER OPTIMIZATION

The method for minimizing a cost function $F(x_1, x_2, \ldots, x_n)$ with independent variables $x_1, x_2, \ldots, x_n$ is similar to the methods we have used for systems with only one parameter (independent variable). The "relaxation method" minimizes the cost function with respect to n parameters by first estimating a value for each parameter. We then minimize the cost function with respect to the first parameter, assuming that the other $n-1$ parameters are fixed. The cost function is then minimized with respect to the second parameter, fixing the value obtained for the first parameter and holding all the other $n-2$ parameters fixed. The process is repeated for all n parameters. Then we return to the first parameter and repeat the cycle until a predetermined tolerance for all parameters is reached.

Although methods for a single-parameter optimization can be generalized and used as suggested above for multiparameter optimization, the process is much more complex and new problems arise. For example, it is difficult to choose directions for the search. Also, the objective function might not be as well behaved as for single-parameter optimization.

The references direct the reader to more sophisticated methods, such as gradient techniques, Powell's multiparameter technique, alternating variable methods, and pattern searches. The references also cover the simplex method and dynamic programming.

11.5 SUMMARY

In this chapter, we have provided a brief introduction to single-variable optimization methods, focusing on two basic techniques: function comparison and function approximation. Ample references for several more powerful methods for optimization are offered to the reader who is interested in further details.

11.6 REVIEW PROBLEMS

1. What do we mean by optimization? How do we choose a cost function?
2. What are the two techniques of optimization algorithms? Discuss the concepts of each.
3. Show that the evaluation of $F(x)$ by Golden Sections at each iteration is symmetric at two points about the midpoint of the interval, the leftmost being 0.618 the length of the entire interval from the right end point of the interval, and the rightmost point being the same distance from the left end point of the interval.
4. A search method for finding the minimum of a cost function $F(x)$ is as follows:

   (a) A choice of the starting point x_0 is made.
   (b) An increment size Δx is chosen, the cost function is evaluated at $x_0 \pm \Delta x$, and a search direction is chosen.
   (c) x is progressively modified by Δx until the optimum point is passed for the first time, and this value is recorded.
   (d) Δx is then reversed in sign and made half as large as the original Δx. The search is continued until the optimum is passed again.

(e) The costs of steps c and d are compared and, if they are within a tolerance, the optimum value of x is taken to be the value of step d. Otherwise, step d is repeated. Write a computer program for this algorithm.

5. Write a computer program for the Golden Sections method.

6. Write a computer program for Powell's method.

7. Develop a model of an urban area that includes population. Choose a cost function and minimize it with respect to population.

8. Airline overbooking is a problem. If more ticketed passengers show up than there are seats available, airlines refund the price of a ticket, provide free accommodations, and fly the passenger free on the next available flight.

a. Develop a model for the process and optimize, with respect to the percentage of overbooking, if we choose to minimize the cost function, which is equal to the amount paid to passengers.

b. Repeat part (a) if we choose to maximize the profit, which is 20% of the airline ticket price.

9. Assuming a two-nation arms race model, let nation 1's cost function be

$$(x_1 - ax_2)^2 + c(x_1 + x_2)$$

where x_1 = the defense expenditures of nation 1

x_2 = the defense expenditures of nation 2

The objective of nation 1 is to maintain some arms advantage over nation 2 and to minimize the total arms expenditures of the two nations.

a. What is the significance of coefficients a and c?

b. What would be the cost function for nation 2?

c. Optimize the model of Chapter 7, problem 8, with respect to α for nation 1's objectives.

10. Find the minimum value of

$$F(x) = x^3 - 4x - 5$$

using the method of the Golden Sections in the interval (−5,5).

11. Find the maximum value of the function

$$G(x) = -x^2 - 6x + 2$$

using Powell's method in the region (−4,6).

12. Find the values x_1^* and x_2^* that minimize the cost function

$$F(x_1, x_2) = x_1^2 + x_2^2 - x_1x_2 + 4$$

in the region bounded by $-1 \leq x_1 \leq 4$ and $-1 \leq x_2 \leq 4$.

11.7 REFERENCES

1. Beveridge, Gordon S.G., and Schechter, Robert S., *Optimization: Theory and Practice* (New York: McGraw-Hill, 1970).
2. Brent, Richard P., *Algorithms for Minimization without Derivatives* (Englewood Cliffs, NJ: Prentice-Hall, 1973).
3. Bunday, Brian D., *Basic Optimization Methods* (London: Edward Arnold, 1984).
4. Chu, Kong, et al., "Population Policy Simulation Performed with an Econometric Model of Taiwan," *Simulation and Games* (September 1972): 291–307.
5. Daellenbach, Hans G.; George, John A.; and McNickle, Donald C., *Introduction to Operations Research Techniques*, 2nd ed. (Newton, MA: Allyn & Bacon, 1983).
6. Dixon, L.C.W., ed., *Optimization in Action* (New York: Academic Press, 1976).
7. Gill, Philip E.; Murray, Walter; and Wright, Margaret H., *Practical Optimization* (London: Academic Press, 1981).
8. Greis, D.M., *Optimization* (London: Longman, 1980).
9. Kochenburger, Ralph J., *Computer Simulation of Dynamic Systems* (Englewood Cliffs, NJ: Prentice-Hall, 1972).
10. Mastascusa, E.J., *Computer-assisted Network and Systems Analysis* (New York: Wiley, 1988).
11. Murray, W., ed., *Numerical Methods for Unconstrained Optimization* (New York: Academic Press, 1972).
12. Powell, M.J.D., "An Interactive Method for Finding Stationary Values of a Function of Several Variables," *Computer Journal* 5 (April 1962–January 1963): 147–51.
13. Powell, M.J.D., "An Efficient Method of Finding the Minimum of a Function of Several Variables without Calculating Derivatives," *Computer Journal* 7 (April 1964–January 1965): 155–62.
14. Rasmussen, David, *Urban Economics* (New York: Harper & Row, 1973).
15. Sivazlian, B.D., and Stanfel, L.E., *Optimization Techniques in Operations Research* (Englewood Cliffs, NJ: Prentice-Hall, 1975).
16. Walsh, G.R., *Methods of Optimization* (New York: Wiley, 1981).
17. Wolfe, A.B., "On the Criterion of the Optimum Population," *American Journal of Sociology* 39, no. 4 (March 1934): 585.
18. Zahradnik, Raymond L., *Theory and Techniques of Optimization for Practicing Engineers* (New York: Barnes & Noble, 1971).

12 Computer Simulation Methodology

12.1 INTRODUCTION

Simulation, the process of model construction and model use, is a tool for solving problems in almost all disciplines that attempt to explain or predict observable phenomena. However, we still lack prescriptive methodologies for the development of simulations. This is because engineering systems, physical systems, and social systems are complex and often not well understood. For example, these systems may exhibit nonlinear behavior, and are affected by random environmental disturbances. Thus, we can see that there is inherent uncertainty in model development, and therefore must interpret the results of our simulation experiments cautiously.

Development of proper simulation methodologies presents a challenge to simulation researchers. We briefly discuss several approaches to simulation from the software engineering field, including the structured system approach, the top-down technique, the Hierarchical Input Process Output (HIPO), and the Warnier method. All of these approaches require us to identify the primary system we are simulating, then identify its subsystems.

In this chapter, through a cardiovascular system example, we present a development of a simulation. We include a list of simulation design pitfalls and ways to avoid them.

12.2 SIMULATION METHODOLOGY: CARDIOVASCULAR SYSTEM EXAMPLE[1]

Women have been included as mission specialists in the U.S. space program. Dr. Sally K. Ride, a physicist who became America's first female astronaut in space, performed a variety of tasks during the seventh Space Shuttle mission. More women have since participated in America's space effort. The first space flight of a woman, Valentina Tchereshkova, occurred on June 16, 1963. The flight lasted for 70 hours and 50 minutes. According to Russian reports, no major cardiovascular problems were encountered. With longer space flights, however, deconditioning has been observed consistently in both U.S. and U.S.S.R. space crews.

Deconditioning, the resultant state of prolonged exposure to weightlessness, is measured by a lowered tolerance to centrifugation, tilt, and reduced lower body negative pressure (LBNP).

Bed rest has been used to simulate weightlessness, because it causes shifts in hydrostatic fluid columns similar to that seen in space flight. LBNP has been used to study cardiovascular response and tolerance to orthostatic stress, since it induces fluid shifts similar to those seen in going from a supine to a sitting position and can be employed in a flight (zero-G) environment. Responses to LBNP following bed rest have been used to study cardiovascular deconditioning following long-term exposure to weightlessness. LBNP exposure to a level of −40mmHg to −50mmHg vacuum has been found to simulate gravitational exposure analogous to 70° passive tilt. Figure 12.1 shows how LBNP is applied. Data from a study by Harold Sandler and David Winter showing typical cardiovascular responses of female subjects to −50mmHg LBNP are shown in Figure 12.2.[17] The data were gathered using the following procedure.

Twelve female nurses (age 23–24 years) volunteered to serve as subjects. Two were active-duty Air Force nurses and 10 were reservists. All were in prime physical condition. Eight of the volunteers served as bed-rest subjects and 4 as ambulatory controls. The subjects' vital statistics are shown in Table 12.1. All subjects were housed throughout the investigation in the Ames Research Center Human Research Facility, which provided a highly regulated environment with regard to temperature and a fixed photoperiod.

The study consisted of a 14-day control period, 17 days of absolute bed rest, and 5 days of recovery. During the bed-rest period, each bed-rest subject was required to refrain from excessive muscular movement, was furnished one pillow, and was allowed to rise up on one elbow during meals. All emunctory functions and bathing were performed in the horizontal position. All subjects were given medical and gynecological examinations before the study was begun and again at

FIGURE 12.1 Application of LBNP

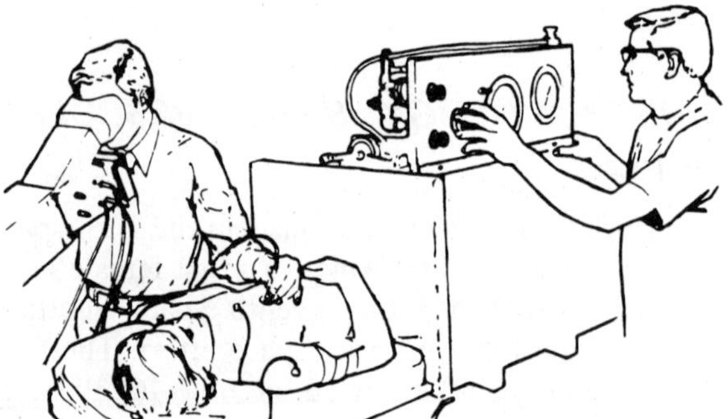

FIGURE 12.2 Typical Cardiovascular Responses to LBNP[17]

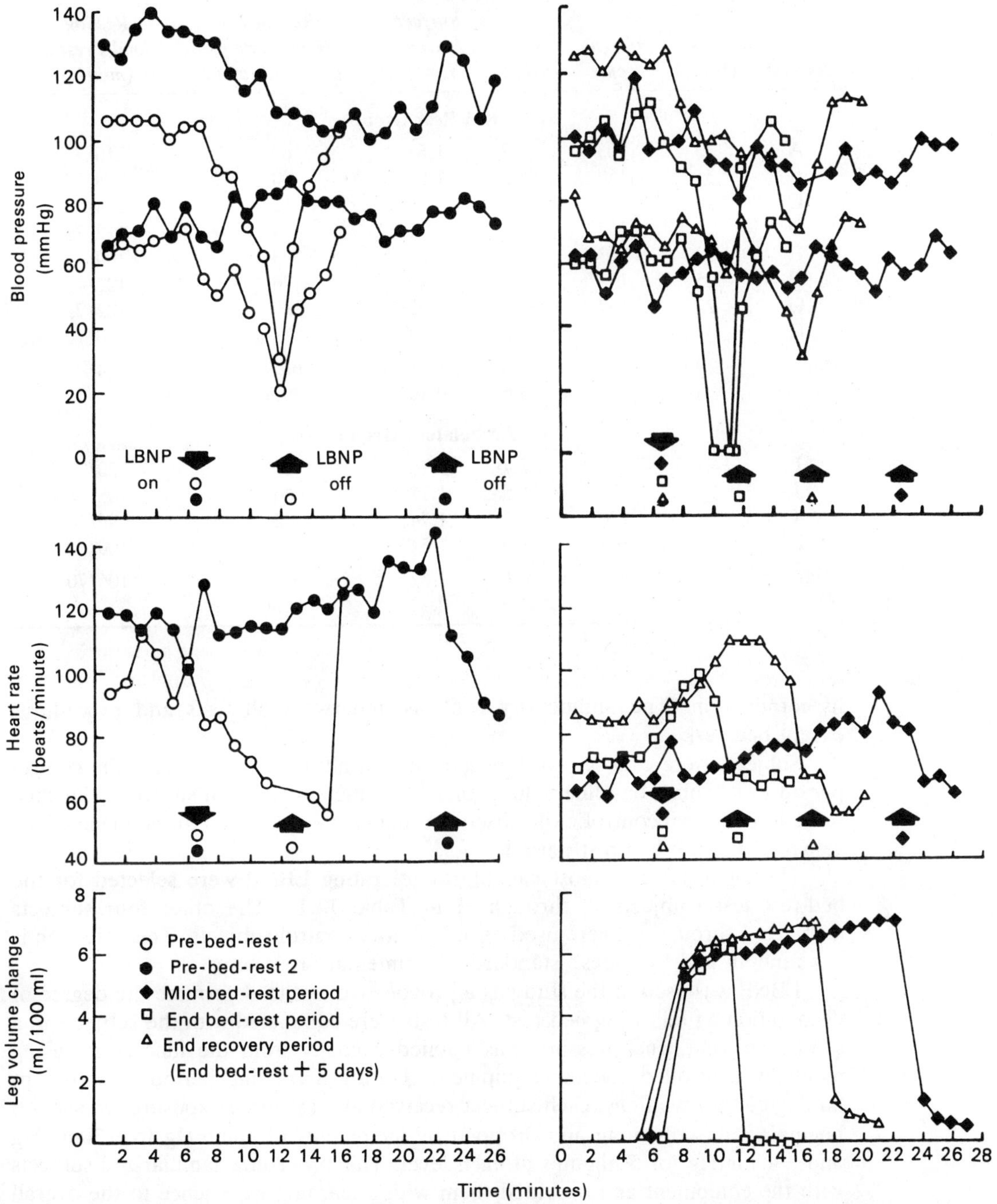

TABLE 12.1 Vital Statistics

Subject	Age (Years)	Height (cm)	Weight (kg)	Body Surface Area (m^2)	Resting Heart Rate (Beats/Minute)	Resting Blood Pressure (mmHg)
			Bed-Rest Group			
A	33	163	60.8	1.65	60	116/72
B	35	159	55.3	1.56	80	108/72
C	26	173	63.4	1.76	65	124/72
D	24	157	53.1	1.52	74	112/78
E	32	163	58.2	1.62	100	100/74
F	25	157	50.4	1.49	76	122/72
G	23	157	53.0	1.52	80	122/72
H	32	164	66.5	1.73	78	110/70
Mean	29	162	57.6	1.61	66.9	114/73
SD	5	5	5.6	0.10	29.5	
			Ambulatory Group			
I	27	159	47.7	1.47	67	120/78
J	29	170	61.5	1.71	60	108/76
K	27	170	64.1	1.74	62	96/64
L	25	164	51.0	1.54	72	100/60
Mean	27	166	56.1	1.62	65.3	106/70
SD	2	5	8.0	0.13	5.4	

its termination. The ambulatory controls underwent all tests and procedures except bed rest.

Subjects were exposed to duplicate runs of all procedures during the control period to establish baseline values for all parameters for each subject, who then served as her own control. All subsequent data were compared to the mean of the values of the pre-bed-rest periods.

The eight subjects most capable of tolerating LBNP were selected for the bed-rest test (subjects A through H in Table 12.1). The other four subjects (subjects I through L) were used as ambulatory control subjects. This selection is not random and "violates" standard experimental procedures.

LBNP was used in the study as a provocative test to determine the degree of deconditioning following bed rest. All tests were conducted with the subjects in a supine position, and pressure was applied caudally from the iliac crests using manually controlled vacuum equipment. During a training period prior to obtaining control readings, each subject received at least three exposures to suction beginning at −20mmHg and dropped by increments of 10mmHg to −30mmHg and −40mmHg for 5 minutes at each level. This procedure familiarized subjects with the equipment and provided them with a learning experience to the overall procedures. Subsequent exposures before and after bed rest lasted 15 minutes at −50mmHg or until subjects exhibited the first symptoms of syncope. Control data were collected and averaged for 5 minutes before the onset of LBNP and for 3 minutes after the release of pressure. Heart rate (HR) was recorded continuously from sternal leads and averaged over 30-second intervals to adjust for respiratory variations.

All subjects were able to tolerate the complete 15-minute exposure to LBNP during the control period. No one was able to complete the test following the bed-rest period; post-bed-rest tolerance times varied from 4.5 to 14 minutes. Tolerance to LBNP had improved toward pre-bed-rest levels by five days of recovery. Summaries of average values for HR obtained before and after bed rest are shown in Table 12.2. In this table, "Test" indicates data from the eight bed-rested subjects and "Ambulatory" refers to the four ambulatory control subjects. Data were arbitrarily divided into 5-minute intervals representing control, early (first 5 minutes of LBNP), mid (second 5 minutes of LBNP), late (last 5 minutes of LBNP), and recovery periods. Average values over each period were determined from 30-second heart rate averages.

Figure 12.2 presents the graphs of mean HR versus time for one of the 12 subjects during the LBNP procedure at each of the four study periods detailed in Table 12.2.

As we stressed earlier in the text, the first step in the simulation process is to identify the purpose or goal of the simulation. This step is often neglected, leading to the failure of many simulations. Our goal will be to develop a simulation model upon which we can test quantitively a hypothesis about the conditioned or deconditioned state of the subject. In other words, can we determine a set of parameters from our model that will discriminate between the conditioned and deconditioned states of a subject? These parameters will then be used to develop criteria for selecting Space Shuttle passengers.

The second step in the process is to assemble a team with appropriate skills for the simulation. In our case, the group included two medical researchers, an engineer, and a student programmer. One of the medical researchers designed the experiments and collected the data shown in Figure 12.2. The engineer had a system theory background, and the student was a senior computer science major.

Our small group then addressed the issue of which type of model would be most appropriate. The models that we considered included:

A physiological model of the cardiovascular system (CVS) that would model the behavior of each component of the system, similar to PHYSBE[12]

A model using an external description[17,20]

A model using an internal description[1]

Our group arrived at a decision consistent with our goal after answering the

TABLE 12.2 LBNP Cardiovascular Data for Heart Rate

Test Condition		(n)[a]	Pre-Bed-Rest, Mean ± SD	(n)[a]	End Bed-Rest, Mean ± SD	(n)[a]	5 Days Post-Bed-Rest, Mean ± SD
			Heart Rate (Beats/Minute)				
Pre-LBNP (resting), 5-min mean	Test	(80)	75 ± 9[b]	(40)	80 ± 9[b,c]	(40)	81 ± 8[b,c]
	Ambulatory	(30)	64 ± 12	(15)	67 ± 7	(15)	62 ± 5
LBNP elapsed time, 1-min values							
5	Test	(16)	94 ± 13[d]	(7)	129 ± 26[c]	(8)	114 ± 16[b,c]
	Ambulatory	(6)	78 ± 17	(3)	112 ± 17[e]	(3)	92 ± 3
10	Test	(16)	102 ± 16	(3)	136 ± 45[c]	(7)	110 ± 25
	Ambulatory	(6)	86 ± 21	(2)	126 ± 17[e]	(3)	100 ± 9
15	Test	(16)	100 ± 18	—	—	(6)	123 ± 17[d,e]
	Ambulatory	(6)	89 ± 20	(2)	132 ± 8[e]	(2)	34 ± 17
Maximum 1-min value during LBNP	Test	(16)	105 ± 18[b]	(2)	160 ± 16	(6)	127 ± 16
	Ambulatory	(6)	89 ± 20	(2)	133 ± 18	(2)	105 ± 4
Recovery, min							
+1	Test	(16)	81 ± 12[d]	(7)	95 ± 29	(8)	84 ± 18
	Ambulatory	(6)	67 ± 9	(3)	68 ± 9	(3)	68 ± 9
+3	Test	(16)	70 ± 11[b]	(8)	71 ± 11	(8)	71 ± 8[b]
	Ambulatory	(6)	55 ± 6	(3)	58 ± 12	(3)	52 ± 3

[a]Total number of test observations.

[b]$p < 0.01$ difference from control subjects.

[c]$p < 0.01$ difference from pre-bed-rest.

[d]$p < 0.05$ difference from control subjects.

[e]$p < 0.05$ difference from pre-bed-rest.

Source: Harold Sandler and David L. Winter, *Physiological Responses of Women to Simulated Weightlessness*, NASA Report, NASA SP-430, 1978, p. 18.

following question: What are the important noninvasive measurable variables of the system? The important variables are heart rate (HR), stroke volume (SV), mean arterial pressure (MAP), pulse pressure (PP), total peripheral resistance (TPR), cardiac output (CO), and central and peripheral blood volume (BV). TPR cannot be measured directly by noninvasive methods but can be approximated by MAP/CO or MAP/(HR × SV). PP is defined as systolic pressure minus diastolic pressure.

FIGURE 12.3 Causal Diagram of the Cardiovascular System

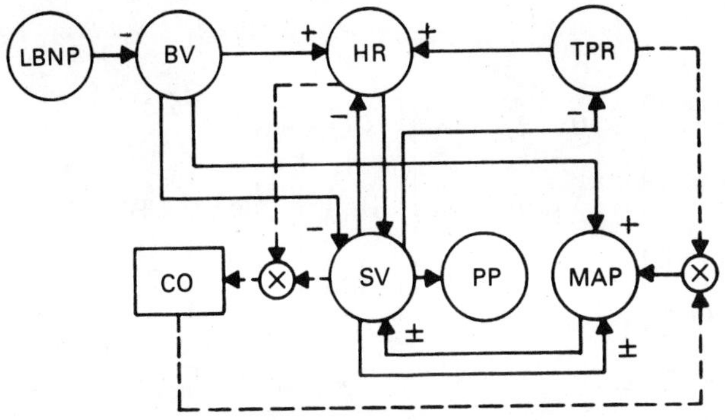

Figure 12.3 is a causal diagram of the CVS. From this diagram, we developed the following internal or state variable description:

$$\begin{bmatrix} \dfrac{d\text{HR}}{dt} \\ \dfrac{d\text{SV}}{dt} \\ \dfrac{d\text{MAP}}{dt} \\ \dfrac{d\text{PP}}{dt} \\ \dfrac{d\text{TPR}}{dt} \\ \dfrac{d\text{BV}}{dt} \end{bmatrix} = \begin{bmatrix} 0 & a_{12} & 0 & 0 & a_{15} & a_{16} & \text{HR} \\ a_{21} & 0 & a_{23} & 0 & 0 & a_{26} & \text{SV} \\ 0 & a_{32} & 0 & 0 & 0 & a_{36} & \text{MAP} \\ 0 & a_{42} & 0 & 0 & 0 & 0 & \text{PP} \\ 0 & a_{52} & 0 & 0 & 0 & 0 & 0 \\ 0 & 0 & 0 & 0 & 0 & 0 & a_{66} \end{bmatrix} \begin{bmatrix} 0 \\ 0 \\ 0 \\ 0 \\ 0 \\ b_{61} \end{bmatrix} \text{LBNP} \qquad \textbf{(12.1)}$$

The model equations have the form

$$\dot{\mathbf{x}} = [A]\mathbf{x} + \mathbf{b}u \qquad \textbf{(12.2)}$$

At this point, from the observable data, the parameters a_{ij} and b_{61} should have been identified using the methods of Chapter 11. However, since TPR cannot be measured noninvasively, no observable data existed; therefore, the TPR state was considered a nonlinear combination of the three states MAP, HR, and SV. Diastolic pressure varies minimally with inducement of LBNP,[13] and PP can be derived from MAP. Also, central blood volume (BV) variations occur rapidly with a step inducement of LBNP and therefore can be considered to have a step response to LBNP. These considerations led us to the reduced state model described by

$$\begin{bmatrix} \dfrac{d\text{HR}}{dt} \\ \dfrac{d\text{SV}}{dt} \\ \dfrac{d\text{MAP}}{dt} \end{bmatrix} = \begin{bmatrix} 0 & f_{12} & f_{13} \\ f_{21} & 0 & f_{23} \\ f_{31} & f_{32} & 0 \end{bmatrix} \begin{bmatrix} \text{HR} \\ \text{SV} \\ \text{MAP} \end{bmatrix} + \begin{bmatrix} g_{11} \\ g_{21} \\ g_{31} \end{bmatrix} \text{BV} \quad \textbf{(12.3)}$$

$$\dot{\mathbf{x}} \quad = \quad [F] \quad \mathbf{x} \quad + \quad \mathbf{g} \quad u$$

In this model, note that the forcing function of the system described in equation (12.3) is the central blood volume. This is a third-order system and can be represented as

$$\dot{\mathbf{x}} = [F]\mathbf{x} + \mathbf{g}u \quad \textbf{(12.4)}$$

As we can see, this model is a reduced form of equation (12.1).

Our next step was to determine the values of the parameters f_{ij} and g_{i1}. This was done using a maximum likelihood parameter identification method similar to the methods presented in Chapter 10. The technique, often referred to as the model reference parameter identification scheme, is shown in Figure 12.4 (see Section 12.8). Parameter values, along with the standard deviation, in equation

FIGURE 12.4 Model Reference Parameter Identification Method

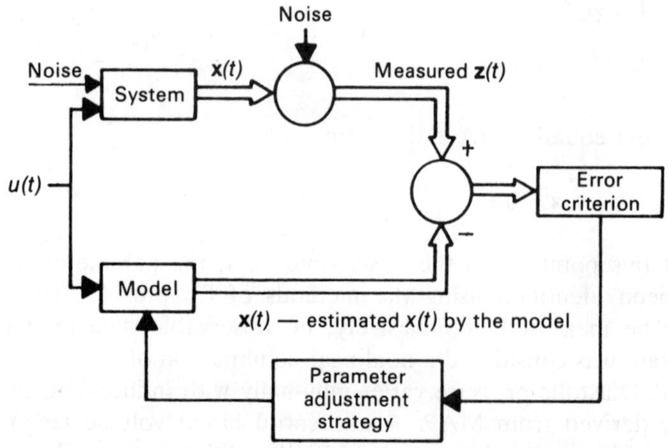

(12.3) for the pre-bed-rest period were calculated to be

$$F = \begin{bmatrix} 0 & (0.489) \pm 0.175 & (-1.375) \pm 0.372 \\ (0.125) \pm 0.028 & 0 & (0.404) \pm 0.082 \\ (0.026) \pm 0.020 & (0.016) \pm 0.026 & 0 \end{bmatrix}$$

$$g = \begin{bmatrix} (1.813) \pm 0.478 \\ (-0.882) \pm 0.168 \\ (-0.062) \pm 0.565 \end{bmatrix}$$

Parameters f_{12} (0.489) and f_{21} (0.125) have opposite signs from those implied by the causal diagram of Figure 12.3. This can be explained by noticing that CO drops during the initial phase of LBNP.

CVS responses predicted by the model and measured by experiment are shown in Figure 12.5. The long-term behavior of the model shows that HR increases exponentially, and SV and MAP decrease until syncope. These findings correspond with the observable data.

This model is an approximation of the unregulated model of CVS response during LBNP and will simulate response only during LBNP. We note that the values of parameters in F and g change when LBNP is removed.

We applied the same technique to the post-bed-rest period. From the data, we were unable to determine parameters that would discriminate between the conditioned and deconditioned states of a subject. Simulation of the model was implemented using the continuous simulation language ACSL.

These results led us to further reduce the model and study only the heart rate response to LBNP, since previous investigators had shown that such changes are a sensitive indicator of CVS deconditioning following bed rest. This decision illustrates the value and importance of the advice, "Build your similation analysis work on the work of others to the extent possible."

The new reduced-order model became

$$\frac{d\text{HR}(t)}{dt} = a\text{HR}(t) + b\text{LBNP}(t) \qquad \textbf{(12.5)}$$

where $\text{HR}(t)$ = the change in HR defined by $\text{HR}(t)$ minus the baseline HR

$$\text{LBNP}(t) = \begin{cases} -50\text{mmHg when the subject is exposed to LBNP} \\ 0 \text{ otherwise} \end{cases}$$

The baseline HR is defined as the average HR for the five-minute period before application of LBNP.

Parameters a and b were determined using the model reference technique presented earlier. The computed values of a, b, and b/a for each subject are shown in Table 12.3. Table 12.4 presents the values of a, b, and b/a computed from the average HR values of all subjects. Figure 12.6 shows a comparison of $\text{HR}(t)$ for four bed-rested subjects to the $\text{HR}(t)$ of the model, using the values of a and b during the pre-bed-rest exposures ($a = 0.25$, $b = -0.14$). Figure 12.7

FIGURE 12.5 Comparison of Model and Actual Data

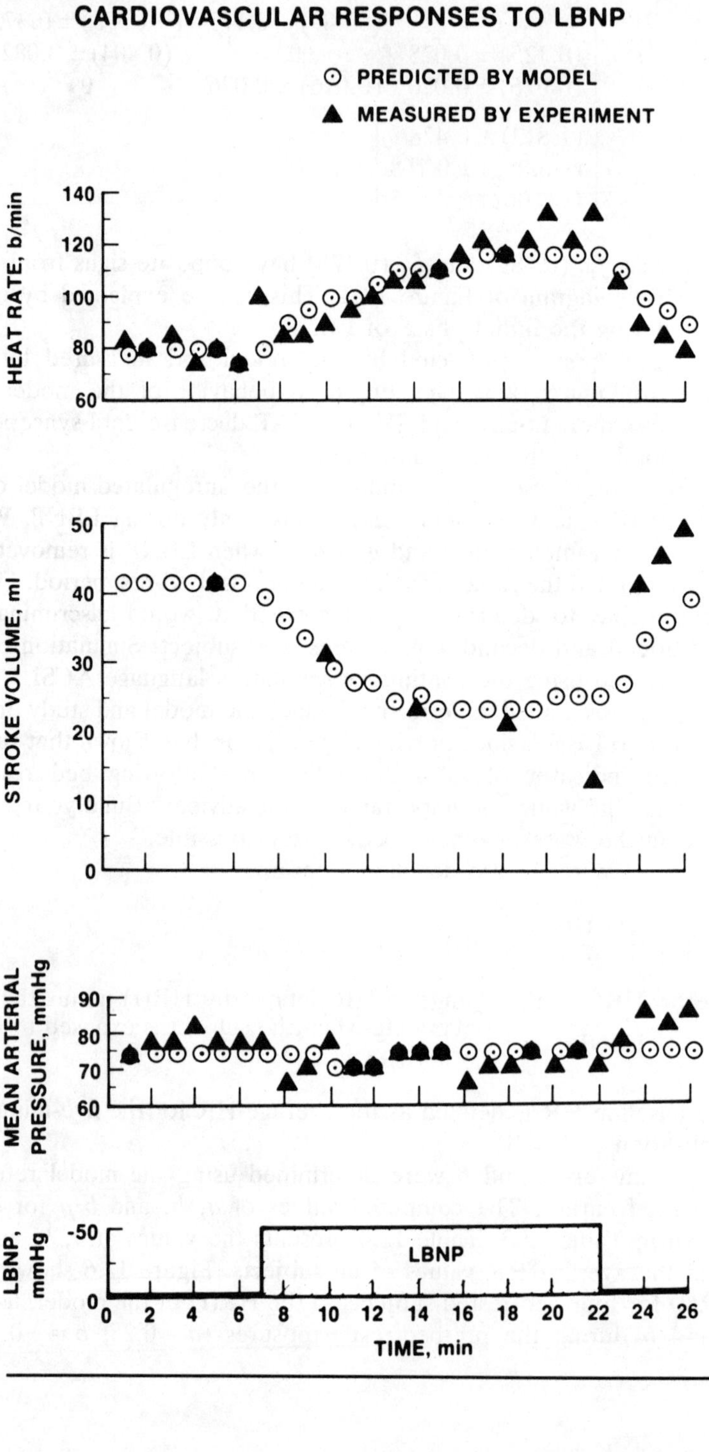

TABLE 12.3 Estimated Parameter Values for Heart Rate Model

Subject	*Exposure*	*a*		*b*		*b/a*
A	1	−2.00	0.09	−0.25	0.09	0.13
	2	−0.12	0.07	−0.02	0.01	0.17
	3	−0.28	0.09	−0.43	0.07	1.54
	4	−0.20	0.06	−0.12	0.02	0.60
B	1	−0.21	0.06	−0.11	0.02	0.55
	2	−1.72	0.42	−0.73	0.18	0.42
	3	−0.31	0.07	−0.41	0.06	1.32
	4	−0.11	0.04	−0.10	0.02	0.90
C	1	−0.74	0.20	−0.59	0.15	0.80
	2	−0.42	0.09	−0.24	0.04	0.57
	3	−2.22	0.01	−1.50	0.11	0.67
	4	−0.50	0.08	−0.58	0.09	1.16
D	1	−0.21	0.06	−0.12	0.02	0.57
	2	−0.16	0.13	−0.04	0.02	0.25
	3	−0.13	0.16	−0.32	0.10	2.44
	4	−0.21	0.05	−0.18	0.03	0.85
E	1	−1.86	0.29	−0.68	0.10	0.36
	2	−1.73	0.46	−0.56	0.14	0.31
	3	−0.44	0.12	−0.58	0.11	1.32
	4	−1.56	0.17	−1.22	0.14	0.78
F	1	−2.25	0.02	−0.33	0.05	0.15
	2	−1.44	0.42	−0.71	0.21	0.49
	3	−2.60	0.05	−0.73	0.16	0.28
	4	−2.37	0.03	−0.51	0.08	0.22
G	1	−0.10	0.03	−0.02	0.00	0.20
	2	−0.03	0.03	−0.08	0.01	2.63
	3	−1.19	0.10	−1.82	0.10	1.54
	4	−0.15	0.05	−0.28	0.04	1.85
H	1	−0.12	0.03	−0.19	0.02	2.78
	2	−0.22	0.04	−0.21	0.03	0.95
	3	−2.19	0.02	−2.30	0.18	1.05
	4	−0.18	0.03	−0.22	0.03	1.22
I	1					
	2	−0.26	0.09	−0.00	0.00	0.00
	3	−1.22	0.31	−0.65	0.14	0.53
	4	−0.13	0.08	−0.11	0.02	0.85
J	1	−0.16	0.04	−0.15	0.02	0.93
	2	−1.49	0.19	−0.48	0.19	0.99
	3	−1.45	0.12	−1.99	0.17	1.36
	4	−1.64	0.30	−1.22	0.22	0.75
K	1	−0.03	0.08	−0.02	0.01	0.66
	2	−1.50	1.60	−0.30	0.32	0.20
	3	−2.19	0.01	−1.03	0.10	0.47
	4	−0.10	0.05	−0.11	0.02	1.10
L	1	−1.89	0.65	−0.29	0.09	0.15
	2	−2.15	0.03	−0.29	0.04	0.13
	3	−0.73	0.14	−0.58	0.10	0.79
	4	−1.62	0.29	−1.15	0.20	0.71

Note: Subjects A through H are test subjects; subjects I through L are ambulatory controls.

TABLE 12.4 Parameter Identification of Average Heart Rate Values

Exposure	*Test Subjects (Subjects A–H)*		*Ambulatory Controls (Subjects I–L)*		*Overall*	
			a			
1	−0.25	0.05	−1.20	0.37	−0.20	0.05
2	−0.19	0.04	−1.65	0.44	−0.18	0.05
3	−0.29	0.04	−0.31	0.05	−0.51	0.07
4	−0.36	0.05	−1.35	0.27	−0.33	0.05
			b			
1	−0.14	0.02	−0.63	0.23	−0.11	0.02
2	−0.11	0.02	−0.62	0.16	−0.10	0.02
3	−0.47	0.04	−0.38	0.05	−0.61	0.07
4	−0.29	0.03	−0.79	0.16	−0.26	0.03
			b/a			
1	0.56		0.53		0.55	
2	0.58		0.38		0.56	
3	1.62		1.23		1.20	
4	0.81		0.59		0.79	

FIGURE 12.6 Comparison of Actual and Model Data Using Pre-Bed-Rest Values

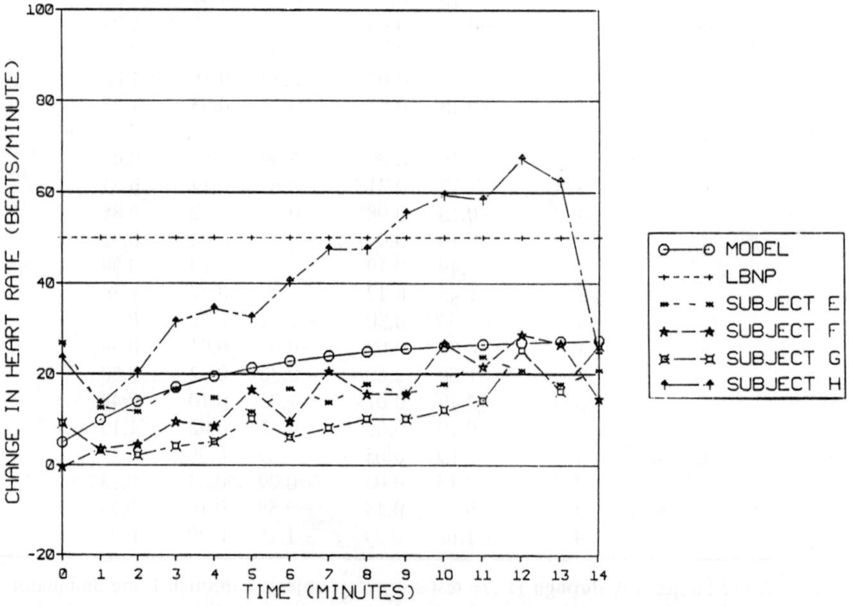

FIGURE 12.7 Comparison of Actual and Model Data Using Post-Bed-Rest Values

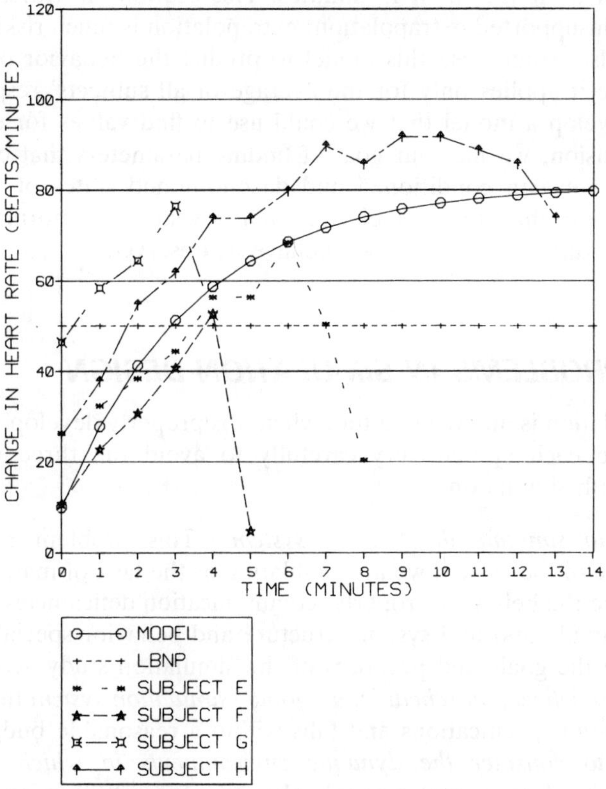

shows corresponding plots for the same four subjects following bed rest ($a = -0.29$, $b = -0.47$). From the tables, we can see that a value of $b/a > 1$ indicates deconditioning.

An advantage of this model is that instantaneous values of a and b can be used to predict deconditioning. The values of a and b can be estimated by exposing a subject to LBNP (−50mmHg) for only a few minutes, which eliminates the need to place the subject under stress for a longer period of time.

In the verification step, we were able to confirm the correctness of the computer program representing the model by looking at the initial and equilibrium conditions. For example, in the simulation program for equation (12.5), the equilibrium HR is given by

$$HR_{equilibrium} = -\frac{b\,LBNP_{equilibrium}}{a}$$

In the validation step, we were able to compare the average HR shown in Figures 12.6 and 12.7 with the individual HR. The results were consistent. We must be careful in using this model, however, because it applies only when LBNP is induced for no longer than 15 minutes. This caution illustrates the importance of avoiding unsupported extrapolation; extrapolation is much riskier than interpolation. We also cannot use this model to predict the behavior of any individual subject, since it applies only for the average of all subjects' responses. We can, however, develop a model that we could use to find values for each subject.

In conclusion, we met our goal of finding parameters that could be used to discriminate between conditioned and deconditioned states of the subject. We must note here that the model does *not* reveal any information about the physiological variables or why they behave as observed.

12.3 MAJOR PROBLEMS IN SIMULATION DESIGN

System simulation is an effective tool when it is properly developed. However, we must analyze each system very carefully to avoid the three major problems associated with simulation:

1. *Failure to simulate the "right" system.* This problem results from our ignorance of the system we are simulating or the lack of mathematical models to capture the behavior properly; communication deficiencies between us (the system simulators) and system structure and behavior specialists; and failure to define the goals and purposes of the simulation study accurately.
2. *Failure to deliver, on schedule, a "good" simulation system* that complies with the required specifications and falls within a reasonable budget.
3. *Failure to consider the dynamic environments in which we develop the simulation.* For example, people change jobs and move away, causing time delays and requiring the introduction of new personnel to the problem.

For a more detailed listing, see Section 12.9.

12.4 TOOLS FOR SIMULATION MANAGEMENT AND METHODOLOGY

The major cost of any system simulation is people, and people cost (a lot of) money. We can improve the management of simulation construction by using the structured system programming approach, which includes

The "team" approach to problem solving (It has been suggested that a two-person team is most creative.[5])

Top-down system simulation design, implementation, and testing

Structured system walk-through with the user of the goals, prototype, and specifications of the simulation

Structured coding, through the use of "good" simulation languages and good tools

Structured walk-throughs of coding by testing, debugging, and verifying the simulation

Structured charts and Hierarchical Input Process Output (HIPO) charts

Use of a librarian for communication and documentation

We can view system simulation development as a sequence of translations that take us step by step from initiation through verification and validation of our simulation model. We can initiate the simulation process by translating *user wants* into *user needs*. User needs are then translated into functional specifications, where we must decide exactly what our simulation should do. Once we have a clear concept of "what," we can translate the functional specifications into design specifications. This is the "how" stage of simulation development. It is here that we consider the details of the system and construct the model by translating the detailed design specifications into program coding. When we have completed the model construction, we test our simulation to demonstrate conclusively the absence of coding mistakes (bugs). Our debugged program is then ready for verification, which we discussed in Chapter 2 and will discuss again in Chapter 13. We then proceed to the validation step and find our results or draw our conclusions. Figure 12.8 shows the above progression of steps diagrammatically.

Information about the system can sometimes get distorted or lost in the translation. Let us use the children's game of "telephone" as an illustration. All the players sit in a circle. One player starts the game by whispering a message in the ear of the player directly to his right. That player then whispers the message to the player on his right. The message is whispered from player to player until it reaches the last player, who then announces the message for all to hear. Usually, the message in its final form has little or no resemblance to the original, and this is the source of real amusement. However, in the "real world," we do not find much humor in distortion and translation errors, which are particularly troublesome in simulation. We classify translation errors into the following categories:

FIGURE 12.8 Steps in System Simulation

User wants ⟶ User needs ⟶ (Analysis) Functional specifications (What?) ⟶

(Construction) Detailed design specification (How?) ⟶ (Construction) Code ⟶ Testing ⟶

Verification ⟶ Validation ⟶ Results or conclusions

Lack of communication (The information was disclosed but was not properly digested or understood, i.e., "You said it but I didn't get it.")

Lack of completeness (Only part of the information was heard, i.e., "You said it but I didn't get all of it.")

Lack of consistency (All the information was heard, but there were conceptual errors, i.e., "You said it, I thought I got it, but something's wrong.")

Lack of full disclosure (i.e., "You didn't say it all.")

Clerical errors

12.5 TOP-DOWN DESIGN

One of the software development methodologies that can be used for the simulation of large-scale systems is top-down design. If design decisions with the greatest impact are done early and design decisions with little scope or impact (implementation details) are saved until later, we are virtually forced to design top-down. "Critical part first" is another top-down design strategy. The distinguishing features of this method are

Formal specification of inputs, outputs, and the functional or structural description of each system

Isolation of the internal variables of the subsystems, which assumes that the transfer of information, material, or energy is by inputs and outputs only

Postponement of details in developing the lower-level subsystem until boundaries of each subsystem are defined

The top-down simulation method is a philosophical approach that breaks down the simulation problem into smaller subproblems until the subproblem is simple enough to solve. This method is illustrated by the organizational diagram in Figure 12.9, often called a Hierarchical Input Process Output (HIPO) diagram.

An important simulation program design concept is that each subproblem be nearly independent with respect to

Program parameters or arguments

Algorithms

Internal variables, tables, or constants

Structure and format of data

This design concept is analogous to the "black box" design approach as shown in Figure 12.9(b).

FIGURE 12.9 HIPO Diagram (a) Organizational Diagram of the Top-Down Method (b) Formal Specifications of Inputs, Structure, and Outputs

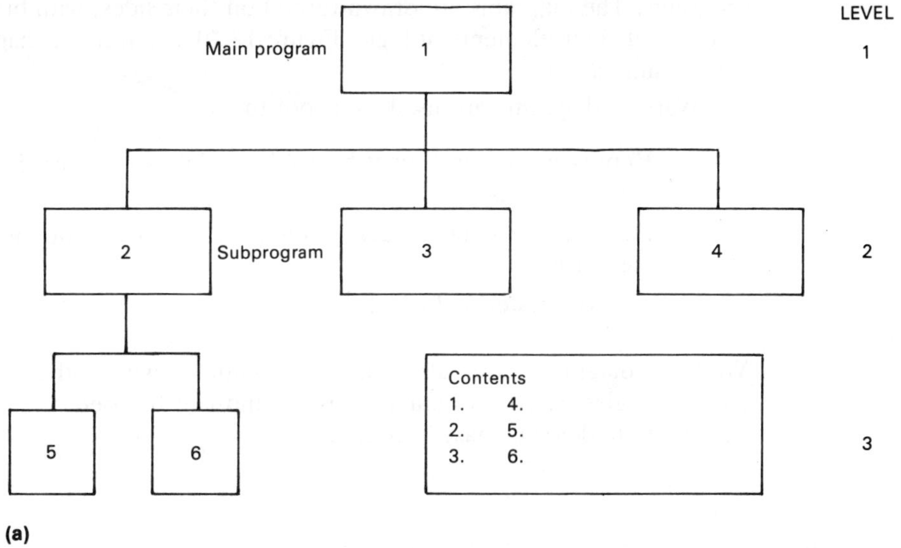

In Figure 12.9(a), level 1 of the simulation program is the main program. Level 2 corresponds to the components of the subprograms that interface with the main program. Similarly, each successive level interfaces with subprograms of the previous level. We must define the inputs, the outputs, and the simulation algorithm. When we use simulation languages, these levels are transparent to us. Indeed, this transparency is the basis for inventing simulation languages. Chattergy and Pooch present a program in FORTRAN for simulating a single-queue system using the top-down method.[7]

The bottom-up method, on the other hand, builds a more complex simulation program until the main program (which is designed last) is powerful enough to simulate the system easily. In this approach, most of the effort is spent on designing subprograms that are tested separately, and the complete simulation is performed last.

12.6 WARNIER DIAGRAMS

Warnier diagrams[22] are ordered hierarchical structure diagrams for data and programs. The diagrams are drawn turned on their sides, with braces but without boxes, and show elements of logic. Figure 12.10 is a Warnier diagram of a simple system simulation.

Warnier diagrams are used as a tool to

> Represent and document hierarchical data structures (i.e., reports, input data, data bases, and programs)
>
> Develop the detailed design for programs using output, input, or data base structures
>
> Analyze systems (logical design)

Warnier diagrams are also fully compatible with other program design methodologies based on data structures, and can be used to supplement system design methodologies based on data flow.

FIGURE 12.10 Warnier Diagram and Definitions

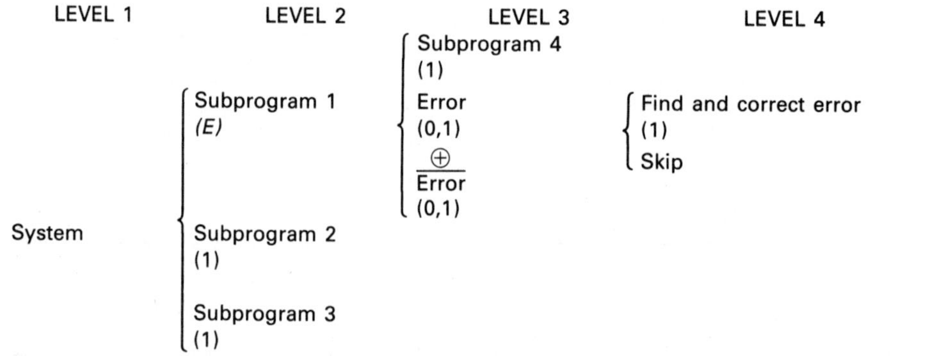

Subprogram 1 (*E*)	Means	The module "Subprogram 1" is executed *E* times at this *point.*
Subprogram 2 (1)	Means	The module "Subprogram 2" is executed 1 time at this point.
Error (0,1)	Means	The module "Error" is executed 0 or 1 times, depending on whether there was an error.
$\overline{\text{Error}}$ (0,1)	Means	The module "Not Error" is executed 0 or 1 times, depending on whether there was an error.
Error (0,1) ⊕ $\overline{\text{Error}}$ (0,1)	Means	The modules "Error" and "Not Error" are mutually exclusive and exhaustive; only one of them will be executed.

The basic ideas for the logical construction of programs (Warnier) are that

Programs, input data, and results (reports) are all information files, and thus have the same basic structures

All data sets can be subdivided so that the sets and subsets form a tree

All programs can be built up in hierarchical order from block, if-then-else, and do-while control constructs (structured coding)

The structure of a program should be derived from the structure of the output and the input and/or logical data base

Simplicity and logic are the strong points of Warnier diagrams. By breaking down the system into separate components (inputs, outputs, etc.), we can easily see the structure of the system and its model. The diagrams provide us with a graphical model of the software for simulation or program, similar to a flow chart, progressing from one step to the next in a logical fashion. In particular, this is an attractive feature for nontechnical users (nonprogrammers), who can gain an understanding of the system model through this visual aid.

12.7 SUMMARY

Realizing that no prescriptive methodologies for system simulation exist, we are forced to rely in large part on our common-sense experience, and logic. In this chapter, we have presented software engineering methodologies that might be useful for simulation construction. In particular, we focused on the top-down and Warnier methods for the logical construction of programs.

Warnier diagrams can be used as a simple tool for representing hierarchical data structures, programs, and systems; in systems analysis to help define system functions and logical data bases; and in a program construction procedure based on input data structures. The construction procedure involves first identifying output data structures (reports) and then determining necessary input data. Recall that we defined simulation as the process of model construction and model use. Using this definition, we can view the construction of the software programs as the model construction stage. The "performance" of the software programs can be tested under various input conditions, and this constitutes model use.

12.8 APPENDIX: PARAMETER IDENTIFICATION USING MAXIMUM LIKELIHOOD ESTIMATION

The model reference parameter identification technique is shown schematically in Figure 12.4. The objective of this technique is to minimize the error criterion. The maximum likelihood criterion minimizes the function

$$F = \frac{1}{2}\sum_{i=0}^{n-1} (\mathbf{z}(t_i) - \mathbf{x}(t_i))B^{-1}(\mathbf{z}(t_i) - \mathbf{x}(t_i)(\mathbf{z}(t_i) - x(t_i))^{\mathrm{T}} + \frac{N}{2}\log|[\mathrm{B}]|$$

where the minimizing value of $[B]$ is given by

$$[B] = \frac{1}{n}\sum_{i=0}^{n-1} (\mathbf{z}(t_i) - \mathbf{x}(t_i)(z(t_i) - \mathbf{x}(t_i))^{\mathrm{T}}$$

where T = the transpose of the matrix
n = the number of sampled points
t_i = the observation time instants
$\mathbf{z}(t_i)$ = the values of the observed heart rates at time instants t_i
$\mathbf{x}(t_i)$ = the values obtained from the model at time instants t_i

It has been shown that minimizing F is equivalent to minimizing the determinant of B.[9] In this procedure, measurement noise and system noise are considered to be white-Gaussian.

12.9 APPENDIX: THE TEN FATAL MISTAKES AND HOW TO AVOID THEM*

1. *Failure to define an achievable goal.* The goal of a simulation project should never be "To model the. . . ." Modelling itself is not a goal; it is a means of achieving a goal. An essential ingredient for successful simulation is a clearly articulated and agreed-upon set of realizable objectives. These depend on answers to questions like What is to be learned about the system under study? What decisions will be based on the simulation results?

The objectives cannot be correctly defined without the active participation of the end user, and they must of course be realizable. Setting the goals is the first step in any simulation project and perhaps the one most commonly bypassed.

2. *Incomplete mix of essential skills.* A successful simulation project calls for a combination of at least four areas of knowledge and experience:

Project leadership—the ability to motivate, lead and manage the simulation team.

Modelling—the ability to design a conceptual model that imitates the system under study at the required level of detail.

Programming—the ability to transform the conceptual model into a readable, modifiable, working computer program.

Knowledge of the modelled system—sufficient understanding of the system to guide the modelling and to judge the validity of the simulation results.

---

* *The Ten Most Frequent Causes of Simulation Analysis Failure . . . and How to Avoid Them!* (Los Angeles, Calif.: CACI, Incorporated, 1985), pp. 4–9. Reprinted with permission.

Teams have typically lacked specialists whose expertise and professional interests lie in modelling and simulation over and above programming. In addition, people knowledgeable about the system together with those who will use the results of the simulation study have typically not tracked the development in sufficient detail to assure that the end product satisfies their needs.

A simulation team cannot succeed if it does not possess all of the requisite skills.

3. *Inadequate level of user participation.* All too often, model developers simply go off by themselves for a year and then proudly drop the "completed," never-to-be-used model on the sponsor's desk.

The model-building team must work with the user organization in order for both to have the confidence and understanding necessary to make and encourage effective use of the completed work.

Provision must be made for regularly scheduled briefings, progress reports, and technical discussions with expected users of the model.

The end user is also the only one who can inform the team about realistic considerations such as politics, bureaucracy, unions, budget, and changes in the sponsoring organization. These will determine the success of the project as much as will the quality of the technical work.

4. *Inappropriate level of detail.* A model is a simplified representation of a system, and it should incorporate only those features of the system thought to be important for the user's purpose.

In modelling a complex system, difficult questions must be addressed—often for the first time. There is a tendency to spend a great deal of effort modelling in unnecessary detail those portions of the system that are well understood, while glossing over poorly defined portions that may be more important. This approach creates the illusion that great progress is being made, until it comes time to produce valid, usable, results.

The goals of the project determine the appropriate level of detail, which must be consistent with the availability of data and other resources.

5. *Poor communication.* It has been said that if you can clearly define what the problem is, you are already halfway toward its solution. To gain understanding of a problem, everyone who can contribute to its definition should be able to do so in a disciplined but natural way.

A conceptual framework and language for communication between the team members is essential. The framework, or world view, should relate closely to the language of the programmed model.

Simulation languages that are problem-oriented and readable, can dramatically simplify model design and programming. They provide a vocabulary and related concepts with which system elements and their interactions can be conveniently described and discussed.

6. *Wrong computer language.* Opinions differ regarding the advantage of simulation programming languages. Some believe that computer languages should be English-like and problem-oriented, while others feel that FORTRAN, and FORTRAN based tools are adequate.

Our view is that the programming language should be English-like, self-documenting, and readable by the user, who is primarily interested in the actual system, not the details of computer programming.

High-level simulation languages have been shown to substantially reduce both programming man-hours and project elapsed time. By design, they offer language, program, and data structures that make models much easier to develop and modify.

7. *Obsolete or nonexistent documentation.* Over the years, we have observed numerous unsuccessful simulation projects that had no documentation except a FORTRAN listing. Many of these listings contained few explanatory comments.

Even a thoroughly commented FORTRAN listing is difficult to decipher for anyone other than the person who wrote it. Often, even the programmer has difficulty understanding it after a short time.

We have also seen great amounts of money often wasted on manuals and flowcharts intended to make it easier to develop, maintain, modify, and enhance the model.

This waste is a consequence of the realities of model development. Most models evolve over a long period of time because of new and increased understanding of the system, changing goals, and availability of new data.

Because of the evolutionary changes, flowcharts, prose documentation, detailed descriptions of routines and variables, and program comments are invariably incomplete, incorrect, and almost always out of date. The longer the model is around, and many models in use today were developed five or more years ago, the more this type of documentation deteriorates.

The only dependable program documentation in a changing environment is the source program listing. The quality and usefulness of this documentation is determined by the model design and the simulation language.

8. *Using an unverified or invalid model.* Verification involves comparing the programmed computer model with the conceptual model. Does the program correctly implement the model as designed? Validation involves comparing the model as designed to the actual system. Does the model adequately represent the real system?

The most effective verification-validation technique is a walk-through, with the programmer explaining the code to someone who is familiar with the system under study. This technique frequently turns up design and coding errors that can be corrected at a fraction of the cost and time that would be required after the model is executing on a computer.

To use the powerful walk-through technique, the program must be readable. Hence, again we see the importance of using an English-like simulation language.

9. *Failure to use modern tools and techniques to manage the development of a large complex computer program.* Practically all large computer program developments are late. Three major reasons for this are:

Premature coding—the irresistible urge to begin coding before the program is designed.
Optimistic scheduling—underestimating the time required for known tasks and neglecting to allocate time for the inevitable, unanticipated problems.
Confusing of effort with progress—forgetting that on some days a great deal of effort will be expended but progress will consist of locating and correcting a single error.

There are tools and techniques that can help in overcoming these problems. Computer based design tools, guide the top-down development of models through statements of objectives, functional requirements, and data structures.

These tools also provide a means for capturing project management information such as progress estimates, responsibility assignments, problem areas, and design revisions.

Software engineering tools and related principles such as structured and modular programming make program development orderly and manageable. They provide a way of stating programs so they clearly reflect the organization and logic of the system being studied. As a result, programs are more reliable; documentation is improved, and coding, debugging, and testing are simplified.

Lack of complete understanding of these methods and failure to use them virtually dooms any effort to develop a large, complex model.

10. *Poor presentation of results.* The results from simulation studies are often presented in a way that the user finds incomprehensible.

A model that gives unexpected or illogical results may do so because certain parameters turn out to be far more significant or insignificant than expected, or because unanticipated interactions between system elements greatly affect system performance. Insight into hidden problems of this sort is typically gained from simulation.

However, unexpected or unusual simulation results that cannot be explained are usually caused by errors, invalid assumptions, or lack of understanding of the real system. Someone familiar with the system under study must examine the simulation results to determine whether they are reasonable. Simulation results must be presented in a way that the user can easily relate to the system under study because that is his world.

12.10 REFERENCES

1. Aburdene, M.F.; Sandler, H.; and Goldwater, D., "Modeling of the Heart Rate Response to LBNP in Women," in *Modeling and Simulation*, edited by William G. Vogt and Marlin H. Mickle, vol. 17, *ISA*, 1986, pp. 1229–34.
2. Bach, Ralph E., *A User's Manual for Ames* (a parameter estimation program), Final Report, NASA 2-7397, NASA-Ames Research Center, Moffett Field, California, June 1974.
3. Baker, T.F., and Mill, H.D., "Chief Programmer Teams," *Datamation* 19, no. 12 (December 1973): 58–61.

4. Bohm, C., and Jacopini, G., "Flow Diagrams, Turing Machines, and Languages with Only Two Formulation Rules," *Communications of the ACM* 9, no. 5 (May 1966): 366–71.
5. Brooks, F.P., *The Mythical Man-Month* (Reading, MA: Addison-Wesley, 1982).
6. "Challenges to Control: A Collective View," *IEEE Transactions on Automatic Control*, vol. AC-32, no. 4 (April 1987): 275–85.
7. Chattergy, R., and Pooch, U.W., "Integrated Design and Verification of Simulation Programs," *IEEE Computer* 10, no. 4 (April 1977): 40–45.
8. Constantine, L.L., *Concepts in Program Design* (Cambridge, MA: Information and Systems Press, 1967).
9. Grove, R.D., et al., *A Procedure for Estimating Stability and Control Parameters from Flight Test Data by Using Maximum Likelihood Methods Employing a Real-Time Digital System*, NASA TN D-6735, May 1972.
10. Heckel, P., *The Elements of Friendly Software Design* (New York: Warner Books, 1984).
11. McGowan, C.L., and Kelly, J.R., *Top-down Structured Programming Techniques* (Princeton, NJ: Petrocelli/Charter, 1975).
12. McLeod, J., "PHYSBE . . . A PHYsiological Simulation Benchmark Experiment," *Simulation* 7, no. 6 (December 1966): 324–29.
13. Musgrave, F.S.; Zechman, F.W.; and Mains, R.C., "Comparison of the Effects of 70° Tilt and Several Levels of Lower Body Negative Pressure on Heart Rate and Blood Pressure in Man," *Aerospace Medicine* 42 (1971): 1065–69.
14. Myers, G.J., *Reliable Software through Composite Design* (New York: Mason/Charter, 1975).
15. Sandler, H., "Cardiovascular Effects of Weightlessness," in *Progress in Cardiology*, edited by Paul N. Yu and John F. Goodwin (Philadelphia: Lea & Febiger, 1976), pp. 227–70.
16. Sandler, H., "Effects of Bedrest and Weightlessness on the Heart," in *Hearts and Heart-like Organs*, edited by Geoffrey H. Bourne, vol. 2 (New York: Academic Press, 1980), pp. 435–524.
17. Sandler, Harold, and Winter, David L., *Physiological Responses of Women to Simulated Weightlessness*, NASA Report, NASA SP-430, 1978.
18. Shannon, Robert E., *Systems Simulation: The Art and Science* (Englewood Cliffs, NJ: Prentice-Hall, 1975).
19. Sommerville, I., *Software Engineering* (Reading, MA: Addison-Wesley, 1985).
20. Steven, P.M.; Miller, P.B.; Gilbert, C.A.; Lynch, T.N.; Johnson, R.L.; and Lamb, L.E., "Influence of Long Term Lower Body Negative Pressure on the Circulatory Function of Man during Prolonged Bed Rest," *Aerospace Medicine* 37 (1966): 466–74.
21. Stevens, W.P.; Meyers, G.J.; and Constantine, L.L., "Structured Design," *IBM Systems Journal* 12, no. 2 (1974): 115–39.
22. Warnier, Jean D., *Logical Construction of Programs* (New York: Van Nostrand Reinhold, 1981).
23. Wirth, N., *Systematic Programming* (Englewood Cliffs, NJ: Prentice-Hall, 1973).
24. Yeh, R.T., ed., *Current Trends in Programming Methodology*, vol. 1: *Software Specification and Design* (Englewood Cliffs, NJ: Prentice-Hall, 1977).
25. Yeh, R.T., ed., *Current Trends in Programming Methodology*, vol. 2: *Program Validation* (Englewood Cliffs, NJ: Prentice-Hall, 1977).

13 Verification and Validation

13.1 THE END OF THE SIMULATION PROCESS: VERIFICATION AND VALIDATION

In Chapter 1, we described the simulation process steps. Here we expand the steps to include verification and validation, shown in Figure 13.1. The inputs to the system, the base model, the lumped model, and the computer implementation given by $\boldsymbol{u}(t)$ are assumed to be known. The outputs of the system and the base model are $\boldsymbol{y}_s(t)$. Because it is difficult to obtain base models of realistic systems, we often use the lumped-parameter models.

After we develop our computer program, we test it to confirm that it is an accurate translation of the lumped model. If it is not, we initiate an iterative process of revision and debugging the program to minimize $e_{mc}(t)$, the difference between the outputs of the lumped model and our computer program. This procedure is called verification and is crucial to simulation. Here we may need all the software engineering, statistical, and system theory tools available to us.

Once we have verified our lumped model, we need to test it against the real-world system we are modelling to confirm its accuracy. This process, called validation, compares the outputs of the system being simulated, $\boldsymbol{y}_s(t)$, with the outputs generated from the computer program, $\boldsymbol{y}_c(t)$.

Our goal is to minimize $e_{sc}(t)$, the difference between $\boldsymbol{y}_s(t)$ and $\boldsymbol{y}_c(t)$, by changing our lumped model or its parameters. Once this is achieved, we can have confidence in our lumped model and any inferences derived from it. However, this validation step is often impossible, as in the case of new systems (e.g., the design of new planes, cars), which offer no observable data. Also, in many instances, we make certain unrealistic assumptions (e.g., that the mathematical model of the system is known), which limits the use of and confidence in our lumped model.

There is no one test that we can use to show the validity of a simulation. The tests we use depend on the system being simulated and our knowledge of the system. If a real system exists, we might check the validity of the simulation by comparing historical data with the output data from the simulation program. This measure of validity, however, does not guarantee that future behavior can be predicted or explained. Therefore, the validity of a simulation is closely related to the objectives and purpose of the simulation, and must be judged on that basis.

FIGURE 13.1 Modelling and Simulation Steps

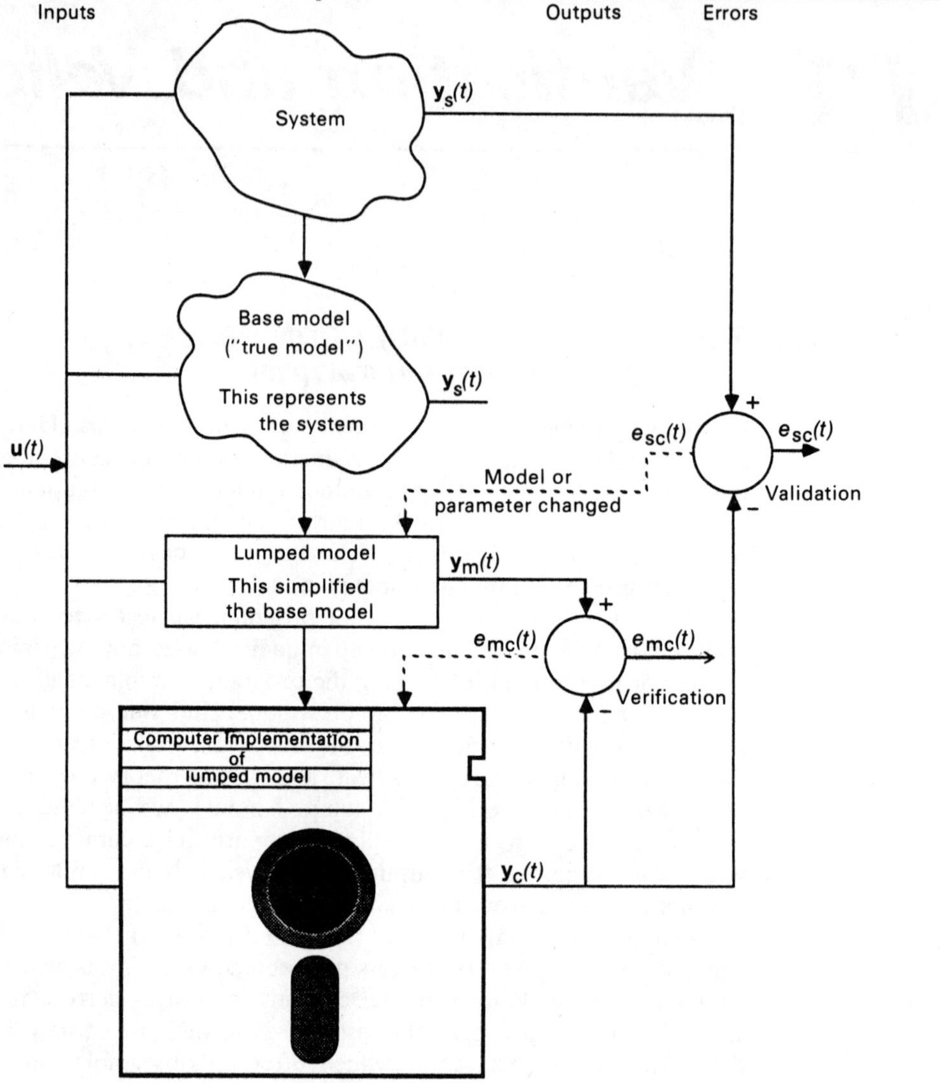

Throughout the development of the simulation, from its inception as a base model through its computer implementation, we have played the role of advocate, believing in the model and program and supporting its "truth." In the validation stage, however, we must change roles and become the critic, challenging the assumptions and premises on which we have constructed the model and asking the question, "Is our model adequate?" We must play this dual role in order to develop the best possible simulation for a system.

J. Gruhl and D.O. Wood believe that human understanding is of little use in the validation process of existing complex large-scale simulation models.[7] They suggest a variety of techniques, including the use of an independent assessor (one not involved in building the model). In addition, they conclude that modellers should acknowledge that

> Some problem areas may not be suitable for simulation modelling, but are best served by evaluation by experts
>
> Each validation of a model cannot be based on a single set of criteria but must be judged independently according to its own circumstances
>
> Model builders are in the best position to discuss the adequacy of the model's predictions (Deeter and Hoffman indicate that 90% of the literature does not address the validity or adequacy of the models used[5])
>
> Without validation by model builders, misuse of results is possible and likely
>
> Lack of necessary funding for manpower and resources often limits validation attempts
>
> One of the strategies for validating models is to report errors in outputs of models regardless of the reasons for the errors

As a warning to us, D.B. Lee indicates that, in general, the goals of large-scale simulation models have not been achieved. Many simulations could have been done at a lower cost, and were not effectively used for long-range planning.[8]

As the philosopher Bertrand Russell said, we believe that "the truth . . . can be discovered by those who are willing to combine the hopefulness, patience, and open mindedness of science with something of the Greek feeling for beauty in the abstract word of logic and for the ultimate intrinsic value in the contemplation of truth."[9]

13.2 REFERENCES

1. Akaike, H., "A New Look at Statistical Model Identification," *IEEE Transactions on Automatic Control*, vol. AC-19, no. 6 (December 1974): 716–23.
2. Bekey, G.A., "Simulability of Dynamical Systems," *Transactions of the Society for Computer Simulation* 2, no. 1 (May 1985): 57–71.
3. Box, G.E.P., and Tiao, G.C., *Bayesian Inference in Statistical Analysis* (Reading, MA: Addison-Wesley, 1973).
4. Chattergy, R., and Pooch, U.W., "Integrated Design and Verification of Simulation Programs," *IEEE Computer* 10, no. 4 (April 1977): 40–45.

5. Deeter, C.R., and Hoffman, A.A.J., *An Annotated Bibliography of Mathematical Models Classified by Validation and Error Analysis Methods*, Texas Christian University Report for NSF, April 30, 1978.

6. Grayeal, W., and Pooch, U.W., *Simulation: Principles and Methods* (Boston: Winthrop, 1980).

7. Gruhl, J., and Wood, D.O., "Independent Assessment of Complex Methods" (Paper presented at NSF/Texas Christian University sponsored Workshop on Validation of Mathematical Models of Energy Related Research and Development, June 21–23, 1978).

8. Lee, D.B., Jr., "Requiem for Large Scale Models," *SIMULETTER* 1, no. 3 (January–April 1975): 16–29.

9. Russell, B., *Our Knowledge of the External World* (London: Allen & Unwin, 1972), p. 40.

10. Sargent, R.G., "Statistical Analysis of Simulation Output Data," *SIMULETTER* 7, no. 4 (July 1976): 39–50.

11. Shannon, Robert E., *Systems Simulation: The Art and Science* (Englewood Cliffs, NJ: Prentice-Hall, 1975).

Index

ACSL simulation language, 248, 260–263
commands in, 267–268
communication interval in, 249
documentation for, 268
model definition in, 267
operators in, 267
plotting in, 268
statements in, 248–250
vs. TOTAL and BUCOP, 267
Algorithms, 3
in top-down simulation, 338
Antithetic series, 109

Balk, 133
Base model, 3, 5, 7
Bernoulli's differential equation, 25n
Bottom-up simulation, 339
BUCOP, 266–267

CACSD systems (computer-aided control system design systems), 265–269
ACSL, 267–268
BUCOP, 266–267
comparison of, 265
NDTRAN, 268–269
TOTAL, 265–266
uses of, 269
Causal loop diagrams, 63
Central limit theorem, 87–91
Chi-square distribution, as random number test, 105–106
Closed-loop feedback system (See Feedback.)
Computers:
program development in, 3
random number generation in, 102
and simulation, 2, 3, 8, 323–345 (See also Simulation methods.)
statistical packages for, 95
Conceptual systems, 45
Continuous systems, 47–48 (See also Dynamic systems.)
time delays in, 230–233
Correlation analysis, 95–96
Cost function minimization, 309–320
function approximation methods, 318–319
function comparison methods, 313–318
multiparameter optimization, 320
Powell method, 319
single-variable optimization, 313–320
CSSLs (continuous-system simulation languages), 246–269
ACSL, 248–254, 260–263
CSMP, 255
DYNAMO, 255–256
NDTRAN, 256–260, 263–264
principles of, 246–248

Delay factor, 29
Deterministic model, 10
Difference equations, 55, 211–212
Differential equations, 213
Bernoulli's, 25n
Discrete-event simulation languages, 129–165 (See also GASP, GPSS, SLAM, SIMAN, SIMSCRIPT, SIMULA.)
of computer communication network, 145–152
defined, 131
of distributed computing system, 141–145
of five-node computer network, 152–158
fixed-increment (synchronous), 132
of manufacturing process, 158–165
next-event (asynchronous), 133
Discrete systems, 47
and Markov processes, 190–207 (See also Markov processes.)
Distribution functions, 73–74
exponential distribution, 79–82
integrals in, 76
normal distribution, 83–85
Poisson distribution, 82–83
and random variables, 108
uniform distribution, 78–79
Domain, 70–71

Dynamic continuous systems, and information diffusion problem, 22–27
Dynamic systems, 22
  critically and overdamped damped, 222–223
  damped oscillatory, 221
  equilibrium of, 27–30, 219, 325
  explosive oscillatory and explosive, 223–224
  feedback loops in, 210
  free vs. forced, 220
  linear vs. nonlinear, 228–229
  model sensitivity in, 236
  nonlinear, 226–228
  oscillatory (undamped), 220–221
  phase plane in, 224
  time-varying, 226
  trajectories in, 224–225

Endogenous variables, 60
Equations:
  as form of model, 55
  recursive, 23
Equilibrium, 27–29
Euler integration, 214–216
Exogenous variables, 60
Expected value, 76–77
Experimental frame, 3
Exponential distribution, 79–82
  mean and variance in, 80
  probability density functions in, 79

Feedback:
  elements of, 62
  loops, 60, 210, 213–214
  system, 61
Functions, 70–71
Function comparison methods, 313–318
  bisection and Fibonacci, 317–318
  Golden Sections, 313–317

GASP simulation language, 114, 130, 137
Gaussian distribution, 83–85, 87
GPSS simulation language, 10, 114, 130, 140–141

HIPO diagram, 338–339

Industrial dynamics (See System dynamics.)
Integrand, 214
Integration, 211
  choice of increments in, 234
  Euler, 214, 215–216, 244, 248, 256, 268 (See Rectangular integration.)
  and exponentiation, 211–214
  rectangular, 27
  Runge–Kutta, 16, 31, 244, 255, 256, 268
  in system dynamics, 210–236
Internal state, 47

Least squares analysis, 92, 294
Levels, 210
Limiting values, 229–230
Linear regression, 92–93
Linear system theory, 63
Lumped model, 3, 5

Machine, 137
Markov process, 190–192
  Markov chain, 190, 192
  models of, 193–197
  probability transition matrix in, 197–200
  probability transition matrix of, 191
  simulation of, 200–206
  states of, 190–191, 197
Material, 137
Mean:
  in exponential distribution, 80
  in normal distribution, 84
  in Poisson distribution, 83
  of samples vs. populations, 86–87
  in sampling distribution, 88
  in uniform distribution, 78
Mixed congruential method, of random number generation, 103–104
Model construction, 337
  as function of simulation, 3
  system description in, 64
Modelling, 7
Models:
  in equation form, 55
  sensitivity of behavior in, 236
  use of, 3
  variables in, 9–10
Monte Carlo simulation, 102–117 (See also Random numbers.)
  antithetic series in, 109, 114
  queuing application, 112–114
  variance reduction in, 113

NDTRAN simulation language, 263–264, 268–269
  plotting in, 258–260, 269
  statements in, 256–257, 268–269
  time-delay function in, 264
  variables in, 258, 268
Negative feedback, 61
Next-event simulation languages, 134–135
Normal distribution, 83–85, 87
Numerical integration, 55 (See also individual techniques.)

Open-loop system, 60 (See also Feedback.)

Parameter estimation, 293–304
  from average value, 295
  graphical technique, 295
  of growth and decay processes, 293–295
  by least squares method, 294
  likelihood function estimation in, 297–303
  logistics curve coefficient in, 295–296
  by regression equation, 295
  of state variable systems, 296–297
  of stochastic systems, 297–304
Parameter optimization, 309–320 (See also Cost function minimization; Function comparison methods.)
Parameters, in top-down simulation, 338
Period, 102, 220–221
Personal computers (PCs), and discrete-event simulation languages, 129, 130
Physical and social systems, 1
Poisson distribution, 82–83
  in parameter estimation, 300
  of random variables, 110–111
  from uniformly distributed numbers, 111
Polynomial regression, 93
Positive feedback, 61
Probability density functions, 74–76
  for chi-square distribution, 105
  continuous, 76
  discrete, 74–75
  in exponential distribution, 79
  histogram, 75
  in normal distribution, 84
  in Poisson distribution, 83
  in uniform distribution, 78
Probability theory, 69–96
  axioms of, 70
  certain event, 70
  relative frequency, 69
Program, 3
Program event, in discrete-event simulation, 131
Pseudorandom numbers, 102

Queue, 137

Random numbers:
  and chi-square distribution, 105–106
  generation of, 102, 103
  and scatter diagrams, 107
  in stochastic simulation, 18
  tests of, 104–107
  uniformly distributed, 103–104
Random variables, 70–72 (See also Distribution functions.)
  discrete vs. continuous, 72
  and distribution functions, 73, 106, 108
  exponential distribution of, 81, 109
  interoccurrence times in, 111
  mean of repetitions, 88
  Poisson distributed, 110–111
  and probability density functions, 76
  statistical calculation of, 95
Rates, 210
  and equilibrium, 219
  as policy statements, 234
Real systems, 3, 17, 27
Rectangular integration, 27 (See also Euler integration.)
Recursive equations, 23
Reference model, 3, 5, 7
Regression, linear, 92–93
Regression analysis, 92–94
  vs. correlation analysis, 95
  as description, 96
  form of equation in, 97
  in nonlinar systems, 226
Runge–Kutta, 16, 31, 244, 255, 256, 268
Run time, 131

Sample size, 21
  and distribution of means, 92
Sampling, 86–88
Scatter diagrams, 107
Second-order systems, 218–219
Seed, 1–2
Server, 137
SIMAN simulation language, 10, 114, 130, 160–165
  model verification in, 164
  summary report, 164
SIMSCRIPT simulation language, 10, 114, 130, 137, 140
SIMULA simulation language, 10, 114, 130, 140
Simulation, 1–11
  of complex systems and long-term events, 6
  and computers, 2, 3, 8
  confidence intervals in, 116–117
  limitations and misapplications of, 7–9
  mathematical models in, 2
  prediction in, 5, 6
  procedure, 17, 210
  in science and engineering, 2
  of social systems, 234–235
  of theoretical events, 6
  uses of, 6
  validation of, 9
Simulation languages, 129–130 (See also CSSLs; individual languages.)

activity method, 140
choice of, 130
classification of, 137
continuous-system, 246–269
definition of terminology, 137
design of, 129
process technique, 140
STELLA, 281–282
types of, 10–11
Simulation methods, 323–345
structured system programming, 336–337
top-down method, 338–339
Sink, 137
SLAM simulation language, 114, 130, 140, 145, 152–157
comparison of models, 146–150
Source, 137
Standard deviation, 77, 84–86
State variable approach, 64
State variable systems, 47, 48, 55
appropriateness of, 64
parameter estimation in, 296–297
Stochastic model, 10
Stochastic systems, 17–21
vs. deterministic systems, 62–63
parameter estimation in, 297–304
random numbers in, 18
statistics from 87
System dynamics, and integration, 210–236
System events, 131, 135
Systems, 1 (See also individual systems.)
base model in, 3
boundaries in, 45
description in model construction, 64
dynamic vs. static, 62
external description of, 45–47
feedback in, 62
inputs and outputs, 45
internal description of, 47
internal states of, 45
linear vs. nonlinear, 63
mathematical relationships in, 8
memory in, 62
natural vs. man-made, 7
open-loop vs. feedback, 60–62
open vs. closed, 60
prediction of future state of, 5
real, 3
state variable, 47, 48, 55
stochastic vs. deterministic, 62–63
structure identification in, 293
time-invariant vs. time-varying, 63
System simulation, 336–337
Warnier diagrams in, 340–341

Tabular simulation, 226–229
Theoretical mean (See Expected value.)
Time delays, 230–233
Top-down simulation, 338–339
TOTAL, 265–266
Trajectories, 224–225
Transaction, 137
Translation errors, 337–338

Uniform distribution, 78–79

Validation, 9, 17, 347–349
Variables:
functional relations of, 92
in top-down simulation, 338
Variance, 74
in exponential distribution, 80
in Monte Carlo simulation, 109
in Poisson distribution, 83
reduction in, 113, 114–115
of samples vs. populations, 87
in sampling distributions, 88
in uniform distribution, 78
Verhalst's equation, 25n
verification, 17, 347–349

Warnier diagrams, 340–341